Geometry of Chemical Graphs

Polycycles and symmetric polyhedra appear as generalizations of graphs in the modeling of molecular structures, such as the Nobel prize winning fullerenes, occurring in chemistry and crystallography. Chemistry has inspired and informed many interesting questions in mathematics and computer science, which in turn have suggested directions for synthesis of molecules.

Here the authors give access to new results in the theory of polycycles and two-faced maps together with the relevant background material and mathematical tools for their study. Organized so that, after reading the introductory chapter, each chapter can be read independently from the others, the book should be accessible to researchers and students in graph theory, discrete geometry, and combinatorics, as well as to those in more applied areas such as mathematical chemistry and crystallography.

Many of the results in the subject require the use of computer enumeration; the corresponding programs are available from the author's website.

MICHEL DEZA is retired Director of Research at CNRS, Director of the Laboratoire interdisciplinaire de géométrie appliquée, and a Professor at Ecole Normale Supérieure, Paris. He is Editor-in-chief of the *European Journal of Combinatorics* and this is his 12th book.

MATHIEU DUTOUR SIKIRIĆ is a Researcher of Mathematics at Institut Rudjer Bošković, Zagreb. His research interests include enumeration and extremal problems, in relation to plane graphs and discrete structures; polyhedral enumeration, lattices, Delaunay polytopes, and dual description problems.

All the titles listed below can be obtained from good booksellers or from Cambridge University Press. For a complete series listing visit http://www.cambridge.org/uk/series/sSeries.asp?code=EOM

Geometry of Chemical Graphs: Polycycles and Two-faced Maps

MICHEL DEZA
École Normale Supérieure, Paris,

MATHIEU DUTOUR SIKIRIĆ
Rudjer Bošković Institute, Zagreb

CAMBRIDGE
UNIVERSITY PRESS

CAMBRIDGE UNIVERSITY PRESS
Cambridge, New York, Melbourne, Madrid, Cape Town, Singapore, São Paulo, Delhi

Cambridge University Press
The Edinburgh Building, Cambridge CB2 8RU, UK

Published in the United States of America by Cambridge University Press, New York

www.cambridge.org
Information on this title: www.cambridge.org/9780521873079

First published 2008

Printed in the United Kingdom at the University Press, Cambridge

A catalogue record for this publication is available from the British Library

ISBN 978-0-521-87307-9 hardback

Contents

v

Preface

Platonic solids have been studied since antiquity and in a multiplicity of artistic and scientific contexts. More generally, "polyhedral" maps are ubiquitous in chemistry and crystallography. Their properties have been studied since Kepler. In the present book we are going to study classes of maps on the sphere or the torus and make a catalog of properties that would be helpful and useful to mathematicians and researchers in natural sciences.

In particular, we are studying here two new classes of maps, interesting for applications, especially in chemistry and crystallography (on the sphere or the torus) generalizing Platonic polyhedra. *Polycycles* are 2-connected plane graphs having prescribed combinatorial type of interior faces and the same degree q for interior vertices, while at the most q for boundary vertices. *Two-faced maps* are the maps having at most two types of faces and the same degree of vertices. Many examples and various generalizations are given throughout the text. Pictures are given for many of the obtained graphs, especially when a full classification is possible. A lot of the presentation is necessarily compact but we hope to have made it as explicit as possible.

We are interested mainly in enumeration, symmetry, extremal properties, face-regularity, metric embedding and related algorithmic problems. The graphs in this book come from broad areas of geometry, graph theory, chemistry, and crystallography. Many new interesting spheres and tori are presented.

The book is organized as follows. Chapters 1 and 2 give the main notions. After reading them, other chapters can be read almost independently.

Chapters 4–8 present the theory of polycycles. In Chapter 4, we explain the general notion of the (r, q)-polycycle, present the cases where classification is possible and the cell-homomorphism into the regular tiling $\{r, q\}$. In Chapter 5, the problem of how the boundary of an (r, q)-polycycle determines it, or not, is addressed. In Chapter 6, we consider the possible symmetries of (r, q)-polycycles and how we can classify these with a symmetry group transitive on faces and/or vertices.

Chapter 7 presents a way to decompose a generalized polycycle into elementary components. This powerful technique is used in Chapters 8, 12, 13, 14, and 18.

The second main subject – k-valent two-faced maps – is treated in Chapter 3 and Chapters 9–19. Chapter 3 deals with our main example, fullerenes, while Chapter 9 classifies strictly face-regular maps on sphere or torus. In Chapters 10–18, we consider a weaker notion of face-regularity. Chapter 19 treats 3-valent two-faced maps with icosahedral symmetry.

Many simple questions (some, possibly, easy) are raised; we hope that this book will be instrumental in their solutions. Much of the results have been obtained (and could only have been obtained) though computer enumeration; the corresponding programs are available from [Du07].

We are grateful to many people for their help with this book, especially, to Jacques Beigbeder, Gunnar Brinkmann, Olaf Delgado Friedrichs, Maja Dutour Sikirić, Patrick Fowler, Jack Graver, Marie Grindel, Viatcheslav Grishukhin, Gil Kalai, Stanislav Jendrol, and Mikhail Shtogrin.

We thank also École Normale Supérieure of Paris, Hebrew University of Jerusalem, Rudjer Bošković Institute of Zagreb, Institute of Statistical Mathematics of Tokyo, and Nagoya University for continued support.

Michel Deza and Mathieu Dutour Sikirić

1

Introduction

In this chapter we introduce some basic definitions for graphs, maps, and polyhedra. We present here the basic notions. Further definitions will be introduced later when needed. The reader can consult the following books for more detailed information: [Grü67], [Cox73], [Mun71], [Cro97].

1.1 Graphs

A *graph* G consists of a set V of *vertices* and a set E of *edges* such that each edge is assigned two vertices at its ends. Two vertices are *adjacent* if there is an edge between them. The *degree of a vertex* $v \in V$ is the number of edges to which it is incident. A graph is said to be *simple* if no two edges have identical end-vertices, i.e. if it has no loops and multiple edges. In the special case of simple graphs, automorphisms are permutations of the vertices preserving adjacencies. For non-simple graphs (for example, when 2-gons occur) an *automorphism* of a graph is a permutation of the vertices and a permutation of the edges, preserving incidence between vertices and edges. By $Aut(G)$ is denoted the group of automorphisms of the graph G; a synonym is *symmetry group*.

For $U \subseteq V$, let $E_U \subseteq E$ be the set of edges of a graph $G = (V, E)$ having end-vertices in U. Then the graph $G_U = (U, E_U)$ is called the *induced subgraph* (by U) of G.

A graph G is said to be *connected* if, for any two of its vertices u, v, there is a path in G joining u and v. Given an integer $k \geq 2$, a graph is said to be *k-connected*, if it is connected and, after removal of any set of $k - 1$ vertices, it remains connected.

Let $G_1 = (V_1, E_1)$ and $G_2 = (V_2, E_2)$ be two graphs. Their *Cartesian product* $G_1 \times G_2$ is the graph $G = (V_1 \times V_2, E)$ with vertex-set:

$$V_1 \times V_2 = \{(v_1, v_2) : v_1 \in V_1 \text{ and } v_2 \in V_2\}$$

1

and whose edges are the pairs $((u_1, u_2), (v_1, v_2))$, where $u_1, v_1 \in V_1$ and $u_2, v_2 \in V_2$, such that either $(u_1, v_1) \in E_1$, or $(u_2, v_2) \in E_2$.

A subset E' of edges of a graph is called a *matching* if no two edges of E' have a common end-vertex. A *perfect matching* is a matching such that every vertex belongs to exactly one edge of the matching.

The following graphs will be frequently used:

- The *complete graph* K_n is the graph on n vertices $v_1, \ldots, v_n$ with v_i adjacent to v_j for all $i \neq j$.
- The *path* $P_n = P_{v_1, v_2, \ldots, v_n}$ is the graph with n vertices $v_1, \ldots, v_n$ and $n - 1$ edges (v_i, v_{i+1}) for $1 \leq i \leq n - 1$.
- The *circuit* $C_n = C_{v_1, v_2, \ldots, v_n}$ (or *n-gon*) is the path $P_{v_1, v_2, \ldots, v_n}$ with additional edge (v_1, v_n).

A *plane graph* is a connected graph, together with an embedding on the plane such that every edge corresponds to a curve and no two curves intersect, except at their end points. A graph is *planar* if it admits at least one such embedding. It is known that any planar graph admits a plane embedding with the edges being straight lines (see [Wa36, Fa48, Tut63]). A *face* of a plane graph is a part of the plane delimited by a circuit of edges. A plane graph defines a partition of the plane into faces. If a is a vertex, edge, or face and b is a edge, face, or vertex, then a is said to be *incident* to b if a is included in b or b is included in a. Two vertices, respectively, faces are called *adjacent* if they share an edge. We will call *gonality* or *covalence* of a face the number of its vertices. A face is *exterior* if it is non-bounded. Bounded faces are called *interior*. Any finite plane graph has exactly one exterior face. An infinite plane graph can have any number, from zero to infinity, of exterior faces. A planar 3-connected graph admits exactly one plane embedding on the sphere, i.e. the set of faces is determined by the edge-set.

The *v-vector* $v(G) = (\ldots, v_i, \ldots)$ of a graph G enumerates the numbers v_i of vertices of degree i. A plane graph is *k-valent* if $v_i = 0$ for $i \neq k$. The *p-vector* $p(G) = (\ldots, p_i, \ldots)$ of a plane graph G counts the numbers p_i of faces of gonality i. For a connected plane graph G, denote its *plane dual graph* by G^* and define it on the set of faces of G with two faces being adjacent if they share an edge. Clearly, $v(G^*) = p(G)$ and $p(G^*) = v(G)$.

1.2 Topological notions

We present in this section the topological notions for the surfaces which will be used. Topology is concerned with continuous structures and invariants under continuous deformations. Since we are working with vertices, edges, and faces, the classical definitions will be adapted to our context.

No proofs are given but we hope to compensate for this by giving some geometrical examples. More thorough explanations are available in basic algebraic topology textbooks, for example, [Hat01] and [God71].

1.2.1 Maps

A *map M* is a family of vertices, edges, and faces such that every edge is contained in at least one and at most two faces. An edge, contained in exactly one face, is called a *boundary edge*; all such edges form the *boundary*. A map is called *closed* if it has no boundary. A map is called *finite* if it has a finite number of vertices, edges, and faces. See below plane graphs related to $Prism_5$ (see Section 1.5) with same vertex- and edge-sets but different face-sets; their boundary edges are boldfaced:

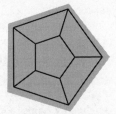

A closed map, cell-complex of a polyhedron. It is a $5R_0$, $4R_2$ plane graph (see Chapter 9)

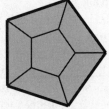

A map with boundary edges. It is a $(\{4, 5\}, 3)$-polycycle (see Chapter 7)

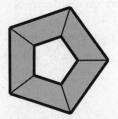

A map with boundary edges; not simply connected. It is a $(4, 3)_{gen}$-polycycle (see Section 4.5)

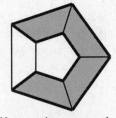

Not a map because two edges are not contained in a face. It is not considered

If M is a closed map, then we can define its dual map M^* by interchanging faces and vertices. See Section 4.1 for some related duality notions for non-closed maps. A map is called a *cell-complex* if the intersection of any two faces, edges, or vertices is a face, edge, vertex, or ∅. Maps with 2-gons are not cell-complexes; they are *CW-complexes* (see, for example, [Rot88]).

Denote by $\mathbb{S}^2$ the 2-dimensional sphere defined by $\{x = (x_1, x_2, x_3) \in \mathbb{R}^3 : x_1^2 + x_2^2 + x_3^2 = 1\}$. For a point A of $\mathbb{S}^2$, let A' be its opposite and the plane $\mathcal{H}_A$ be the plane orthogonal to AA' passing through A'. In the following, take $A = (0, 0, -1)$; then $\mathcal{H}_A$ is the set of all $x \in \mathbb{R}^3$ with $x_3 = 1$. If $B \in \mathbb{S}^2 - \{A\}$, then the intersection of the line AB with the plane $\mathcal{H}_A$ defines a point $f_A(B)$. This establishes a bijection, called a *Riemann map*, between $\mathbb{S}^2 - \{A\}$ and the plane $\mathcal{H}_A$; we can extend f_A to A by defining $f_A(A)$ to be the "point at infinity" of the plane $\mathcal{H}_A$.

Let G be a finite plane graph on $\mathbb{R}^2 \simeq \mathcal{H}_A$ and let $f_A^{-1}(G)$ denote its image in $\mathbb{S}^2$. The vertices of G correspond to points of the sphere $\mathbb{S}^2$, the edges of G correspond to non-intersecting curvilinear lines on $\mathbb{S}^2$ and the faces of G correspond to domains of $\mathbb{S}^2$ delimited by circuit of those lines. The exterior face of G corresponds to a domain of $\mathbb{S}^2$ containing A. Reciprocally, if we have a map M on the sphere, then we can find a point A, which does not belong to M and the corresponding plane $\mathcal{H}_A$, the image of M on $\mathcal{H}_A$ is a finite plane graph. So, by abuse of language, we will use the term "sphere" not only for the surface $\mathbb{S}^2$ but also for any combinatorial map on it, i.e. a finite plane graph.

A reader who is interested only in plane graphs, our main subject, can move now directly to Section 1.4. But, for full understanding of the toroidal case, we need maps in all their generality. For reference on Map Theory; see, for example, [BoLi95] and [MoTh01].

We will also work with maps having an infinite number of vertices, edges, and faces. The vertex-degrees will always be bounded by some constant; however, faces could have an infinity of edges. Amongst plane drawing of those maps, we will allow only *locally finite* ones, i.e. those admitting an embedding such that any bounded domain contains a finite number of vertices. Consider, for example, the map $\widetilde{\mathbb{Z}^2}$ obtained as the quotient map of square tiling $\mathbb{Z}^2$ by the translation operation $(x, y) \mapsto (x + 10, y)$. The map $\widetilde{\mathbb{Z}^2}$ is an infinite cylinder made of consecutive rings of ten 4-gons. We can draw those rings concentrically on the plane, but the resulting plane graph will not be locally finite.

1.2.2 Orientability and classification of surfaces

A *flag of a map* is a triple (v, e, f), where v is a vertex contained in the edge e and e is contained in the face f. Given a flag $F = (v, e, f)$, the flags differing from F only in v, e or f are denoted by $\sigma_0(F)$, $\sigma_1(F)$ and $\sigma_2(F)$, respectively. $\sigma_0(F)$ and $\sigma_1(F)$ are always defined over the flag-set $\mathcal{F}(M)$ of M but $\sigma_2(F)$ is not always defined if M is not closed.

The map M is called *oriented* if there exist a bipartition $\mathcal{F}_1$, $\mathcal{F}_2$ of $\mathcal{F}(M)$ such that, for any $(v, e, f) \in \mathcal{F}_1$, the flag $\sigma_i(v, e, f)$, if it exists, belongs to $\mathcal{F}_2$. We will be almost exclusively concerned with oriented maps.

The notion of orientation is easy to define algebraically but difficult to visualize, because closed non-orientable maps cannot be represented by a picture. Fortunately, this is easier for maps with boundaries; see Figure 1.1 for a non-orientable map, called *Möbius strip*. The non-orientability can be seen in the following way: moving along one side of the strip and doing a full circuit, we arrives at the other side of the strip. All boundary edges of a Möbius strip belong to a unique cycle; after adding a face to this cycle, we obtain the *projective plane* $\mathbb{P}^2$. The projective plane can also be obtained by taking a map on the sphere (like Dodecahedron) and identifying the opposite vertices, edges, or faces, i.e. taking the *antipodal quotient*.

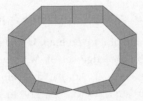

Figure 1.1 A Möbius strip

Given a surface S, we can add to it a *handle*:

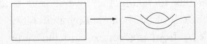

or a *cross-cap*.[1] The handle and cross-cap can be seen as a cylinder and a Möbius strip, respectively.

Consider now the classification of finite maps:

Theorem 1.2.1 *Any finite closed map is one of the following:*

1 *the sphere* $\mathbb{S}^2$ *(orientable),*
2 *the sphere* $\mathbb{S}^2$ *with g handles (orientable),*
3 *the sphere* $\mathbb{S}^2$ *with g cross-caps (non-orientable).*

---

[1] See, for example, http://mathworld.wolfram.com/Cross-Cap.html for pictures of cross-caps.

Theorem 1.2.1 is proved, for example, in [Mun71]. The number g above is called the *genus* of the map. All finite closed maps, that occur below, are:

1 the sphere $\mathbb{S}^2$ with $g = 0$ (orientable),
2 the *torus* $\mathbb{T}^2$ with $g = 1$ (orientable),
3 the *projective plane* $\mathbb{P}^2$ with $g = 1$ (non-orientable),
4 the *Klein bottle* $\mathbb{K}^2$ with $g = 2$ (non-orientable); one way to obtain the Klein bottle is to take the quotient of a torus $\mathbb{R}^2/\mathbb{Z}^2$ by the fixed-point-free automorphism $f(x, y) = (x + \frac{1}{2}, -y)$.

If M is a finite non-closed map, then we can add some faces along the boundary edges and obtain a closed map. So, finite non-closed maps are obtained by removing some faces of closed ones.

1.2.3 Fundamental groups, coverings, and quotient maps

Fix an orientation on every edge of a given map M and define the free group $G(\mathcal{M})$ with generators g_e indexed by the edge-set of M (see, for example, [Hum96] for relevant definitions in Group Theory). An *oriented path* $\mathcal{O}P = (v_1, v_2), \ldots, v_m$ is a sequence of vertices with v_i adjacent to v_{i+1}. For an edge $e_i = (v_i, v_{i+1})$, denote by $g(v_i, v_{i+1})$ the group element g_{e_i} if e_i is oriented from v_i to v_{i+1}, and $g_{e_i}^{-1}$ otherwise. Associate to the oriented path the product $g(\mathcal{O}P) = g(v_1, v_2)g(v_2, v_3)\ldots g(v_{m-1}, v_m)$.

Denote by $\mathcal{Z}_v(M)$ the set of all $g(\mathcal{O}P)$ with $\mathcal{O}P$ being the set of oriented closed paths starting and finishing at a given *base vertex* v. It is a group; reversing orientation corresponds to taking the inverse and product to concatenating closed paths. Given a face F of M, bounded by a circuit of vertices $(v_1, \ldots, v_{|F|})$, and an oriented path $\mathcal{O}P$ from the vertex v to the vertex v_1, consider a group element $g(\mathcal{O}P)g(v_1, v_2, \ldots, v_{|F|}, v_1)g(\mathcal{O}P)^{-1}$. Denote by $\mathcal{B}_v(M)$ the subgroup of $G(\mathcal{M})$ generated by all such elements. The *fundamental group* $\pi_1(M)$ is the quotient group of the group $\mathcal{Z}_v(M)$ by the normal subgroup $\mathcal{B}_v(M)$. Two oriented closed paths having a common vertex v are called *homotopic* if they correspond to the same element in the group $\pi_1(M)$. The group $\mathcal{B}_v(M)$ is the group of all elements homotopic to the *null path*, i.e. the path from v to v with 0 edges. If we replace the base vertex v by another base vertex w, then, for any oriented path $\mathcal{O}P$ from v to w, we have $\mathcal{Z}_v(M) = g\mathcal{Z}_w g^{-1}$ and $\mathcal{B}_v(M) = g\mathcal{B}_w g^{-1}$ with $g = g(\mathcal{O}P)$. So, the fundamental group depends on the base vertex, but only up to conjugacy. A map is called *simply connected* if $\pi_1(M)$ is trivial, i.e. every path is homotopic to the null path. This is equivalent to saying that every two paths with the same beginning and end can be continuously deformed one to the other.

See below three homotopic paths in the same map:

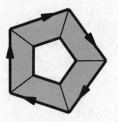 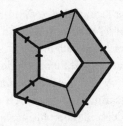 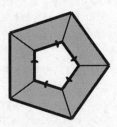

See below again three plane maps and a closed path represented in it:

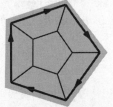 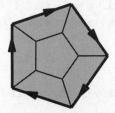 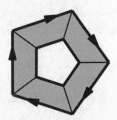

M_1 is simply connected M_2 is simply connected M_3 is not simply connected

In M_1 and M_2, the closed path is homotopic to a null path. In M_1, this cycle is the boundary of a face, while in M_2, the closed path is the boundary of all faces put together. More generally, a plane graph and a finite plane graph minus a face are simply connected. But the closed path in M_3 is not homotopic to a null path. Actually, this closed path is a generator of the fundamental group $\pi_1(M_3) \simeq \mathbb{Z}$.

Given two maps M and M', a *cell-homomorphism of maps* $\phi : M \to M'$ is a function that maps vertices, edges, and faces of M to the ones of M', while preserving the incidence relations. An *isomorphism* is a cell-homomorphism that is bijective. If $M = M'$, it is called an *automorphism*; the set of all automorphism of a map M is called the *symmetry group* of M. An automorphism f of a map M is called *fixed-point-free* if f is the identity or, for every vertex, edge, face of M, its image in f is different from it. If G is a group of fixed-point-free automorphisms of a map M, then M/G is the *quotient map* of M by G. Its vertices, edges, and faces are formed by orbits of vertices, edges, and faces of M (by G) with the incidence relations being induced by the ones of M.

The quotient of a map can be a map with loops and multiple edges. Consider, for example, the 4-valent plane tiling $\{4, 4\}$ (see Section 1.5) formed by 4-gons and the group $\mathbb{Z}^2$ acting by translations on it. There is one orbit of vertices, two orbits of edges, and one orbit of faces under $\mathbb{Z}^2$; so the quotient $\{4, 4\}/\mathbb{Z}^2$ is a torus represented by a single vertex and two loops.

For a vertex v (or edge e, or face f), the *standard neighborhood* $N(v)$ is the set of all vertices, edges, and faces incident to v. A *local isomorphism* is a continuous mapping $\phi: M \rightarrow M'$ such that, for any vertex $v \in M$ (or edge, or face), the mapping from $N(v)$ to $N(\phi(v))$ is bijective. A *covering* is a local isomorphism such that for every vertex $v' \in M'$ (or edge, or face) and $w' \in N(v')$, if $\phi^{-1}(v') = (v_i)_{i \in I}$, we have an element $w_i \in N(v_i)$ such that $w_i \neq w_j$ if $i \neq j$ and $\phi^{-1}(w') = (w_i)_{i \in I}$.

If $\mathcal{O}P' = (v'_1, \ldots, v'_m)$ is an oriented path in M', ϕ is a covering and v_1 is a vertex in M with $\phi(v_1) = v'_1$, then there exists a unique oriented path $\mathcal{O}P = (v_1, \ldots v_m)$ in M such that $\phi(\mathcal{O}P) = \mathcal{O}P'$. A *deck automorphism* is an automorphism u of M such that $\phi \circ u = \phi$, u is necessarily fixed-point-free. If $v' \in M'$, then, for any two $v_1, v_2 \in \phi^{-1}(v')$, there exists a deck automorphism u such that $u(v_1) = v_2$.

Given a map M, its *universal cover* is a simply connected map $\widetilde{M}$ (unique up to isomorphism) with a covering $\phi : \widetilde{M} \rightarrow M$. The map $\widetilde{M}$ is finite if and only if M and $\pi_1(M)$ are finite. The fundamental group $\pi_1(M)$ is isomorphic to the group of deck automorphisms of ϕ. If H is a subgroup of a group G, then its *normalizer*, denoted by $N_G(H)$, is defined as:

$$N_G(H) = \{x \in G \ : \ xhx^{-1} \in H \text{ for all } h \in H\}.$$

The group $Aut(M)$ of automorphisms of M is identified with the quotient group:

$$N_{Aut(\widetilde{M})}(\pi_1(M))/\pi_1(M).$$

The simplest and most frequently used case is when M is a closed finite map on the sphere. In this case π_1 is trivial and we can represent the map nicely on the plane with a face chosen to be exterior. An infinite locally finite closed simply connected map can be represented on the plane. In this case, there is no exterior face and the map fills completely the plane.

A closed torus M can be represented as a 3-dimensional figure projected on to the plane, but this is not very practical. We represent its universal cover $\widetilde{M}$ as a plane having two periodicity directions, i.e. a *2-periodic plane graph*. The group $\pi_1(M)$ is isomorphic to $\mathbb{Z}^2$ and it is represented on $\widetilde{M}$ as a group of translation symmetries.

By choosing a *finite index subgroup* H of the group G (i.e. such that there exist $g_1, \ldots, g_m \in G$ with $G = \cup_i g_i H$) of deck transformations and taking the quotient, we can obtain a bigger torus; such tori have a translation subgroup, which is isomorphic to the quotient G/H.

On the other hand, given a torus with non-trivial translation group, there exists a unique *minimal torus* with the same universal cover and trivial translation subgroup. Those minimal tori correspond, in a one-to-one way, to periodic tilings of the plane.

1.2.4 Homology and Euler–Poincaré characteristic

Given a map M, assign an orientation on each of its edges and form a $\mathbb{Z}$-module $C_1(M)$ using this set of oriented edges as basis. The $\mathbb{Z}$-module $Z_1(M)$ is the submodule of $C_1(M)$ generated by the set of closed oriented cycles of M. Given any face of M, associate to it the set of incident edges in clockwise orientation; the generated $\mathbb{Z}$-module is denoted by $B_1(M)$. It is easy to see that $B_1(M)$ is a submodule of $Z_1(M)$.

The *homology group* $H_1(M)$ is the quotient of $Z_1(M)$ by its subgroup $B_1(M)$. Again, we refer to Algebraic Topology textbooks for details. If M is a torus, then $H_1(M)$ is isomorphic to $\pi_1(M)$.

If M is an orientable finite closed map, then $H_1(M)$ is isomorphic to $\mathbb{Z}^{2g}$, where g is the genus of M. The *Euler–Poincaré characteristic* of a finite map M is defined as $\chi(M) = v - e + f$ with v the number of vertices, e the number of edges, and f the number of faces.

Theorem 1.2.2 *For a finite closed map M of genus g it holds:*

(i) if M is orientable, then $\chi(M) = 2 - 2g$,
(ii) if M is non-orientable, then $\chi(M) = 2 - g$.

This theorem is the main reason why we are able to use topology in dimension two to derive non-trivial combinatorial results.

Theorem 1.2.3 *Let G be a k-valent closed map on a surface M; then:*

(i) the following Euler formula *is valid:*

$$\sum_{j \geq 2} p_j(2k - j(k-2)) = 2k\chi(M), \tag{1.1}$$

where p_i is the number of i-gonal faces.
(ii) If G has no 2-gonal faces, then $k \leq 5$ if M is a sphere and $k \leq 6$ if M is a torus.

Proof. (i) The relation $2e = kv$ allows us to rewrite the Euler–Poincaré characteristic as:

$$\chi(M) = \left(\frac{2}{k} - 1\right)e + \sum_{i \geq 2} p_i .$$

Using that $2e = \sum_{i \geq 2} ip_i$ in the above equation, yields the result.

If $j \geq 3$, then $2k - j(k-2) \leq 0$ for $k \geq 6$ and $2k - j(k-2) < 0$ for $k \geq 7$. Assertion (ii) is deduced by noticing that $\chi = 2, 0$ for sphere, and torus, respectively. $\qquad\square$

1.3 Representation of maps

A *polytope P* is the convex hull of a finite set of points in $\mathbb{R}^n$; its *dimension* is the dimension of the smallest affine space containing it. We assume it to be full-dimensional. A linear inequality $f(x) \geq 0$ is called *valid* if it holds for all $x \in P$. A *face* of P is a set of the form $\{x \in P : f(x) = 0\}$ with $f \geq 0$ being a valid inequality.

We will consider only 3-dimensional polytopes; they are called *polyhedra*. Their 0-dimensional faces are called *vertices* and the 2-dimensional faces are called just *faces*. Two vertices are called *adjacent* if there exist an *edge*, i.e. a 1-dimensional face containing both of them. The *skeleton* of a polyhedron is the graph formed by all its vertices with two vertices being *adjacent* if they share an edge. This graph is 3-connected and admits a plane embedding.

Given a polyhedron P, its skeleton $skel(P)$ is a planar graph. Furthermore, for any face F of P, we can draw $skel(P)$ on the plane so that F is the exterior face of the plane graph. Those drawings are called *Schlegel diagrams* (see, for example, [Zie95]). Steinitz proved that a finite graph is the skeleton of a polyhedron (and so, an infinity of polyhedra with the same skeleton) if and only if it is planar and 3-connected (see [Ste22], [Zie95, Chapter 4] and [Grü03] for a clarification of the history of this theorem).

A *Riemann surface* is a 2-dimensional compact differentiable surface, together with an infinitesimal element of length (see textbooks on differential and Riemannian geometry, for example, [Nak90]). The *curvature* $K(x)$ at a point x is the coefficient α in the expansion:

$$Vol(D(x,r)) = \pi r^2 - \alpha r^4 + o(r^4)$$

with $D(x,r)$ being the disc consisting of elements at distance at most r from x. The curvature of a Riemann surface S satisfies the *Gauss–Bonnet formula*:

$$\int_S K(x)dx = 2\pi(1 - g).$$

All Riemann surfaces, considered in this section, will be of constant curvature. If a surface has constant curvature, then, for any two points x and y of it, there exist two neighborhoods N_x and N_y and a local isometry ϕ mapping x to y and N_x to N_y. Hence, Riemann surfaces of constant curvature do not have local invariants and the only invariants they have are global (see, for example, [Jos06]). For genus zero, the curvature has positive integral. Up to rescaling, we can assume that this curvature is 1. There is only one such Riemann surface: the sphere $\mathbb{S}^2$. For genus 1, the curvature has integral 0 and so it is 0. The Teichmüller space T_1 has dimension 2, which means that Riemann surfaces of genus 1 are parametrized by two real parameters. Geometrically, they are very easy to depict: take $\mathbb{R}^2$ and quotient it by a group $v_1\mathbb{Z} + v_2\mathbb{Z}$. For higher genus, the situation is much more complicated.

Given a map M, its *circle-packing representation* (see [Moh97]) is a set of disks on a Riemann surface Σ of constant curvature, one disk $D(v, r_v)$ for each vertex v of M, such that the following conditions are fulfilled:

1 the interior of disks are pairwise disjoint,
2 the disk $D(u, r_u)$, $D(v, r_v)$ touch if and only if uv is an edge of M.

Simultaneous circle-packing representations of a map M and its dual M^* are called *primal-dual circle representation* of M if it holds:

1 If $e = (u, v)$ is an edge of M and u^*, v^* are the corresponding vertices in M^*, then the disks $D(u, r_u)$, $D(v, r_v)$ corresponding to e touch at the same point as the disks $D(u^*, r_{u^*})$, $D(v^*, r_{v^*})$.
2 The disks $D(u, r_u)$, $D(u^*, r_{u^*})$ cross at that point perpendicularly.

See Figure 1.2 for an illustration of this feature and an example of a primal-dual circle representation.

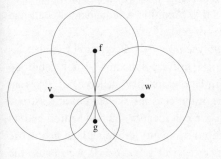

The local picture of a primal-dual circle
representation

The edges, vertex circles and face circles of
a primal-dual representation

Figure 1.2 Illustration of primal-dual circle representations

A map M is called *reduced* (see [Moh97, Section 3]) if its universal cover is 3-connected and is a cell-complex. It is shown in [Moh97, Corollary 5.4] that reduced maps admit unique primal-dual circle packing representations on a Riemann surface of the same genus; moreover, a polynomial time algorithm allows one to find the coordinates of those points relatively easily. This means that the combinatorics of the map determines the structure of the Riemann surface.

The primal-dual representations allow us to demonstrate the uniqueness of the representation of a given map. But, actually, for finite plane graphs, i.e. spheres, we use the program CaGe ([BDDH97]), which does not apply primal-dual representations. CaGe draws a Schlegel diagram of finite plane graphs on the plane, which we consider (see, Subsection 1.2.1); sometimes, in order to show the symmetry, it is a good idea to have one vertex or edge of the graph at infinity (see, for example, Figure 2.1, Section 9.1, Chapter 10).

For tori, we take their universal covers on the plane and use the primal-dual representation obtained from the program TorusDraw ([Dut04b]). For the projective plane $\mathbb{P}^2$, we take its universal cover, which is the sphere, and draw a circular frame, where antipodal boundary points are to be identified (see Figure 3.1). For the Klein bottle $\mathbb{K}^2$, we draw a rectangle with boundary identifications (see Figure 3.1).

1.4 Symmetry groups of maps

For finite closed maps on the sphere, there is a complete classification of possible symmetry groups. For finite closed maps on the torus, we can describe the possible symmetry groups of their universal covers.

We remind that an *automorphism* of a simple graph is a permutation of the vertices preserving adjacencies between vertices. For plane graphs, we require also that faces are sent to faces but for 3-connected graphs this condition is redundant. Recall that $Aut(G)$ denotes the group of automorphisms of G.

The *automorphism group $Aut(P)$ of a polyhedron P* is the group of isometries preserving P. This group of isometrics $Aut(P)$ of a polyhedron P is a subgroup of the group of symmetries $Aut(G)$ of the plane graph (the skeleton) of P. Mani [Man71] proved that any 3-connected plane graph G is the skeleton of at least one polyhedron P with $Aut(G) = Aut(P)$. So, we can identify the polyhedron and its skeleton, as well as the algebraic (permutation) symmetry group and the geometric (isometry) point group. For closed maps on surfaces of genus $g > 0$, we can use the primal-dual representation of the preceding section to prove that the symmetry group of the map can be realized as the isometry group of the surface.

A *point group* is a finite subgroup of the group $O(3)$ of isometries of the space $\mathbb{R}^3$, fixing the origin. Those groups have been classified a long time ago. They are described, for example, in [FoMa95], [Dut04a] using the Schoenflies notation, which is used here (for the Hermann–Maugin notation, see [OKHy96, Chapter 3], [Dut]). Every symmetry group of a finite plane graph is identified with a point group.

The list of point groups is split into two classes: seven infinite families and seven sporadic cases. Every point group contains a normal subgroup formed by its rotations.

We now list the infinite series of point groups:

1 The group C_m is the cyclic group of rotations by angle $\frac{2\pi}{m}k$ with $0 \leq k \leq m - 1$ around a fixed axis Δ.

2 The group C_{mh} is generated by C_m and a reflection of plane P with P being orthogonal to Δ.

3 The group C_{mv} is generated by C_m and a reflection of plane P with P containing Δ.

4 The group D_m is generated by C_m and a rotation by angle π, whose axis is orthogonal to Δ.

5 The group D_{mh} is generated by C_{mv} and a rotation by angle π, whose axis is orthogonal to Δ and contained in a reflection plane.

6 The group D_{md} is generated by C_{mv} and a rotation by angle π, whose axis is orthogonal to Δ and going between two reflection planes.

7 For any even positive integer m, the group S_m is the cyclic group generated by the composition of a rotation by angle $\frac{2\pi}{m}$ with axis Δ and a reflection of plane P with P being orthogonal to Δ.

The particular cases C_1, $C_s = C_{1h} = C_{1v}$ and $C_i = S_2$ correspond to the *trivial group*, the *plane symmetry group*, and the *central symmetry inversion group*, respectively.

The point groups T_d, O_h, and I_h are the respective symmetry group of Tetrahedron, Cube, and Icosahedron; the point groups T, O, and I are their respective normal subgroup of rotations. The point group T_h is generated by T and the central symmetry inversion of the centre of the Isobarycenter of the Tetrahedron.

We now list the *strip groups* (also called *frieze groups*); this part follows [Cla, Dut]. In their representation given below, the lines correspond to line reflections, while the dashed lines correspond to glide reflection, i.e. reflection followed by translation. For $n \geq 3$ the center of rotations of order n is represented by a regular n-gon, while centers of rotation of order 2 are represented by small parallelogram. Furthermore, we indicate a translation vector to give the translational symmetries of the group. There are seven strip groups, since every such group corresponds to one of the infinite series of symmetry groups of plane graphs (imagine a graph of symmetry $C_m, \ldots, S_m$ and let m go to infinity, the figure becoming a strip).

1 $p111$ ($=C_\infty$); it has only translational symmetry:

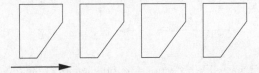

2 $p1m1$ ($=C_{\infty h}$); it has a horizontal mirror symmetry:

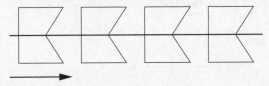

3 $pm11$ ($=C_{\infty v}$); it has vertical mirror symmetries:

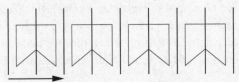

4 $p112$ $(=D_\infty)$; it has only 2-fold rotations, spaced at half the translation distance:

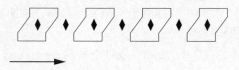

5 $pmm2$ $(=D_{\infty h})$; it has vertical mirror symmetries, horizontal mirror symmetry, and 2-fold rotations where the mirror intersect:

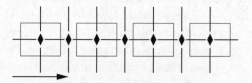

6 $pma2$ $(=D_{\infty d})$; it has a glide reflections, with alternating vertical mirror and 2-fold rotations:

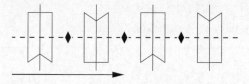

7 $p1a1$ $(=S_\infty)$; it has glide reflections, half the length of the translation distance:

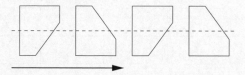

For 2-periodic plane maps, there are 17 possible symmetry groups, called *wallpaper groups* (or *plane crystallographic groups*); this part follows [OKHy96, Chapter 1]. In the representation given below, we give the fundamental domain as a parallelogram (or rectangle, or square), which tiles the plane under translation. We also give the rotation axis, reflection, and glide reflections with the same conventions as for strip groups:

1 Wallpaper groups without rotations:

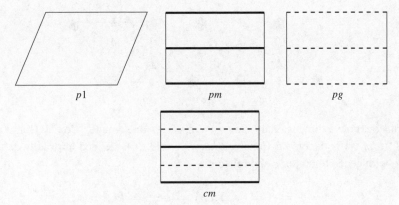

2 Wallpaper groups with rotations of order 2:

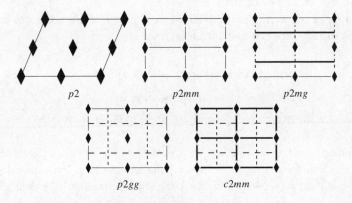

3 Wallpaper groups with rotations of order 3:

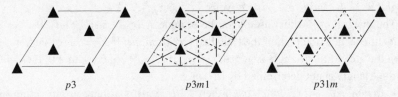

4 Wallpaper groups with rotations of order 4:

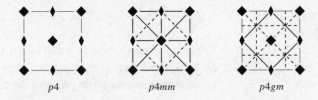

5 Wallpaper groups with rotations of order 6:

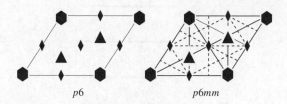

p6   *p6mm*

Consider now 2-dimensional *Coxeter groups* (see, for example, [Cox73, Hum90]). By $T^*(l, m, n)$ is denoted a Coxeter *triangle group*. It is defined abstractly as the group with generators a, b, c and relations:

$$a^2 = b^2 = c^2 = 1 \text{ and } (ab)^l = (ac)^m = (bc)^n = 1.$$

Denote by $\alpha(l, m, n)$ the number $\frac{1}{l} + \frac{1}{m} + \frac{1}{n} - 1$.

The group $T^*(l, m, n)$ can be realized as a group of isometries of a simply connected surface X of constant curvature, where:

- $X = \mathbb{S}^2$, i.e. the 2-dimensional sphere, if $\alpha(l, m, n) > 0$,
- $X = \mathbb{R}^2$, i.e. the Euclidean plane, if $\alpha(l, m, n) = 0$,
- $X = \mathbb{H}^2$, i.e. the hyperbolic plane, if $\alpha(l, m, n) < 0$ (see [Cox98]).

For a group G acting on a surface X, a *fundamental domain* is a closed set $\mathcal{D}$, the orbit of which under G tiles X, i.e. every point belongs to at least one image of $\mathcal{D}$ under G and the intersection of any two domains in the orbit have empty interior. For the group $T^*(l, m, n)$ acting on X, we can find a fundamental domain, which is a triangle ABC, so that $a, b,$ and c are reflections along the sides BC, AC, AB. The angles at $A, B,$ and C are, respectively, $\frac{\pi}{n}, \frac{\pi}{m},$ and $\frac{\pi}{l}$.

The integral of the curvature over the triangle ABC is $\pi\alpha(l, m, n)$. If $X = \mathbb{S}^2$ of curvature 1, then the area of the triangle is $\pi\alpha(l, m, n)$. If $X = \mathbb{H}^2$ of curvature -1, then the area of the triangle is $-\pi\alpha(l, m, n)$. If $X = \mathbb{R}^2$ of curvature 0, then the area of the triangle is not determined by $\alpha(l, m, n)$.

If $\alpha(l, m, n) > 0$ (*elliptic case*), then $T^*(l, m, n)$ is a finite group acting on the sphere $\mathbb{S}^2$; the only possibilities for (l, m, n) are:

1 $(2, 2, n)$ for $n \geq 2$ with $T^*(2, 2, n) \simeq D_{nh}$ being the automorphism group of the regular n-gon,
2 $(2, 3, 3)$ with $T^*(2, 3, 3) \simeq T_d$ being the automorphism group of the Tetrahedron,
3 $(2, 3, 4)$ with $T^*(2, 3, 4) \simeq O_h$ being the automorphism group of the Octahedron,
4 $(2, 3, 5)$ with $T^*(2, 3, 5) \simeq I_h$ being the automorphism group of the Icosahedron.

If $\alpha(l, m, n) = 0$ (*parabolic case*), then $T^*(l, m, n)$ is a wallpaper group acting on the plane $\mathbb{R}^2$; the only possibilities for (l, m, n) are:

1. $(2, 4, 4)$ with $T^*(2, 4, 4) \simeq p4mm$ being the automorphism group of the square tiling $\mathbb{Z}^2$ of $\mathbb{R}^2$,
2. $(2, 3, 6)$ with $T^*(2, 3, 6) \simeq p6mm$ being the automorphism group of the triangular tiling of $\mathbb{R}^2$.

If $\alpha(l, m, n) < 0$ (*hyperbolic case*), then there is an infinity of possibilities for (l, m, n).

By $T(l, m, n)$ is denoted the normal subgroup of index two of $T^*(l, m, n)$ formed by rotations of X (i.e. orientation-preserving elements of $T^*(l, m, n)$); see, for example, [Mag74].

There are the following relations with strip groups:

$$T^*(2, 2, \infty) = pmm2 \text{ and } T(2, 2, \infty) = p112 \approx pm11 \approx pma2.$$

Remark that $p1m1$ also has index two in $T^*(2, 2, \infty)$, but it is not isomorphic to $T(2, 2, \infty)$. Recall also that $T(2, 3, \infty) \approx PSL(2, Z)$ (the *modular group*) and $T^*(2, 3, \infty) \approx SL(2, Z)$. The groups $T(l, m, n)$ contain a large number of non-isomorphic subgroups of finite index, which renders futile any hope of classification of possible symmetry groups of maps on orientable surfaces of genus $g \geq 2$. However, in many cases considered here, the groups $T(l, m, n)$ and $T^*(l, m, n)$ are sufficient for our purposes.

Remark 1.4.1 *In Chapters 2, 4, and 7, a pair (r, q) will be called* elliptic, parabolic, *or* hyperbolic *if $rq < 2(r + q)$, $rq = 2(r + q)$, $rq > 2(r + q)$, respectively. This is equivalent to $\alpha(2, r, q) > 0$, $\alpha(2, r, q) = 0$, $\alpha(2, r, q) < 0$, respectively. In Chapter 4, the link will be made direct, since every (r, q)-polycycle has a cell-homomorphism into the tiling $\{r, q\}$ (see Section 1.5), whose symmetry group is $T^*(2, r, q)$.*

In Chapter 7, there will be no such link; however, the pair (r, q) such that there is a countable number of elementary (r, q) polycycles are exactly the elliptic ones; see Theorem 7.2.1, 7.4.1, 7.5.1, 7.6.1. But we have no general explanation of this fact.

A k-valent sphere, whose faces have gonality a or b, is called a $(\{a, b\}, k)$-sphere (see Chapter 2). We call the parameters $(\{a, b\}, k)$ elliptic, parabolic, hyperbolic, according to the sign of $\alpha(2, b, k)$. This sign has a consequence for the finiteness and growth of the number of graphs in the class of $(\{a, b\}, k)$-spheres. Here, the link is provided by the Euler formula (1.1).

1.5 Types of regularity of maps

We list here some classification results for maps on the sphere or on the plane.

A map is *regular* if its automorphism group act *transitively* on flags, i.e. if, for any two flags f and f', there is an automorphism ϕ with $\phi(f) = f'$.

We are now in position to define formally the *regular tiling* $\{r, q\}$.

Definition 1.5.1 *The triangle group $T^*(2, r, q)$ act on X (with $X = \mathbb{S}^2$, $\mathbb{R}^2$, $\mathbb{H}^2$) if $\alpha(2, r, q) > 0$, $=0 < 0$, respectively. The fundamental domain $\mathcal{D}$ (triangle ABC) has angles $\frac{\pi}{q}$, $\frac{\pi}{r}$, and $\frac{\pi}{2}$ at A, B, C, respectively.*

To every point B' in the orbit of B under $T^*(2, r, q)$, associate an r-gon formed by all $2r$ triangles containing B'. The tiling $\{r, q\}$ is the set of all those r-gons; it satisfies:

- $\{r, q\}$ *is a q-valent tiling of X by r-gons.*

- *The group $T^*(2, r, q)$ act regularly on $\{r, q\}$, i.e. any two flags of $\{r, q\}$ are equivalent under $T^*(2, r, q)$.*

- *The curvature of those r-gons is $2r\alpha(2, r, q)$.*

See below the Platonic (regular) polyhedra P with their groups $Aut(P)$:

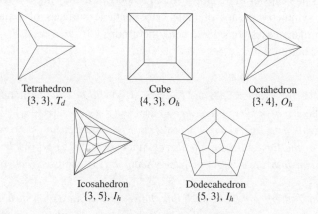

Tetrahedron	Cube	Octahedron
$\{3, 3\}$, T_d	$\{4, 3\}$, O_h	$\{3, 4\}$, O_h
Icosahedron	Dodecahedron	
$\{3, 5\}$, I_h	$\{5, 3\}$, I_h	

One has the duality $Cube = (Octahedron)^*$ and $Icosahedron = (Dodecahedron)^*$, while *Tetrahedron* is self-dual.

Denote by $Bundle_m$, $m \geq 2$, the plane graph with two vertices and m edges between them (so, m 2-gonal faces). The plane graph $Bundle_m$, which is dual to m-gon, has the symmetry group $D_{mh} = T^*(2, 2, m)$ and it is a regular map, which is not a cell-complex.

Three regular tilings of the plane $\mathbb{R}^2$ are:

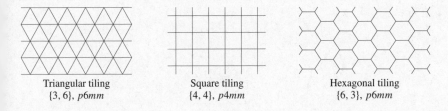

Triangular tiling	Square tiling	Hexagonal tiling
$\{3, 6\}$, *p6mm*	$\{4, 4\}$, *p4mm*	$\{6, 3\}$, *p6mm*

The triangular and hexagonal tilings are dual to each other, while the square tiling is self-dual. For other parameters (r, q), the tiling $\{r, q\}$ lives in hyperbolic space $\mathbb{H}^2$ (see many pictures in [Eps00]) but we will not need it.

Given two circuits $U = (u_1, \ldots, u_m)$ and $V = (v_1, \ldots, v_m)$, an *Prism$_m$* (*m-sided prism*), for $2 \leq m \leq \infty$, is a 3-valent plane graph, where each u_i is joined to v_i by an edge. Its symmetry group is D_{mh} if $m \neq 4$ and O_h if $m = 4$.

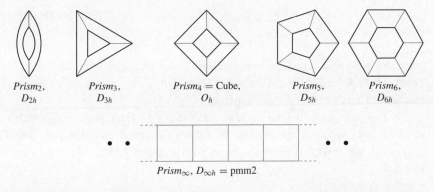

Prism$_2$, D_{2h}	*Prism$_3$*, D_{3h}	*Prism$_4$* = Cube, O_h	*Prism$_5$*, D_{5h}	*Prism$_6$*, D_{6h}

Prism$_\infty$, $D_{\infty h}$ = pmm2

An *APrism$_m$* (*m-sided antiprism*), for $2 \leq m \leq \infty$, is a 4-valent plane graph formed by adding to two circuits U and V the cycle $(u_1, v_2, u_2, v_3, \ldots, v_m, u_m, v_1, u_1)$. Its symmetry group is D_{md} if $m \neq 3$ and O_h if $m = 3$.

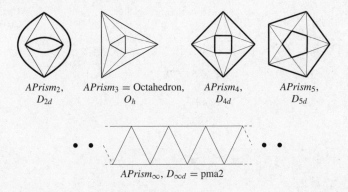

APrism$_2$, D_{2d}	*APrism$_3$* = Octahedron, O_h	*APrism$_4$*, D_{4d}	*APrism$_5$*, D_{5d}

APrism$_\infty$, $D_{\infty d}$ = pma2

For $2 \leq m \leq \infty$, denote by *snub Prism$_m$* a 3-valent plane graph with two m-gonal faces separated by two m-rings of 5-gons. Its symmetry group is D_{md} if $m \neq 5$ and I_h if $m = 5$.

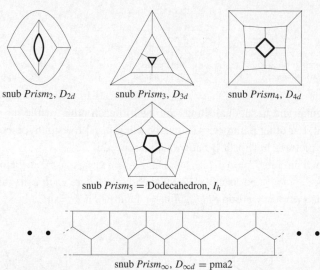

snub *Prism$_2$*, D_{2d} snub *Prism$_3$*, D_{3d} snub *Prism$_4$*, D_{4d}

snub *Prism$_5$* = Dodecahedron, I_h

snub *Prism$_\infty$*, $D_{\infty d}$ = pma2

Snub *Prism$_3$* is also called *Dürer octahedron* (see it on the painting *Melencolia I* by Dürer, 1514, depicting the muse of mathematics at work) and it can be obtained by truncating Cube on two opposite vertices.

For $2 \leq m \leq \infty$, denote by *snub APrism$_m$* a 5-valent plane graph with two m-gonal faces separated by $6m$ 3-gonal faces as in examples below (see [DGS04, page 119], for formal definition). Its symmetry group is D_{md} if $m \neq 3$ and I_h if $m = 3$.

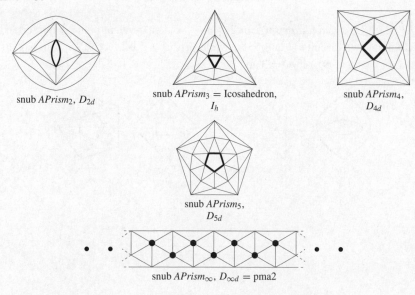

snub *APrism$_2$*, D_{2d} snub *APrism$_3$* = Icosahedron,   snub *APrism$_4$*,
                                I_h D_{4d}

snub *APrism$_5$*,
D_{5d}

snub *APrism$_\infty$*, $D_{\infty d}$ = pma2

Snub $APrism_4$ is one of 92 *regular-faced*[2] polyhedra called *snub square antiprism* (see [Joh66] and [Zal69]).

A map is called *Archimedean* if its symmetry group acts transitively on vertices but the map itself is not regular. Any Archimedean polyhedron belongs either to 13 sporadic examples, or to one of two infinite series $Prism_m$ and $APrism_m$ for $m \geq 3$. Like the Platonic polyhedra, they are known since the antiquity and were rediscovered during the Renaissance; Kepler ([Kep1619]) gave them their modern names. Their duals are called *Catalan polyhedra*. The Archimedean tilings of the plane are also classified. There are eight such maps; their duals are called *Laves tilings*.

All 92 regular-faced polyhedra were found by the work of many people, especially of Johnson and Zalgaller (see, for example, [Joh66, Zal69]). The eight (namely, Tetrahedron, Octahedron, Icosahedron and duals of $Prism_3$ and four ($\{4, 5\}$, 3)-spheres from Figure 2.1) whose faces are regular triangles, are called *deltahedra*. A *mosaic* is a tiling of the Euclidean plane by regular polygons. All (165) mosaics are classified in [Cha89].

1.6 Operations on maps

A *decorated* $\{r, q\}$ is a map obtained by adding some vertices and edges to the regular tiling $\{r, q\}$; see in Tables 9.1 and 9.3 many such decorations.

We list here some operations transforming a map M into another map M'.

Truncation and capping: The *truncation of M at a vertex v* of degree m consists of replacing the vertex v by a m-gonal face. The *capping* is, in a sense, dual to truncation, it consists of adding a new vertex v to a face F of M such that v is adjacent to all vertices of F, i.e. putting a pyramid on F.

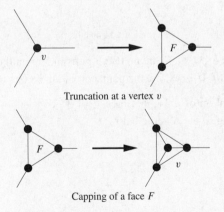

Truncation at a vertex v

Capping of a face F

[2] http://mathworld.wolfram.com/JohnsonSolid.html

The *truncation of M*, respectively, *capping of M* consist of performing truncation, respectively, capping of all vertices, respectively, faces of M. The *t-capping of M* is obtained by the capping of t distinct faces of M. The *b-cap of M* is the map obtained by capping all b-gonal faces of M.

Elongation: Let C be a simple circuit of adjacent vertices in M. An *elongation of M along C* consists of replacing C by a ring of 4-gons (see a related notion of *central circuit* in Chapter 2). The elongation of M along a circuit C, bounding a face F, means adding a prism on F.

m-**halving**: Given an even number m, a *m-halving of M* is obtained by putting a new edge, connecting the mid-points of opposite edges, on each m-gon.

4-triakon: The *4-triakon of a 3-gonal face F* of M is obtained by partitioning F into three 4-gons according to the scheme below:

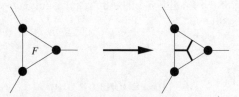

4-triakon of a 3-gonal face F

The *4-triakon of M* consists of performing 4-triakon of all 3-gons of M.

Pentacon: The *pentacon of a 5-gonal face F of M* consists of partitioning F into six 5-gons according to the scheme below:

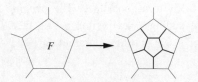

Pentacon of a 5-gonal face F

If S is a set of 5-gons of M such that no two 5-gons in S are adjacent, then denote by $P_S(M)$ the *pentacon of M on S*, i.e. the pentacon of all 5-gons in S.

5-triakon: The 5-triakon of a 3-gonal face F of M consists of partitioning F into nine 5-gons according to the scheme below:

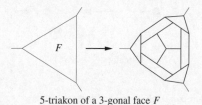

5-triakon of a 3-gonal face F

Actually, the 3-gons to which we will apply this construction come from the truncation of a set of vertices S of a map M. The result of this operation (truncation followed by 5-triakon of new 3-gonal faces) is denoted by $T_S(M)$ and is called 5-*triakon of M on S*.

Note that last two operations amount to replacing a face by $(5, 3)$-polycycles A_5, A_3 (see Chapter 7 and Figure 7.2), respectively.

2

Two-faced maps

Call a *two-faced map* and, specifically, $(\{a, b\}, k)$-*map* any k-valent map with only a- and b-gonal faces, for given integers $2 \leq a < b$. We will also use terms $(\{a, b\}, k)$-*sphere* (moreover, $(\{a, b\}, k)$-*polyhedron* if it is 3-connected) or $(\{a, b\}, k)$-*torus* for maps on sphere $\mathbb{S}^2$ or torus $\mathbb{R}^2$, respectively. Call $(\{a, b\}, k)$-*plane* any infinite k-valent plane graph with a- and b-gonal faces and without exterior faces. More generally, for $R \subset \mathbb{N} - \{1\}$, call (R, k)-*map* a k-valent map whose faces have gonalities $i \in R$.

When presenting a $(\{a, b\}, k)$-sphere in a drawing, we will indicate its number of vertices and its Schoenflies symmetry group (see Section 1.4). The notation (v, p_a, p_b) under picture of a minimal $(\{a, b\}, k)$-torus indicate its number of vertices, its number of a- and b-gonal faces; we also indicate its wallpaper symmetry group (see Section 1.4).

Call *corona* (of a face) the cyclic sequence of gonalities of all its consecutive neighbors. The *corona* of a vertex is the sequence of gonalities of all consecutive faces containing it. Recall that v, e, and f denote the number of vertices, edges, and faces, respectively, of a given finite map. Denote by p_i the number of its i-gonal faces and by e_{a-b}, e_{a-a} the number of $(a$-$b)$-*edges*, $(a$-$a)$-*edges*, i.e. edges separating a- and b-gonal faces or, respectively, a-gonal faces. Euler Formulas (1.1) for $(\{a, b\}, k)$-sphere and $(\{a, b\}, k)$-torus are:

$$p_a(2k - a(k - 2)) + p_b(2k - (k - 2)b) = 4k,$$
$$p_a(2k - a(k - 2)) + p_b(2k - (k - 2)b) = 0.$$

We can interpret the quantity $2k - b(k - 2)$ as the curvature of the faces of gonality b; Euler formula is the condition that the total curvature is a constant, equal to $4k$, for k-valent plane graphs. This curvature has an interpretation and applications in Computational Group Theory, see [Par06] and [LySc77, Chapter 9].

A pair (R, k) is called *elliptic, parabolic, hyperbolic* if $\frac{1}{r} + \frac{1}{k} > \frac{1}{2}, = \frac{1}{2}, < \frac{1}{2}$, where $r = \max_{i \in R} i$, respectively.

24

The $(\{5, 6\}, 3)$-spheres are called *fullerenes* in Organic Chemistry, where fullerenes and other two-faced maps are prominent molecular models. The $(\{5, 7\}, 3)$-spheres are called, in chemical context, *azulenoids* (see Figure 7.1 for *azulene* and [BCC96]).

The $(\{a, b\}, k)$-spheres with elliptic $(\{a, b\}, k)$, i.e. with $b(k - 2) < 2k$, are listed in Figure 2.1.

We will mainly consider in this chapter $(\{a, b\}, k)$-spheres with parabolic $(\{a, b\}, k)$, i.e. with $2k = b(k - 2)$; for those the number p_a of a-gonal faces remains fixed:

$$p_a = \frac{4k}{2k - a(k - 2)}.$$

In other words, every a-gonal face of G has positive curvature and every b-gonal face has zero curvature.

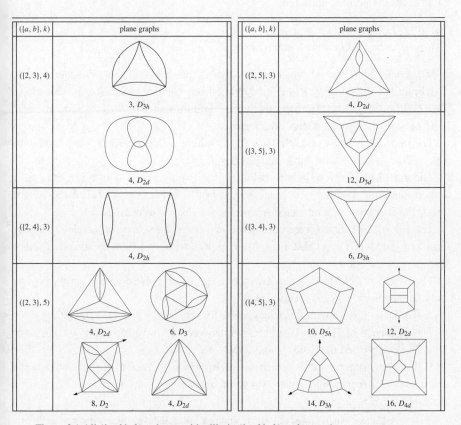

Figure 2.1 All $(\{a, b\}, k)$-spheres with elliptic $(\{a, b\}, k)$ and $p_b \geq 1$

Is is easy to see that the only (seven) solutions are $(\{a, b\}, k) = (\{2, 6\}, 3)$, $(\{3, 6\}, 3)$, $(\{4, 6\}, 3)$, $(\{5, 6\}, 3)$, $(\{2, 4\}, 4)$, $(\{3, 4\}, 4)$, and $(\{2, 3\}, 6)$. [1]

If the number of vertices becomes large, then those plane graphs are formed by a few faces of gonality a in a sea of faces of gonality b. But there is a unique way to have k-valent tiling of the plane with faces of gonality b: the regular 2-periodic tiling $\{b, k\}$ of $\mathbb{R}^2$. The case $(\{5, 6\}, 3)$ of fullerenes alone created an entire industry (see [FoMa95]), the first reference is [Gol35] and even Rev. Kirkman, who studied in 1882 40-vertex fullerenes, was cited in [Gol35].

For the hyperbolic classes $(\{a, b\}, k)$, i.e. those with $2k < b(k - 2)$ (see [BCC96]), things are much more complicated. It is likely that the number of such graphs grows more than exponentially with the increasing number v of vertices (the growth rate for them is unknown, as for many other things) and the combinatorics is so rich that it becomes intractable by the methods exposed here. But Malkevitch ([Mal70]) proved that 3-valent polyhedra, i.e. $(\{a, b\}, 3)$-polyhedra with $b \geq 7$, exist, with finite number of exceptions, if and only if $(6 - a)p_a - (b - 6)p_b = 12$, i.e. Euler formula (1.1) holds. Moreover, he found:

(i) $b \in \{7, 8, 9, 10\}$, if $a = 3$,
(ii) p_b is even, if $2a$ divide b and $a = 4, 5$.

If $p_b = 0$, then the above seven possible classes of $(\{a, b\}, k)$-spheres with parabolic $(\{a, b\}, k)$ give $Bundle_3$, Tetrahedron, Cube, Dodecahedron, $Bundle_4$, Octahedron, and $Bundle_6$, respectively (see definition of $Bundle_m$ in Section 1.5).

If $p_b = 1$, then such spheres do not exist.

Theorem 2.2.1 below gives that $(\{2, 6\}, 3)$-sphere with v vertices exists if and only if $v = 2(k^2 + kl + l^2)$ for some integers $0 \leq k \leq l$.

Theorem 2 of [GrMo63] gives that a $(\{3, 6\}, 3)$-polyhedron with v vertices exists only for any $v \geq 4$ with $v \equiv 0$ (mod 4), except $v = 8$. For $v = 8$, a $(\{3, 6\}, 3)$-sphere exists but it is only a 2-connected sphere T_1; see Proposition 2.0.2.

The $(\{4, 6\}, 3)$-spheres were considered in a chemical setting in [GaHe93]. Theorem 1 of [GrMo63] gives that a $(\{4, 6\}, 3)$-sphere with v vertices exists only for any even $v \geq 8$, except $v = 10$.

Theorem 1 of [GrMo63] gives also that a $(\{5, 6\}, 3)$-sphere with v vertices exists only for any even $v \geq 20$, except $v = 22$.

A $(\{2, 4\}, 4)$-sphere with v vertices exists for any even $v \geq 2$ (see [DeSt03]).

The existence of $(\{3, 4\}, 4)$-spheres with v vertices only for any $v \geq 6$, except $v = 7$, is established in [Grü67, page 282].

A $(\{2, 3\}, 6)$-sphere with v vertices exists for any $v \geq 2$ and the proof (in the same spirit as for other parabolic classes) is given below:

---

[1] We attributed before to those seven classes the following names and notation, respectively: 2_v, 3_v, 4_v, 5_v (in [DeDu05]), 4-hedrites, octahedrites (in [DeSt03, DDS03, DHL02]), $(2, 3)_v$ (in [DeGr01]).

Theorem 2.0.1 *For any $v \geq 2$, there exists a $(\{2, 3\}, 6)$-sphere with v vertices.*

Proof. Consider the regular tiling $\{3, 6\}$ and take the doubly infinite path l of vertices lying on a a straight line in $\{3, 6\}$. If we take another parallel line l' at distance t, then l and l' bound a domain D_t in $\{3, 6\}$.

If we take the group G generated by a translation of three edges along l, then the quotient $\widetilde{D}_t$ of D_t by G is formed of t rings, each of six 3-gons. The domain $\widetilde{D}_t$ has two faces bounded by vertices of degree 4. There are two possible caps to close those structures:

| Incomplete structure | Cap Nr. 1 | Cap Nr. 2 |

Denote by $23_1(t)$ the $(\{2, 3\}, 6)$-sphere with $3t + 3$ vertices, which is formed by closing domain $\widetilde{D}_t$ by two caps Nr. 1. Denote by $23_2(t)$ the $(\{2, 3\}, 6)$-sphere with $3t + 4$ vertices, which is formed by closing domain $\widetilde{D}_t$ by one cap Nr. 1 and one cap Nr. 2. Denote by $23_3(t)$ the $(\{2, 3\}, 6)$-sphere with $3t + 5$ vertices, which is formed by closing domain $\widetilde{D}_t$ by two caps Nr. 2. See below the first two members of those series:

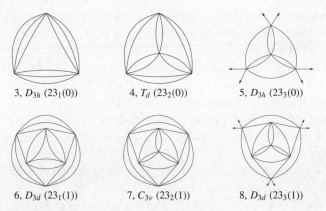

| 3, D_{3h} ($23_1(0)$) | 4, T_d ($23_2(0)$) | 5, D_{3h} ($23_3(0)$) |
| 6, D_{3d} ($23_1(1)$) | 7, C_{3v} ($23_2(1)$) | 8, D_{3d} ($23_3(1)$) |

We conclude by noting that $Bundle_6$ has two vertices. $\qquad\square$

Denote by $(T_n)_{n \geq 1}$ the infinite series of $4(n + 1)$-vertex $(\{3, 6\}, 3)$-spheres, whose first three members are shown below:

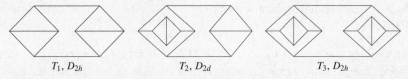

| T_1, D_{2h} | T_2, D_{2d} | T_3, D_{2h} |

The symmetry group of T_n is D_{2d} or D_{2h} if n is even or odd, respectively.

Theorem 2.0.2 *([DeDu05]) For a 3-valent plane graph G with faces of gonality between 3 and 6, it holds:*

 (i) *G is 2-connected.*
 (ii) *If G is not 3-connected, then it belongs to the infinite series T_n of ($\{3, 6\}$, 3)-spheres.*

There is a similar theorem in [DDS03] for 4-valent plane graphs with faces of size 2, 3, or 4. From this it follows that ($\{3, 4\}$, 4)-, ($\{4, 6\}$, 3)- and ($\{5, 6\}$, 3)-spheres are polyhedra.

2.1 The Goldberg–Coxeter construction

The Goldberg–Coxeter construction takes a 3- or 4-valent plane graph G_0, two integers k and l, and returns another 3- or 4-valent plane graph denoted by $GC_{k,l}(G_0)$. This construction occurs in many contexts, whose (non-exhaustive) list (for the main case of G_0 being Dodecahedron) is given below:

1 Every fullerene ($\{5, 6\}$, 3) of symmetry I or I_h is of the form $GC_{k,l}$ (*Dodecahedron*) for some k and l, i.e. is parametrized by a pair of integers $k, l \geq 0$. This result was proved by Goldberg in [Gol37]; see some other proofs in [Cox71] and Theorem 2.2.2.
2 The famous fullerene $C_{60}(I_h)$ (called *buckminsterfullerene* or *soccer ball*) has the skeleton of $GC_{1,1}(Dodecahedron)$. $GC_{k,l}(Dodecahedron)$ constitute a particularly studied class of fullerenes (see [FoMa95, Diu03]).
3 A certain class of virus *capsides* (protein shells of virions) have a spherical structure, that is modeled on dual $GC_{k,l}(Dodecahedron)$ (see [CaKl62, Cox71, DDG98]).
4 Geodesic domes, designed with the method of Buckminster Fuller, are based again on those two parameters k and l (see [Cox71]).
5 In Numerical Analysis on the sphere, we need systems of points that look roughly uniform. The vertices of dual $GC_{k,l}(Dodecahedron)$ provide such a point-set (see [ScSw95]).
6 Some conjectural solutions of many extremal problem on the sphere (Thomson, Tammes, Skyrme problems, etc.) have (solving exactly the problem is almost impossible) the combinatorial structure of $GC_{k,l}(Dodecahedron)$ or its dual (see [HaSl96]).

In Virology, the number $t(k, l) = k^2 + kl + l^2$ (used for icosahedral fullerenes) is called *triangulation number*. In terms of Buckminster Fuller, the number $k + l$ is called *frequency*, the case $l = 0$ is called *Alternate*, and the case $l = k$ is called

Triacon. He also called the Goldberg–Coxeter construction *Breakdown* of the initial plane graph G_0.

The *root lattice* A_2 is defined by $A_2 = \{x \in \mathbb{Z}^3 : x_0 + x_1 + x_2 = 0\}$. The *square lattice* is denoted by $\mathbb{Z}^2$.

The ring $\mathbb{Z}[\omega]$, where $\omega = e^{\frac{2\pi}{6}i} = \frac{1}{2}(1 + i\sqrt{3})$, of *Eisenstein integers* consists of the complex numbers $z = k + l\omega$ with $k, l \in \mathbb{Z}$. The norm of such z is denoted by $N(z) = z\bar{z} = k^2 + kl + l^2$ and we will use the notation $t(k, l) = k^2 + kl + l^2$. If we identify $x = (x_1, x_2, x_3) \in A_2$ with the Eisenstein integer $z = x_1 + x_2\omega$, then it holds that $2N(z) = \|x\|^2$.

The ring $\mathbb{Z}^2 = \mathbb{Z}[i]$ consists of the complex numbers $z = k + li$ with $k, l \in \mathbb{Z}$. The norm of such z is denoted by $N(z) = z\bar{z} = k^2 + l^2$ and we will use the notation $t(k, l) = k^2 + l^2$.

The Goldberg–Coxeter construction for 3- or 4-valent plane graphs can be seen, in algebraic terms, as the scalar multiplication by Eisenstein or Gaussian integers in the parameter space. More precisely, $GC_{k,l}$ corresponds to multiplication by complex number $k + l\omega$ or $k + li$ in the 3- or 4-valent case, respectively.

Let us now build the graph $GC_{k,l}(G_0)$. First consider the 3-valent case. By duality, every 3-valent plane graph G_0 can be transformed into a *triangulation*, i.e. into a plane graph whose faces are triangles only. The Goldberg–Coxeter construction with parameters k and l consists of subdividing every triangle of this triangulation into another set of faces, called *master polygon*, according to Figure 2.2, which is defined by two integer parameters k, l. The obtained faces, if they are not triangles, can be glued with other non-triangle faces (coming from the subdivision of neighboring triangles), in order to form triangles. So, we end up with a new triangulation (see Figure 2.3).

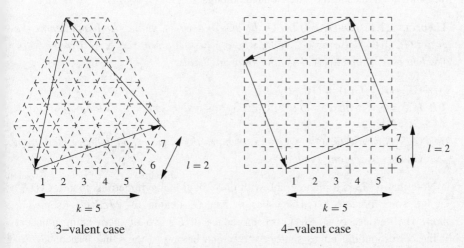

3–valent case   4–valent case

Figure 2.2 The master polygon in 3- or 4-valent case

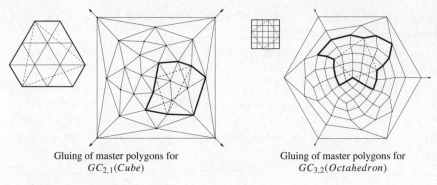

Gluing of master polygons for
$GC_{2,1}(Cube)$

Gluing of master polygons for
$GC_{3,2}(Octahedron)$

Figure 2.3 Two examples of gluing of master polygons

The triangle of Figure 2.2 has area $\mathcal{A}(k^2 + kl + l^2)$ if $\mathcal{A}$ is the area of a small triangle. By transforming every triangle of the initial triangulation in such a way and gluing them, we obtain another triangulation, which we identify with a (dual) 3-valent plane graph and denote by $GC_{k,l}(G_0)$. The number of vertices of $GC_{k,l}(G_0)$ (if the initial graph G_0 has v vertices) is $vt(k, l)$ with $t(k, l) = k^2 + kl + l^2$.

For a 4-valent plane graph G_0, the duality operation transforms it into a quadrangulation and this initial quadrangulation is subdivided according to Figure 2.2, which is also defined by two integer parameters k, l. After merging, the obtained non-square faces, we get another quadrangulation and the duality operation yields the 4-valent plane graph $GC_{k,l}(G_0)$ having $nt(k, l)$ vertices with $t(k, l) = k^2 + l^2$ (see Figure 2.3).

The faces of G_0 correspond to some faces of $GC_{k,l}(G_0)$. If $t(k, l) > 1$, then those faces are not adjacent; they are isolated amongst 6-gons or 4-gons.

Theorem 2.1.1 *([DuDe03]) Let G_0 be a 3- or 4-valent plane graph and denote the graph $GC_{k,l}(G_0)$ also by $GC_z(G_0)$, where $z = k + l\omega$ or $z = k + li$ in the 3- or 4-valent case, respectively. Then the following holds:*

(i) $GC_z(GC_{z'}(G_0)) = GC_{zz'}(G_0)$.

(ii) If $z' = z\alpha^u$ with $\alpha = \omega$ or i, G_0 is 3- or 4-valent and $u \in \mathbb{Z}$, then $GC_z(G_0) = GC_{z'}(G_0)$.

(iii) $GC_{\bar{z}}(G_0) = GC_z(\overline{G_0})$, where $\overline{G_0}$ denotes the plane graph, which differs from G_0 only by a plane symmetry.

Note that if G_0 is a plane graph with only rotational symmetries, then $GC_z(G_0)$ is not isomorphic to $GC_{\bar{z}}(G_0)$. For a map G, denote by $Med(G)$ its *medial map*. The vertices of $Med(G)$ are the edges of G, two of them being adjacent if the corresponding edges share a vertex and belong to the same face of G. So, $Med(G) = Med(G^*)$ and we can check that $Med(G) = GC_{1,1}(G)$. We have

$Med(Tetrahedron) = Octahedron$ and $Med(Cube) = Cuboctahedron$. The skeleton of $Med(G)$ is the line graph of the skeleton of G if G is a 3-valent map.

For any 3-valent map G, the *leapfrog* $leap(G)$ of G is the truncation of G^* (see [FoMa95]). We have $leap(G) = GC_{1,1}(G)$.

If $l = 0$, then $GC_{k,l}(G_0)$ is called *k-inflation* of G_0. For $k = 2, l = 0$, it is called *chamfering of G_0*, because Goldberg called the result of his construction for $(k, l) = (2, 0)$ on Dodecahedron, *chamfered Dodecahedron*. All symmetries of G occur in $GC_{k,l}(G)$ if $l = 0$ or $l = k$, while only rotational symmetries are preserved if $0 < l < k$. The Goldberg–Coxeter construction can be also defined, similarly, for maps on orientable surfaces. While the notions of medial, leapfrog, and k-inflation go over for non-orientable surfaces, the Goldberg–Coxeter construction is not defined on a non-orientable surface. Some examples of Goldberg–Coxeter construction can be found in Chapter 9. It is interesting to study graph parameters of $GC_{k,l}(G)$ for a fixed G; one example of such study concerns *zigzags* and *central circuits* in [DuDe03].

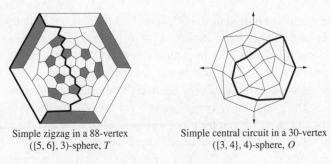

Simple zigzag in a 88-vertex
$(\{5, 6\}, 3)$-sphere, T

Simple central circuit in a 30-vertex
$(\{3, 4\}, 4)$-sphere, O

Figure 2.4 Examples of simple zigzag and central circuit; one orbit of such 12 zigzags and one orbit of six such circuits

2.2 Description of the classes

Here are presented some known theoretical and computer tools for generating $(\{a, b\}, k)$-spheres. We first give the cases for which the description of the Goldberg–Coxeter construction suffices. Then, for $(\{3, 6\}, 3)$- and $(\{2, 4\}, 4)$-spheres simple combinatorial constructions give a full description of those classes. For the remaining classes, we have less possibilities and the number of graphs grows more sharply with the increasing number of vertices.

Theorem 2.2.1 *Every $(\{2, 6\}, 3)$-plane graph comes as $GC_{k,l}(Bundle_3)$; its symmetry group is D_{3h} if $l \in \{0, k\}$ and D_3, otherwise.*

Proof. The description of those spheres is given in [GrZa74] and it is, actually, a Goldberg–Coxeter construction. □

Theorem 2.2.2 *(i) Any ({3, 6}, 3)-sphere with symmetry T or T_d is $GC_{k,l}$ (Tetrahedron),*

(ii) any ({4, 6}, 3)-sphere with symmetry O or O_h is $GC_{k,l}(Cube)$,

(iii) any ({4, 6}, 3)-sphere with symmetry D_6 or D_{6h} is $GC_{k,l}(Prism_6)$,

(iv) any ({5, 6}, 3)-sphere with symmetry I or I_h is $GC_{k,l}(Dodecahedron)$,

(v) any ({2, 4}, 4)-sphere with symmetry D_4 or D_{4h} is $GC_{k,l}(Bundle_4)$,

(vi) any ({3, 4}, 4)-sphere of symmetry O or O_h is $GC_{k,l}(Octahedron)$.

(vii) Let $\mathcal{G}P_m$ (for $m \neq 2, 4$) denote the class of 3-valent plane graphs with two m-gonal faces, m 4-gons, and p_6 6-gonal faces. Every such graph, having a m-fold axis, comes as $GC_{k,l}(Prism_m)$ and has symmetry group D_m or D_{mh}.

(viii) Let $\mathcal{G}F_m$ (for $m \neq 2, 3$) denote the class of 4-valent plane graphs with two m-gonal faces, m 2-gons, and p_4 4-gonal faces. Every such graph, having a m-fold axis, comes as $GC_{k,l}(Foil_m)$ and has symmetry group D_m or D_{mh}.

The above results are proved in [Gol37, Cox71, DuDe03]. We will prove only (ii), the other cases being very similar: Take a ({4, 6}, 3)-sphere of symmetry O or O_h. One 4-fold symmetry axis goes through a 4-gon, say, F_1. After adding p rings of 6-gons around F_1, we find a 4-gon and so, by symmetry, four 4-gons, say, F'_1, F'_2, F'_3, F'_4. The position of the square F'_1 relatively to F_1 defines an Eisenstein integer $z = k + l\omega$. The graph can be completed in a unique way and this proves that it is $GC_{k,l}(Cube)$. Looking back at the proof we can see that we used only the existence of a 4-fold rotation axis to get the result. Therefore, the symmetry groups C_4, D_4, and so on are not possible for ({4, 6}, 3)-spheres, i.e. such symmetry implies some higher symmetries.

We are now giving the general construction of all ({3, 6}, 3)-spheres following [GrMo63] (see also [DeDu05]). A *zigzag* in a plane graph is a circuit of edges, such that any two, but no three, consecutive edges belong to the same face. A zigzag is called *simple* if it has no self-intersection (see, for example, Figure 2.6). A *closed railroad* in a 3-valent plane graph is a circuit of 6-gonal faces, such that any 6-gon

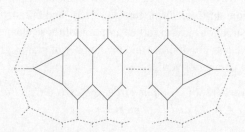

Figure 2.5 A railroad and a closed railroad in a ({3, 6}, 3)-sphere

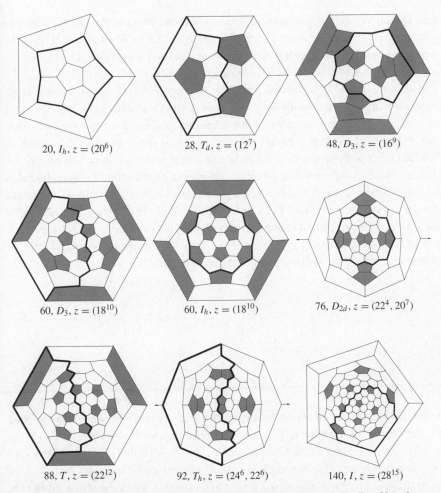

20, I_h, $z = (20^6)$ 28, T_d, $z = (12^7)$ 48, D_3, $z = (16^9)$

60, D_3, $z = (18^{10})$ 60, I_h, $z = (18^{10})$ 76, D_{2d}, $z = (22^4, 20^7)$

88, T, $z = (22^{12})$ 92, T_h, $z = (24^6, 22^6)$ 140, I, $z = (28^{15})$

Figure 2.6 All known tight ($\{5, 6\}$, 3)-spheres with simple zigzags; z is the list of lengths of the zigzags

is adjacent to its neighbors on opposite edges. A 3-valent plane graph is called *tight* if it has no closed railroads. A *railroad* between two non 6-gonal faces F and F' is a sequence of 6-gons, say, $F_1, \ldots, F_l$, such that putting $F_0 = F$ and $F_{l+1} = F'$ we have that any F_i, $1 \le i \le l$, is adjacent to F_{i-1} and F_{i+1} on opposite edges. It is proved in [GrMo63] that closed railroads and railroads of a ($\{3, 6\}$, 3)-sphere do not self-intersect, i.e. a given 6-gon occur only once. Take a ($\{3, 6\}$, 3)-sphere G, a railroad between two triangles T_1, T_2 and denote by s the number of 6-gons of this railroad. Around this structure, we adds ring of 6-gons, which happen to be closed railroads. After adding m such rings, we add a triangle T_3. The structure is then completely determined and we obtain a railroad between T_3 and T_4, which is isomorphic to the one between T_1 and T_2 (see Figure 2.5 for illustration). We get, by direct computation, the equality $p_6 = 2(sm + s + m)$ and $v = 4(s + 1)(m + 1)$, where p_6 is the number of 6-gons and v the number of vertices.

Following [DeSt03], we now explain how to describe all $(\{2, 4\}, 4)$-spheres. In an Eulerian map (i.e. ones with vertices of even degrees), a *central circuit* is a circuit of edges such that any two consecutive edges are not contained in a common face (see Figure 2.4). A central circuit is called *simple* if it has no self-intersection. A central circuit is determined by a single edge, so the edge-set is partitioned by the central circuit. A *railroad* is a, possibly self-intersecting, circuit of 4-gons bounded by two central circuits. A 4-valent graph is called *tight* if it has no railroad. Non-tight 4-valent graphs are obtained from tight ones by duplicating some of their central circuits. It is proved that all central circuits of $(\{2, 4\}, 4)$-sphere do not self-intersect. Let us describe the tight $(\{2, 4\}, 4)$-spheres. They have exactly two central circuits. Below we indicate the starting point of the construction for $v = 5$:

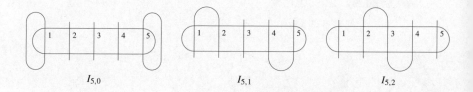

$$I_{5,0} \qquad\qquad I_{5,1} \qquad\qquad I_{5,2}$$

From the scheme drawn above, it is clear that there is only one way of completing the structure so as to obtain a $(\{2, 4\}, 4)$-sphere, which we denote by $I_{v,i}$. Also $I_{v,i}$ is a tight $(\{2, 4\}, 4)$-sphere if and only if $gcd(v, i) = 1$. All tight $(\{2, 4\}, 4)$-spheres are obtained in this way. The non-tight $(\{2, 4\}, 4)$-spheres are obtained by replacing the central circuits by railroads, which do not self-intersect. The description given here of $(\{3, 6\}, 3)$-spheres and $(\{2, 4\}, 4)$-spheres is not the only one possible. We can see them as dual agglomerations of triangles and 4-gons (see, for example, [FoCr97]). Another possibility is to take the quotient of a $(6, 3)$-torus by an involution fixing four 6-gonal faces; see [DGMŠ07], where this is used to determine the eigenvalues of $(\{3, 6\}, 3)$-spheres.

Remark 2.2.3 *The number of zigzags in a tight $(\{a, 6\}, 3)$-sphere is ≤ 3, 3 for $a = 2$, 3 and, conjecturally ≤ 8, ≤ 15 for $a = 4$, 5 (see [DeDu05]). The number of central circuits in a tight $(\{a, 4\}, 4)$-sphere is 2, ≤ 6 for $a = 2$, 3 (see [DeSt03]). The number of tight $(\{a, 6\}, 3)$-spheres with only simple zigzags is 0, ∞, 2 for $a = 2, 3, 4$ and, conjecturally, 9 for $a = 5$ (see Figure 2.6 and [DeDu05]). The number of tight $(\{a, 4\}, 4)$-spheres with only simple central circuits is ∞, 8 for $a = 2$, 3 (including the right one on Figure 2.4. See [DeSt03, DDS03] for details).*

The polyhedra, which are dual to the 3-valent polyhedra without b-gonal faces, $b > 6$, are studied in [Thu98]; they are called there *non-negatively curved triangulations*. Thurston developed there a global theory of parameter space for sphere triangulations with degree of vertices at most 6. The main Theorem 0.1 there describes them as the elements of L_+/G, where L is a lattice in complex Lorenz space $C^{(1,9)}$, G is a group of automorphisms and L_+ is the set of lattice points of positive square-norm. Clearly, our ($\{a, 6\}$, 3)-spheres with $a = 3, 4, 5$ are covered by Thurston's considerations. Let s denote the number of vertices of degree less than 6; such vertices reflect the positive curvature of the triangulation of the sphere $\mathbb{S}^2$. Thurston has built a parameter space with $s - 2$ degrees of freedom (complex numbers). Using this, Theorem 3.4 in [Sah94] (which is an application of a preliminary version of [Thu98]) implied that the number of ($\{3, 6\}$, 3)-, ($\{4, 6\}$, 3)-, ($\{5, 6\}$, 3)-spheres with v vertices grows as $O(v)$, $O(v^3)$, $O(v^9)$. We believe that the hypothesis on degree of vertices (in dual terms, that the graph has no b-gonal faces with $b > 6$) in [Thu98] is unnecessary to his theory of parameter space. Also, his theory can be extended, perhaps, to the case of quadrangulations instead of triangulations.

The graphs, that can be described in terms of Goldberg–Coxeter construction, can be thought of as those expressed in terms of one complex parameter. Considering intermediate symmetry groups (for example, ($\{5, 6\}$, 3)-polyhedra of symmetry D_5), we can use a small number of complex parameters to describe them and this is, actually, done in [Dut02] or in [FCS88] for generating them up to a large number of vertices. See also [Gra05] for some classification results, based on those methods. For general classes of graphs with no hypothesis on symmetry, the best is, probably, to use CPF ([BDDH97]) or ENU ([Hei98]), which generates all 3- or 4-valent graphs with the number of faces of size i being given in advance.

Remark 2.2.4 *The possible symmetries of* ($\{2, 3\}$, 6)-*spheres have not been determined yet. Also, the Goldberg–Coxeter construction has not been defined for 6-valent spheres, although we do not see an obstruction to it. Also it could be interesting to extend remark 2.2.3 on those spheres.*

For the ($\{2, 3\}$, 6)-*spheres, a useful transformation for computer enumeration and for theoretical purposes is the following. Take the dual, which is a plane graph with 6-gonal faces and vertices of degree 2 or 3; then remove the 2-valent vertices. The resulting sphere is a 3-valent one with faces of gonality at most 6. We can thus generate many* ($\{2, 3\}$, 6)-*spheres: just take a 3-valent plane graph with faces of size at most 6, put adequately some vertices on some edges, and then take the dual.*

We now list the known results on symmetry of those spheres:

Theorem 2.2.5 *The symmetry groups are as follows:*

(i) *([GrZa74]) For* ($\{2, 6\}$, 3))-*spheres, it is one of* D_3 *or* D_{3h}.
(ii) *([FoCr97]) For* ($\{3, 6\}$, 3)-*spheres, it is one of* D_2, D_{2h}, D_{2d}, T, T_d.

(iii) ([DeDu05]) For ({4, 6}, 3)-spheres, it is one of 16 groups: C_1, C_s, C_2, C_i, C_{2v}, C_{2h}, D_2, D_3, D_{2d}, D_{2h}, D_{3d}, D_{3h}, D_6, D_{6h}, O, O_h.

(iv) ([FoMa95]) For ({5, 6}, 3)-spheres, it is one of 28 groups: C_1, C_2, C_i, C_s, C_3, D_2, S_4, C_{2v}, C_{2h}, D_3, S_6, C_{3v}, C_{3h}, D_{2h}, D_{2d}, D_5, D_6, D_{3h}, D_{3d}, T, D_{5h}, D_{5d}, D_{6h}, D_{6d}, T_d, T_h, I, I_h.

(v) ([DDS03]) For ({2, 4}, 4)-spheres, it is one of D_{4h}, D_4, D_{2h}, D_{2d}, D_2.

(vi) ([DDS03]) For ({3, 4}, 4)-spheres, it is one of 18 groups: C_1, C_s, C_2, C_{2v}, C_i, C_{2h}, S_4, D_2, D_{2d}, D_{2h}, D_3, D_{3d}, D_{3h}, D_4, D_{4d}, D_{4h}, O, O_h.

In [Kar07] a general method for determining the symmetry groups of ({5, b}, 3)-spheres is given. It is proved in [Kar07] that if a group occurs as a symmetry group of a ({5, b}, 3)-sphere, then it occurs as a symmetry group of an infinity of ({5, b}, 3)-spheres. It is also proved in [Kar07] that a group G occurs as a symmetry group of ({5, b}, 3)-spheres for any $b \geq 7$ if and only if G is a subgroup of I_h.

Another interesting class of spheres, which could be considered, consists of the self-dual ones, whose vertices are of degree 3 or 4 and faces have gonality 3 or 4. The medials (defined in Section 2.1) of such spheres are ({3, 4}, 4)-spheres. In particular, $p_3 + v_3 = 8$, which implies $p_3 = v_3 = 4$. Furthermore, the self-duality becomes an ordinary symmetry of the ({3, 4}, 4)-sphere. (See [ArRi92, SeSe94, SeSe95, SeSe96] for some other constructions of self-dual maps on the sphere.)

2.3 Computer generation of the classes

We now present the general ideas on computer generation of those classes of plane graphs. The main technique in combinatorial construction is the exhaustive search: we build a plane graph, face by face, until it is completed. The main problem is that the number of possibilities to be considered is, usually, very large. Sometimes, we can prove that a group of faces cannot be completed to the desired graph and this yields speedup. But in practice, the benefit, while tremendous, does not change the nature of the problem. Typical examples of this scheme of combinatorial enumeration are presented in Chapters 5 and 10.

The main and fundamental objection is that we need, sometimes, to make huge computations lasting months for finding only a few graphs. Fortunately, for the above classes, there are some other ways to diminish the magnitude of the problem. A simple zigzag or central circuit (see Figure 2.4 for some examples) splits the sphere into two parts, which are much easier to enumerate. Of course, all is not so easy in practice since a zigzag or central circuit is self-intersecting most of the time. But the basic idea remains (see [BHH03]) and is used in the following programs:

1 The program CPF ([BDDH97]) generates 3-valent plane graphs with specified p-vector.

2 The program ENU ([BHH03] and [Hei98]) does the same for 4-valent plane graphs.
3 The program CGF ([Har]) generates 3-valent orientable maps with specified genus and p-vector.

If we consider a larger class, like the plane triangulations, then another strategy is possible. It is known (see [Ebe1891]) that every plane triangulation arises from Tetrahedron by a sequence of the three following operations Op_i:

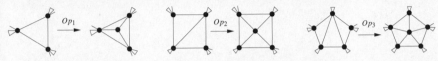

Of course, there is a very large number of triangulations. But the running time of the algorithm is approximately proportional to the number of triangulations to be found and it scales with almost 100% efficiency on parallel computers. Note also that the above algorithm is not limited by memory: for a given plane triangulations T, there are many ways of using the operations Op_i to obtain T from Tetrahedron. But the canonical augmentation scheme (see [McK98]) provides a unique path to get T from the Tetrahedron by applying the operations Op_i, thereby avoiding memory problems.

The enumeration of triangulations, triangulations of minimum degree 4 or 5, Eulerian triangulations, quadrangulations, 3-connected plane graphs, plane graphs, 3-connected plane graphs of minimum degree at least 4 or 5 are done in the program plantri (see [BrMK], [BrMK06], and [BGGMTW05]) using such kinds of elementary operations. There is no reasonable hope of applying this kind of algorithm to the enumeration of $(\{a, b\}, k)$-spheres because there is no simple operation such as the Op_i that would preserve the property of being a $(\{a, b\}, k)$-sphere and generate them all.

All computations used the GAP computer algebra system [GAP02] and the package PlanGraph ([Dut02]) by the second author; the programs are available from [Du07].

The program CaGe ([BDDH97]) was used for most of the graph drawings.

3

Fullerenes as tilings of surfaces

The discovery of the fullerene molecules and related forms of carbon, such as nan-otubes, has generated an explosion of activity in chemistry, physics, and materials science, which is amply documented, for example, in [DDE96] and [FoMa95]. In chemistry, the "classical" definition is that a fullerene is an all-carbon molecule in which the atoms are arranged as a map on a sphere made up entirely of 5-gons and 6-gons, which, therefore, necessarily includes exactly 12 5-gonal faces. We are concerned here with the following generalization: what fullerenes are possible if a fullerene is a finite 3-valent map with only 5- and 6-gonal faces embedded in any surface? This seemingly much larger concept leads only to three extensions to the class of spherical fullerenes. Embedding in only four surfaces is possible: the sphere, torus, Klein bottle, and projective plane. In [DFRR00], the spectral properties of those fullerenes are examined. The usual spherical fullerenes have 12 5-gons, projective fullerenes 6, and toroidal and Klein bottle fullerenes none. Klein bottle and projective fullerenes are the antipodal quotients of centrally symmetric toroidal and spherical fullerenes, respectively. Extensions to infinite graphs (plane fullerenes, cylindrical fullerenes) are indicated. Detailed treatment of the concept of the extended fullerenes and their further generalization to higher dimensional manifolds are given in [DeSt99b].

3.1 Classification of finite fullerenes

Define a 3-*fullerene* as a 3-valent map embedded on a surface and consisting of only 5-gonal and 6-gonal faces. Each such object has, say, v vertices, e edges, and f faces of which p_5 are 5-gons and p_6 are 6-gons.

From Theorem 1.2.3, we know that the Euler characteristic χ satisfies to:

$$p_5 = 6\chi.$$

For a surface, in which a finite 3-fullerene can be embedded, the number χ is, therefore, a non-negative integer. Let us use the topological classification of finite closed maps in Theorem 1.2.1 and recall the expression of χ in Theorem 1.2.2:

$$\chi = 2(1 - g) \quad \text{(for an orientable surface)}$$
$$= 2 - g \quad \text{(for a non-orientable surface).}$$

The cases compatible with non-negative integral solutions for χ are thus exactly four in number. The only surfaces admitting finite 3-fullerene maps are therefore: $\mathbb{S}^2$ (the sphere, orientable with $g = 0$), $\mathbb{T}^2$ (the torus, orientable with $g = 1$), $\mathbb{P}^2$ (the projective plane, non-orientable with $g = 1$), and $\mathbb{K}^2$ (the Klein bottle, non-orientable with $g = 2$). All embeddings are *2-cell-embeddings*, i.e. each face is homeomorphic to an open disk. An immediate consequence of Euler formula is that fullerenes on $\mathbb{S}^2$, $\mathbb{T}^2$, $\mathbb{K}^2$, and $\mathbb{P}^2$ have exactly 12, 0, 0, and 6 5-gons, respectively. Toroidal and Klein bottle fullerenes may also be called toroidal and Klein bottle *polyhexes* ([FYO95, Kir94, Kir97, KlZh97]) as they include no 5-gons.

Figure 3.1 shows the smallest fullerenes from the four classes, drawn as the graph, the map, and its dual triangulation on the appropriate surface. Note that the Petersen and Heawood graphs, which appear naturally here, are, actually, the 5- and 6-cages (a *k-cage* is a 3-valent graph of smallest cycle size k with the largest possible number of edges); their duals in $\mathbb{P}^2$ and $\mathbb{T}^2$, K_6 and K_7, realize the *chromatic number* (i.e. the minimal number of colors a map on the surface can be colored with, so that no two faces of the same color are adjacent; see, for example, [GrTu87, Chapter 5]) of the corresponding surfaces.

Spherical and toroidal fullerenes have an extensive chemical literature, and Klein bottle polyhexes have been considered, for example, in [Kir97, KlZh97].

Note that at least one spherical fullerene with v vertices exists for all even v with $v \geq 20$, except for the case $v = 22$ ([GrMo63]).

3.2 Toroidal and Klein bottle fullerenes

The $(6, 3)$-tori and $(6, 3)$-Klein bottles are related to $\{6, 3\}$ in a straightforward way. The underlying surfaces are quotients of the Euclidean plane $\mathbb{R}^2$ under groups of isometries generated by two translations (for $\mathbb{T}^2$) or one translation and one glide reflection (for $\mathbb{K}^2$). Each point of $\mathbb{T}^2$ and $\mathbb{K}^2$ corresponds to an orbit of the generating group. Note that the groups generated by a single translation or a single reflection, respectively, give, as quotients, the cylinder and the twisted cylinder (the Möbius strip, see Figure 1.1). Construction and enumeration of polyhexes can therefore be envisaged as a process of cutting parallelograms out of the "graphite plane" $\{6, 3\}$ and gluing their edges according to the rules implied in Figure 3.1.

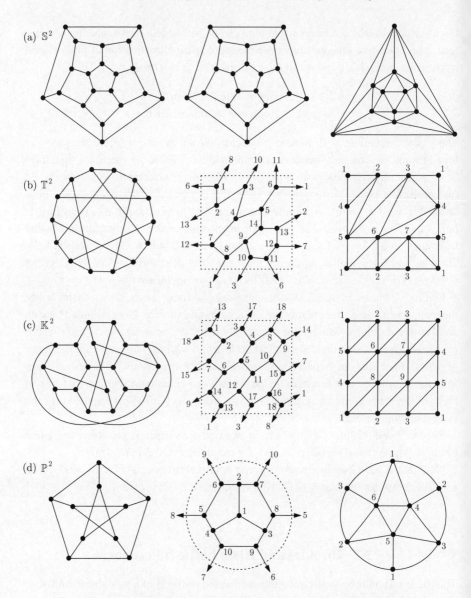

Figure 3.1 Smallest spherical, toroidal, Klein bottle and projective fullerenes. The first column lists the graphs drawn in the plane, the second the map on the appropriate surface and the third the dual on the same surface. The examples are: (a) Dodecahedron (dual Icosahedron), (b) the Heawood graph (dual K_7), (c) a smallest Klein bottle polyhex (dual $K_{3,3,3}$), and (d) the Petersen graph (dual K_6).

Some confusion exists in the mathematical and chemical literature on toroidal polyhexes. Negami ([Neg85]), Altschuler ([Alt73]), and other topological graph theorists define regular 3-valent maps on the torus to mean 2-cell embeddings with all faces 6-gonal, without further qualification. Coxeter and others, working in a group theoretical tradition, use the same term in a more restricted sense of polyhexes with automorphism groups $Aut(G)$ of the maximal possible order, in other words, those that realize the equality in the bound $|Aut(G)| \leq 4e(G) (= 6p_6$ for a polyhex). All such regular maps are: (on $\mathbb{S}^2$) the five Platonic polyhedra, (on $\mathbb{P}^2$) six graphs that include the Petersen graph and its dual, (on $\mathbb{K}^2$) no graphs at all ([Nak96]), and (on $\mathbb{T}^2$) the polyhexes that arise by the Goldberg–Coxeter construction from the 6-gon (see Section 2.1).

In Negami's construction ([Neg85]), a three-parameter code represents any toroidal polyhex (or, equivalently, any 6-regular triangulation of $\mathbb{T}^2$) as a tiling of $\{6, 3\}$. Each graph of this type is denoted $T(p, q, r)$, with integer parameters p, q, and r, where p is the length of a geodesic cycle of edge-sharing 6-gons, r is the number of such cycles, and q is an offset.

At least one toroidal polyhex that is cell-complex exists for all numbers of vertices $v \geq 14$. The unique cell-complex toroidal fullerene at $v = 14$ is a realization of the Heawood graph. It is $GC_{2,1}(hexagon)$ in terms of Goldberg–Coxeter construction and is the dual of K_7, which itself realizes the 7-color map on the torus. This map and its dual are shown in Figure 3.1.

A description of Klein bottle polyhexes can be developed along similar lines ([Nak96]). Each toroidal graph $T(p, 0, r)$ can be used to obtain two Klein bottle 6-regular triangulations (and, hence, by taking dual, 3-fullerenes), the *handle*, and *cross-cap* types $K_h(p, r)$ and $K_c(p, r)$, respectively. The torus is cut along a geodesic of length p. Then the *handle* construction amounts to identification of opposite sides of the resulting parallelogram with reversed direction. In the *cross-cap* construction, the opposite sides are each converted to cross-caps, with slightly different rules for odd and even p. The unique smallest cell-complex Klein bottle polyhex has 18 vertices (9 6-gonal faces) and is the dual of the tripartite $K_{3,3,3}$; the graph and the map and its dual are shown in Figure 3.1.

3.3 Projective fullerenes

The projective plane arises as a quotient space of the sphere, the required group being C_i. It is obtained by identifying antipodal points of the spherical surface; in other words, it is the antipodal quotient of the sphere (see Section 1.2.2). $\mathbb{P}^2$ is the simplest compact non-orientable surface in the sense that it can be obtained from the sphere by adding just one cross-cap.

Clearly, this construction can be carried over to maps: the antipodal quotient of a centrosymmetric map on the sphere has vertices, edges, and faces obtained by identifying antipodal vertices, edges, and faces, thereby halving the number of each type of structural component. For example, the antipodal quotient of Icosahedron is K_6, and that of Dodecahedron is the Petersen graph, famous as a counterexample to many conjectures (see, for example, [HoSh93]). The Petersen graph is not a planar graph, but it is called *projective-planar* in the sense that it can be embedded without edge crossings in the projective plane.

In this terminology, our definition of projective fullerenes amounts to selection of cell-complex projective-planar 3-valent maps with only 5- and 6-gonal faces. As noted above, $p_5 = 6$ for these maps. Thus, the Petersen graph is the smallest projective fullerene. In general, the projective fullerenes are exactly the antipodal quotients of the centrally symmetric spherical fullerenes.

Thus, the problem of enumeration and construction of projective fullerenes reduces simply to that for centrally symmetric conventional spherical fullerenes. The point symmetry groups that contain the inversion operation are C_i, C_{mh}, (m even), D_{mh} (m even), D_{md} (m odd), T_h, O_h, and I_h. A spherical fullerene may belong to one of 28 point groups ([FoMa95]) of which eight appear in the previous list: C_i, C_{2h}, D_{2h}, D_{6h}, D_{3d}, D_{5d}, T_h, and I_h. Clearly, a fullerene with v vertices can be centrally symmetric only if v is divisible by four as p_6 must be even. After the minimal case $v = 20$, the first centrally symmetric fullerenes are at $v = 32$ (D_{3d}) and $v = 36$ (D_{6h}).

3.4 Plane 3-fullerenes

An example of infinite 3-fullerene is given by *plane fullerenes*, i.e. 3-valent partitions of the plane into (combinatorial) 6-gons and p_5 5-gons. Such partitions have $p_5 \leq 6$ (see [DeSt02b] for a reduction of proof to Alexandrov theory in [Ale50] and [Ale48, Chapter VIII]). For $p_5 = 0, 1$ such 3-fullerene is unique; for any $2 \leq p_5 \leq 6$ there is an infinity of them.

4

Polycycles

4.1 (r, q)-polycycles

A (r, q)-*polycycle* is a simple plane 2-connected locally finite graph with degree at most q, such that:

(i) all interior vertices are of degree q,
(ii) all interior faces are (combinatorial) r-gons.

We recall that any finite plane graph has a unique exterior face; an infinite plane graph can have any number of exterior faces, including zero and infinity. Denote by p_r the number of interior faces; for example, Dodecahedron on the plane has $p_5 = 11$.

See in Figure 4.1 some examples of connected simple plane graphs that are not (r, q)-polycycles.

We will prove later (in Theorem 4.3.2) that all vertices, edges, and interior faces of an (r, q)-polycycle form a cell-complex (see Section 1.2.1).

The *skeleton of a polycycle* is the edge-vertex graph defined by it, i.e. we forget the faces. By Theorem 4.3.6, except for five Platonic ones, the skeleton has a unique *polycyclic realization*, i.e. a polycycle for which it is the skeleton.

The parameters (r, q) are called *elliptic* if $rq < 2(r + q)$, *parabolic* if $rq = 2(r + q)$, and *hyperbolic* if $rq > 2(r + q)$; see Remark 1.4.1. Call a polycycle *outerplanar* if it has no interior vertices. For parabolic or hyperbolic (r, q), the tiling $\{r, q\}$ is a (r, q)-polycycle. For elliptic (r, q), the tiling $\{r, q\}$ with a face deleted is an (r, q)-polycycle. Different, but all isomorphic, polycyclic realizations for those five exceptions to the unicity, come from different choices of such deleted (exterior) faces.

The (r, q)-polycycles $\{r, q\}$ with parabolic and hyperbolic parameters (r, q) do not have a boundary.

Recall that an *isomorphism* between two plane graphs, G_1 and G_2, is a function ϕ mapping vertices, edges, and faces of G_1 to the ones of G_2 and preserving incidence: relations. Two (r, q)-polycycles, P_1 and P_2, are *isomorphic* if there is an

Figure 4.1 Some plane graphs that are not (r, q)-polycycles

isomorphism ϕ of their skeletons preserving the set of interior faces. Recall also that the *automorphism group* $Aut(G)$ of a plane graph G is the group of all its *automorphisms*, i.e. isomorphisms of G to G. The *automorphism group* $Aut(P)$ of a polycycle P consists of all automorphisms of plane graph G preserving the set of interior faces.

The notion of duality of plane graphs applies as well for (r, q)-polycycles, but it ignores the exterior faces, which we want to keep unchanged. We will introduce two notions of duality for (r, q)-polycycles and call them *inner dual* (see [BCH02] for some applications in enumeration) and *outer dual*. They are always defined, but the resulting plane graph is not necessarily a (q, r)-polycycle.

The inner dual $Inn^*(P)$ of an (r, q)-polycycle P is the graph obtained by taking the interior faces as vertices and having, as edges, the edges between two adjacent interior faces. The inner dual $Inn^*(P)$ is not necessarily 2-connected. See the two examples below:

A $(5, 3)$-polycycle and its inner dual   A $(3, 5)$-polycycle and its inner dual

The outer dual $Out^*(P)$ of an (r, q)-polycycle P is the graph obtained by taking, as vertex-set, the interior faces and some exterior vertices. Such exterior vertices are taken around every boundary vertex of P, so as to make sure that they correspond to q-gonal faces of $Out^*(P)$. The degree of boundary vertices may be higher than r. Another possible obstruction to $Out^*(P)$ being a (q, r)-polycycle is not having any exterior face, as can happen for the tiling $\{r, q\} - f$ with elliptic (r, q). See the two examples below:

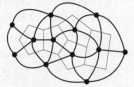   

A $(5, 3)$-polycycle and its outer dual   A $(3, 5)$-polycycle and its outer dual

We have the equality $P = Inn^*(Out^*(P))$ and $P = Out^*(Inn^*(P))$ for an (r, q)-polycycle P, provided that all the maps appearing in those equations are (r, q)- or

(q, r)-polycycles. If all operations are defined correctly and P is an (r, q)-polycycle, then $Inn^*(Inn^*(P))$ is the (r, q)-polycycle P with all r-gons having boundary edges removed, while $Out^*(Out^*(P))$ is the (r, q)-polycycle P with a ring of r-gons being added on its outermost layer, so that all boundary vertices become interior vertices.

Call a polycycle *proper* if it is a partial subgraph of $\{r, q\}$ and a *helicene*, otherwise (this term will be justified later by Theorem 4.3.1). Call a proper (r, q)-polycycle *induced* (moreover, *isometric*) if this subgraph is, in addition, an induced (moreover, isometric) subgraph of $\{r, q\}$. Another interesting possible property of a proper (r, q)-polycycle is being convex in $\{r, q\}$ (see Section 4.4).

For $(r, q) = (3, 3), (4, 3), (3, 4)$, any induced (r, q)-polycycle is isometric but, for example, the path of three 5-gons is an induced non-isometric $(5, 3)$-polycycle.

Consider now the notion of reciprocity, defined for some proper polycycles. Let P be a proper bounded (r, q)-polycycle. Consider the union of all r-gonal faces of $\{r, q\}$ outside of P. Easy to see that this union will be an (r, q)-polycycle; call it then a *reciprocal polycycle* to P if either P is elliptic or P is infinite and has a connected boundary. Call a polycycle *self-reciprocal* if it admits the reciprocal polycycle and is isomorphic to it.

All self-reciprocal (r, q)-polycycles with $(r, q) = (3, 3), (4, 3), (3, 4), (5, 3)$ are: $\{3, 3\} - e$, $\{4, 3\} - v$, $P_2 \times P_4$, $\{3, 4\} - v$, $\{3, 4\} - C_3$, Tr_4 and 9 (out of 11) $(5, 3)$-polycycles with $p_5 = 6$, including 6 chiral ones. An example of self-reciprocal $(3, q)$-polycycle, for any $q \geq 3$, is a $(3, q)$-polycycle on one of two shores of *zigzag* (see definition in Section 2.2), cutting $\{3, q\}$ in two isomorphic halves; it includes $\{3, 3\} - e$, $\{3, 4\} - C_3$ and is infinite for $q \geq 6$.

A general theory of polycycles is considered in [DeSt98, DeSt99a, DeSt02c, DeSt00a, DeSt00b, DeSt00c, DeSt01, DeSt02b, Sht99, Sht00].

4.2 Examples

Call an (r, q)-polycycle *elliptic*, *parabolic*, or *hyperbolic*, if $rq < 2(r + q)$, $rq = 2(r + q)$, or $rq > 2(r + q)$. This corresponds to $\{r, q\}$ being the regular tiling of $\mathbb{S}^2$, $\mathbb{R}^2$, or $\mathbb{H}^2$, respectively.

There is a literature (see, for example, [GrSh87a, Section 9.4], [BGOR99], [BCH02], and [BCH03]) about proper parabolic polycycles (*polyhexes*, *polyamonds*, *polyominoes* for $\{6, 3\}$, $\{3, 6)\}$, $\{4, 4\}$, respectively); the terms come from familiar terms *hexagon*, *diamond*, *domino*, where the last two correspond to the case p_3, $p_4 = 2$.

Polyominoes were considered first by Conway, Penrose, Golomb as tiles (of $\mathbb{R}^2$ etc.; see, for example, [CoLa90]) and in Game Theory; later, they were used for enumeration in physics and statistical mechanics.

Polyhexes are used widely (see, for example, [Dia88, Bal95]) in organic chemistry: they represent *completely condensed PAH (polycyclic aromatic hydrocarbons)* $C_n H_m$ with n vertices (atoms of the carbon C), including m vertices of degree two, where atoms of the hydrogen H are adjoined (see Figure 7.1).

All 39 proper (5, 3)-polycycles were found in [CCBBZGT93] in chemical context, but were already given in [Har90] all 3, 6, 9, 39, 263 proper elliptic (r, q)-polycycles for $(r, q) = (3, 3), (4, 3), (3, 4), (5, 3), (3, 5)$, respectively.

Now, we list all (3, 3)-, (4, 3)-, (3, 4)-polycycles. Clearly, all (3, 3)-polycycles are:

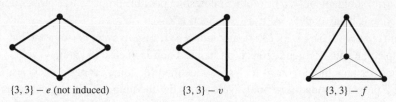

$\{3, 3\} - e$ (not induced) $\{3, 3\} - v$ $\{3, 3\} - f$

Recall that P_n denotes a path with n vertices; denote by $P_{\mathbb{N}}$, $P_{\mathbb{Z}}$ infinite paths in one or both directions. All (4, 3)-polycycles are:

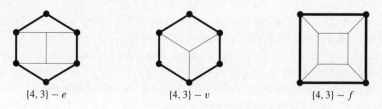

$\{4, 3\} - e$ $\{4, 3\} - v$ $\{4, 3\} - f$

the infinite series $P_2 \times P_n$ (for any $n \geq 2$; see below examples with $n = 2, 3, 4$):

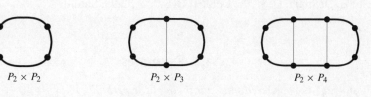

$P_2 \times P_2$ $P_2 \times P_3$ $P_2 \times P_4$

and two infinite ones:

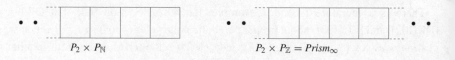

$P_2 \times P_{\mathbb{N}}$ $P_2 \times P_{\mathbb{Z}} = Prism_\infty$

Only $\{4, 3\}$, $\{4, 3\} - v$, $P_2 \times P_2$, $P_2 \times P_3$, $P_2 \times P_4$, $\{4, 3\} - e$ are proper; amongst them only the last two are not induced.

The number of (3, 4)-polycycles is also countable, including two infinite ones (9 of (3, 4)-polycycles are proper and 5 of proper ones are induced).

Namely, all (3, 4)-polycycles are:

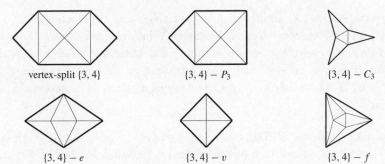

vertex-split {3, 4}   {3, 4} − P_3 {3, 4} − C_3

{3, 4} − e   {3, 4} − v   {3, 4} − f

the infinite series Tr_n (for any $n \geq 1$; see below examples with $n = 1, 2, 3, 4$):

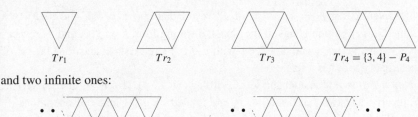

Tr_1 Tr_2 Tr_3 $Tr_4 = \{3, 4\} − P_4$

and two infinite ones:

$Tr_\mathbb{N}$ $Tr_\mathbb{Z} = APrism_\infty$

For all other parameters (r, q), there is a continuum of (r, q)-polycycles and the number of finite ones amongst them is countable.

The *vertex-split Octahedron* and the *vertex-split Icosahedron* are polycycles obtained from Octahedron and Icosahedron, respectively, by splitting a vertex into two vertices and the edges, incident to it, into two parts, accordingly. The vertex-split Octahedron is drawn on Figure 4.2,[1] and both of them are given on Figure 8.3; they are the only, besides five Platonic $\{r, q\} − f$, non-extensible finite (r, q)-polycycles.

(4, 3)-helicene
$P_2 \times P_5$

(5, 3)-helicene
from Dodecahedron

(3, 4)-helicene
vertex-split Octahedron

Figure 4.2 Some small helicenes

[1] It is the *Hexagon*, HSBC logo from 1983; it was developed from bank's nineteenth-century house flag: a white rectangle divided diagonally to produce a red hourglass shape. This flag was derived from the Scottish flag: *saltire* or *crux decussata* (heraldic symbol in the form of diagonal cross); Saint Andrew was crucified upon such a cross and thirteenth-century tradition states that the cross was X-shaped at his own request, out of respect for Jesus.

4.3 Cell-homomorphism and structure of (r, q)-polycycles

Given two maps X and X', recall that a *cell-homomorphism* is a function $\phi : X \to X'$ that transforms vertices, edges, and faces of X into vertices, edges, and faces of X', while preserving the incidence relations. Recall also that a *flag h* of a map X is a sequence $h = (v, e, f)$ with v, e, f being a vertex, edge, and face of X and $v \in e \subset f$. For a flag $h = (v, e, f)$ of a polycycle P, f is necessarily an interior face.

Theorem 4.3.1 *([DeSt98, GHZ02, Gra03]) Any (r, q)-polycycle P admits a cell-homomorphism into $\{r, q\}$ and such homomorphism is defined uniquely by a flag and its image.*

Proof. Given a flag $h = (v, e, f)$ of P, recall that $\sigma_0(h)$, $\sigma_1(h)$, and $\sigma_2(h)$ (see Section 1.2.2) are unique flags, if they exist, differing from h on v, e, and f, respectively. Given a vertex v and an edge e with $v \in e$, there exists at least one face f of P such that (v, e, f) is a flag.

Given a vertex v, consider the set $\mathcal{F}_v$ of flags (v, e, f). If v is an interior vertex, then, clearly, any two flags $h, h' \in \mathcal{F}_v$ are related by a sequence of operations σ_1 and σ_2. If v belongs to the boundary, then, since v cannot disconnect the graph, the vertex corona of v contains only one exterior face. This means that, again, any two flags h, h' in $\mathcal{F}_v$ are related by operations σ_1 and σ_2. Since the graph, underlying the polycycle, is connected, any two flags are related by a sequence of operations σ_0, σ_1, and σ_2.

Given a cell-homomorphism ϕ, the operations on flags should satisfy to $\phi(\sigma_i(h)) = \sigma_i(\phi(h))$ for any $i = 0, 1, 2$. Consider a flag h_0 in P and k_0 in $\{r, q\}$. Any other flag h in P is related to h_0 by a sequence of operations σ_i. Hence, if $\phi(h_0) = k_0$, then $\phi(h)$ is completely determined. This proves the uniqueness of ϕ. But this also gives a way to prove the existence of an homomorphism ϕ with $\phi(h_0) = h'_0$. In fact, define $\phi(h) = \sigma_{i_1} \ldots \sigma_{i_t}(k_0)$ with $h = \sigma_{i_1} \ldots \sigma_{i_t}(h_0)$ and $0 \leq i_j \leq 2$. Since tiling $\{r, q\}$ has no boundary, all faces are interior and the operations σ_i are always defined in $\{r, q\}$.

But, in order for this construction to work, we should prove that $\phi(h)$ is independent of the expression $\sigma_{i_1} \ldots \sigma_{i_t}$ chosen to express h in terms of h_0. Consider two expressions of h:

$$h = \sigma_{i_1} \ldots \sigma_{i_t}(h_0) = \sigma_{i'_1} \ldots \sigma_{i'_{t'}}(h_0).$$

Associate to it the flag paths:

$$P'' = (h_0, h_1, \ldots, h_t = h) \text{ with } h_j = \sigma_{i_j}(h_{j-1}),$$
$$P' = (h'_0 = h_0, h'_1, \ldots, h'_t = h) \text{ with } h_j = \sigma_{i'_j}(h'_{j-1}).$$

Those sequences correspond to vertex sequences, i.e. a path in the (r, q)-polycycle P. Locally, there is no obstruction to the coherence of the definition of $\phi(h)$. Around a boundary vertex there is no problem, and around an interior vertex, the condition of having degree q provides coherence. Also, if we consider flags around a face f, we do not encounter coherence problems. The simple connectedness of P allows us to modify the path P'' into the path P' by changing around faces and vertices. So, no ambiguity will appear. $\qquad \square$

All properties, defining (r, q)-polycycles, were used in the proof of Theorem 4.3.1. In particular, (r, q)-polycycles have to be simply connected; see some examples on Figure 4.4.

Clearly, the above cell-homomorphism is an isomorphism if and only if P is a proper polycycle, i.e. there is no pair of vertices or edges having the same image. In view of Theorem 4.3.1, any improper (r, q)-polycycle is called (r, q)-*helicene* (see Figure 4.2). It is easy to check that an (r, q)-helicene exists if and only if $(r, q) \neq (3, 3)$ and $p_r \geq (q - 2)(r - 1) + 1$ with equality only for the helicene being a ring of r-gons, going around an r-gon.

A natural parameter to measure an (r, q)-helicene, will be the degree of the corresponding homomorphism into $\{r, q\}$ (on vertices, edges, and faces). For $q \geq 4$, helicenes appear with vertices, but not edges, having same homomorphic image. The vertex-split Octahedron is a unique such maximal helicene for $(r, q) = (3, 4)$ (two 2-valent vertices are such; see Figure 4.2). There is a finite number of such helicenes for $(r, q) = (3, 5)$; one of them is the vertex-split Icosahedron.

Theorem 4.3.2 *([DeSt05]) The vertices, edges, and interior faces of any (r, q)-polycycle form a cell-complex.*

Proof. To prove that it is a cell-complex, we shall prove that the intersection of any two cells (i.e. vertices, edges, or interior faces) of an (r, q)-polycycle P is again a cell of P or $\emptyset$. For the intersection of vertices with edges or faces this is trivial. For the intersection of edges or faces, we will use the cell-homomorphism ϕ from P to $\{r, q\}$.

If two interior faces F and F' of P intersect in several cells (for example, two edges or two vertices), then their images in $\{r, q\}$ also intersect in several cells. It is easy to see that this cannot occur in $\{r, q\}$. So, F and F' intersect in an edge, vertex, or $\emptyset$. The same proof works for other intersections of cells. $\qquad \square$

Theorem 4.3.3 *If a finite (r, q)-polycycle has a boundary vertex whose degree is less than q, then the total number of these vertices is at least two.*

Proof. Take a flag f in this (r, q)-polycycle P and a flag f' in $\{r, q\}$. Let us now use the cell-homomorphism ϕ as in Theorem 4.3.1. The boundary $\mathcal{B}$ of P consists of vertices $v_1, \ldots, v_k$. The image of this boundary in $\{r, q\}$ is also a cycle $\phi(\mathcal{B})$.

Whenever $\mathcal{B}$ has a vertex v_i of degree q, the edges $e_i = v_{i-1}v_i$ and $e_{i+1} = v_iv_{i+1}$ belong to a common face F. Assume now, in order to reach a contradiction, that we have only one vertex of degree different from q. This implies that all e_i are incident to the same face F. In particular, the length of the boundary is a multiple of r. But if we turn around to face F, it must always be in the same direction, i.e. rightmost turn or leftmost turn. So, a vertex of degree different from q is impossible. □

The *girth* of a graph is the length of a minimal edge circuit in it.

Theorem 4.3.4 *It holds that for the skeleton of $\{r, q\}$:*

(i) its girth is r,
(ii) its minimal edge circuits are the boundaries of faces of $\{r, q\}$.

Proof. This statement can be easily verified for elliptic (r, q). Two edges of Tetrahedron always belong to the same triangle. Two adjacent edges of the same face of Octahedron (or Icosahedron) enter the boundary triangle of the face, while if they do not belong to the same face, then they enter an edge circuit of length at least four. Three successive edges of Cube (or Dodecahedron), belonging to the same face, enter the boundary quadrangle (pentagon) of the face, while if they do not belong to the same face, then they enter an edge circuit of length at least six.

The verification in the parabolic or hyperbolic case proceeds as follows. Take a simple edge circuit in the tiling $\{r, q\}$. By the Jordan theorem, this circuit bounds a finite domain in the plane; this domain contains at least one 2-dimensional r-gon of $\{r, q\}$. Draw the rays from the center of this circuit though its vertices. These rays divide the central angle into r sectors, each sector based on its own side of the r-gon. Any line connecting two external points of the boundary rays of a sector, is longer than the side of the r-gon. Therefore, the number of edges in any edge circuit containing the r-gon and not coinciding with it is greater than r. Hence, the girth of the skeleton of the tiling is r. Assertion (ii) follows in the same way. □

Corollary 4.3.5 *It hold that for any (r, q)-polycycle P:*

(i) P has girth r, and the length of its boundaries is at least r.
(ii) If a boundary of P has length r, then either P is an r-gon, or (r, q) is elliptic and P is $\{r, q\} - f$.

Proof. The assertion (i) is trivial using Theorem 4.3.4 and 4.3.1. Let us now prove (ii). If the length of a boundary of P is r, then P has only one boundary of finite length and so, P is a finite polycycle. Let us consider the mapping ϕ from P to $\{r, q\}$. The image of the boundary of P is a face of $\{r, q\}$. If the image lies inside of this r-gon, then P is an r-gon. Otherwise, all boundary vertices of P have degree q.

Since all vertices of the image of P in $\{r, q\}$ have degree q in $\{r, q\}$, we obtain that $\phi(P)$ covers completely $\{r, q\}$ except for one face. So, the parameters (r, q) are elliptic and $P = \{r, q\} - f$. □

Theorem 4.3.6 *Given a graph G that is the skeleton of an (r, q)-polycycle different from (one of 5) elliptic $\{r, q\}$, then the polycyclic realization is completely determined by G.*

Proof. From Corollary 4.3.5, we know that the boundary has length greater than r. Every face of a polycyclic realization of G yields a cycle of length r in G. Take a polycyclic realization P of G and consider the homomorphism ϕ from P to $\{r, q\}$. Every cycle of length r in G determines, under the mapping ϕ, a face F in $\{r, q\}$. Since any boundary has length greater than r, the face F corresponds to an interior face in P. Therefore, we have a one-to-one correspondence between interior faces of a polycyclic realization of G and cycles of length r in G. So, the graph G determines completely the polycyclic realization. □

Theorem 4.3.6 is an analog of the Steinitz theorem for 3-connected planar graphs.

4.4 Angles and curvature

Recall that the regular tiling $\{r, q\}$ lives on $X = \mathbb{S}^2$, $\mathbb{R}^2$, or $\mathbb{H}^2$ according to whether parameters (r, q) are elliptic, parabolic, or hyperbolic. The r-gons are regular in X and their curvature is $2r\alpha(2, r, q)$.

A set D in X is called *convex* if, for any two points $x, y \in D$, the geodesic, i.e. shortest path, joining x to y is contained in D. (Note that there are other definitions of convexity in hyperbolic space; see, for example, [GaSo01].) An (r, q)-polycycle is called *convex* if its image in $\{r, q\}$ is convex. See Figure 4.3 for a convex (r, q)-polycycle that is not proper. The only convex $(r, 3)$-polycycles are r-gon and $\{r, 3\}$. This is because, if a vertex $v \in \{r, 3\}$ is contained in two r-gons F_1 and F_2, then we can find two vertices $v_1 \in F_1$ and $v_2 \in F_2$ such that the geodesic between v_1 and v_2 pass by the third r-gon.

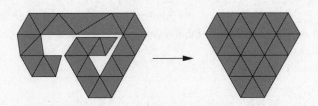

Figure 4.3 A convex non-proper $(3, 6)$-polycycle and its image in $\{3, 6\}$

Theorem 4.4.1 *Let P be an outerplanar (r, q)-polycycle. Then P, seen as an $(r, 2q - 2)$-polycycle, is convex and proper.*

Proof. Consider a face F in P. Starting with $Dec(P)_0 = F$, the finite (r, q)-polycycle $Dec(P)_{n+1}$ is formed by $Dec(P)_n$ and all faces of P sharing an edge with $Dec(P)_n$. $Dec(P)_n$ and P are outerplanar (r, q)-polycycles; denote by ϕ the cell-homomorphism of P into $\{r, 2q - 2\}$.

Every vertex of $Dec(P)_n$ is contained in at most $q - 1$ r-gons; hence, the interior angle in the image $\phi(Dec(P)_n)$ is at most π. This is a necessary and sufficient condition for the image $\phi(Dec(P)_n)$ to be convex (see, for example, [GaSo01, Lemma 3.1]). Therefore, the image of the boundary of $Dec(P)_n$ in $\{r, 2q - 2\}$ is not a self-intersecting curve and $Dec(P)_n$, considered as a $(r, 2q - 2)$-polycycle, is proper.

If x and y are two points in $\phi(P)$, then there exists an integer n_0 such that $x, y \in \phi(Dec(P)_{n_0})$. The geodesic d between x and y is included in $\phi(Dec(P)_{n_0})$ and, therefore, just as well in $\phi(P)$. If P is not proper, then there exist two distinct vertices v, v' (or edges, faces) of P, whose images in $\{r, q\}$ coincide. There exists an integer n_0 such that $v, v' \in Dec(P)_{n_0}$. But $Dec(P)_{n_0}$ is proper; so their images in $\{r, q\}$ do not coincide. □

In the proof of the above theorem, we need to use finite polycycles because infinite polycycles can have an infinity of boundaries.

Theorem 4.4.2 *Let P be an outerplanar $(3, q)$-polycycle; then P is a proper $(3, q + 2)$-polycycle.*

Proof. By the proof of Theorem 4.4.1, we can assume, without loss of generality, that P is finite. Denote by ϕ the cell-homomorphism in $\{3, q+2\}$ and assume further that P is not a proper $(3, q + 2)$-polycycle. Then we can find two vertices v, v' on the boundary of P with $\phi(v) = \phi(v')$ and the image of the boundary path $\mathcal{P} = (v_0 = v, v_1, \ldots, v_{p-1}, v_p = v')$ by ϕ being not self-intersecting. The Image $\phi(\mathcal{P})$ defines a finite $(3, q + 2)$-polycycle, denoted by P'. Since P was originally a $(3, q)$-polycycle, the boundary vertices of P' are of degree at least 4, except, possibly, the vertex $\phi(v) = \phi(v')$. Denote by v_i the number of boundary vertices of P' of degree i different from $\phi(v)$ and by v_{int} the number of interior vertices. Denote by p_3 the number of 3-gons of P'.

The number e of edges satisfies to:

$$2e = 1 + \sum_{i=4}^{q+2} v_i + 3p_3 = (deg\ \phi(v)) + \sum_{i=4}^{q+2} i v_i + (q + 2)v_{int},$$

which implies $3p_3 - (q + 2)v_{int} = (deg \, \phi(v)) - 1 + \sum_{i=4}^{q+2}(i - 1)v_i$. Then Euler formula $v - e + f = 2$ with $v = 1 + v_{int} + \sum_{i=1}^{q+2} v_i$ and $f = 1 + p_3$ gives:

$$p_3 = q v_{int} + \sum_{i=4}^{q+2} \left(\frac{i}{2} - 1 \right) v_i + \frac{deg \, \phi(v)}{2}.$$

Eliminating p_3, we get:

$$0 = (2q - 2)v_{int} + \sum_{i=4}^{q+2} \left(\frac{i}{2} - 2 \right) v_i + \frac{deg \, \phi(v)}{2} + 1 > 1,$$

which is impossible. $\qquad\qquad\qquad\qquad\qquad\qquad\qquad\qquad\qquad\qquad\square$

Note, that for $p_3 = 7$, there are outerplanar $(3, 4)$- and $(3, 5)$-polycycles, which remain helicenes in $\{3, 5\}$ and $\{3, 6\}$, respectively. A *fan of* $(q - 1)$ *r-gons* with q-valent common (boundary) vertex, is an example of outerplanar (r, q)-polycycle, which is a proper non-convex $(r, 2q - 3)$-polycycle.

We now consider another geometric viewpoint on (r, q)-polycycles. In the above consideration, the curvature was uniform and the triangles were viewed as embedded into a surface of constant curvature. Consider now the curvature to be constant, equal to zero, in the triangle itself, and to be concentrated on the vertices, where r-gons meet.

The r-gons are now regular r-gons and the angle at its vertices is $\frac{r-2}{r}\pi$. Consider a point A where q r-gons meet. The curvature of A is the difference between 2π and the sum of the angles of the r-gons, i.e. $2\pi - q\frac{r-2}{r}\pi$. The total curvature of the (r, q)-polycycle is then:

$$v_{int} \left(2\pi - q\frac{r - 2}{r}\pi \right) = v_{int} \frac{\pi}{r}(2(r + q) - rq).$$

It is different from the curvature defined earlier in this section, since boundary vertices contribute differently to it. If (r, q) is elliptic, parabolic, and hyperbolic, then the curvature of interior vertices is positive, zero, and negative, respectively.

We will use this curvature only for non-extensible polycycles in Section 8.2, and it will be only a counting, i.e. combinatorial, argument. However, the above curvature, concentrated on points, can be made geometric. This is the subject of Alexandrov theory (see, for example, [Ale50]).

4.5 Polycycles on surfaces

We now present an extension of the notion of a (r, q)-polycycle that concerns maps on surfaces. Given integers $r, q \geq 3$, an $(r, q)_{gen}$-*polycycle* is a 2-dimensional surface pasted together out of r-gons, so that the degrees of interior vertices are equal to q and the degrees of boundary vertices are within $[2, q]$.

A formal definition is the following: an $(r, q)_{gen}$-polycycle is a non-empty 2-connected map on surface S with faces partitioned in two non-empty sets F_1 and F_2, so that it holds that:

(i) all elements of F_1 (called *proper faces*) are combinatorial r-gons;
(ii) all elements of F_2 (called *holes*) are pairwisely disjoint, i.e. have no common vertices;
(iii) all vertices have degree within $\{2, \ldots, q\}$ and all *interior* (i.e. not on the boundary of a hole) vertices are q-valent.

Condition (ii) is here to forbid a vertex or an edge to belong to more than one hole. This condition is not necessary for (r, q)-polycycle, since the simple connectedness and the 2-connectedness imply it. An example of a map that does not satisfy this condition is shown below:

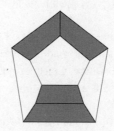

An $(r, q)_{gen}$-polycycle, which is simply connected, is, in fact, an (r, q)-polycycle, i.e. it can be drawn on the plane and the holes become exterior in this drawing. Some $(r, q)_{gen}$-polycycles can be drawn on the plane, for example, half of those in Theorem 4.5.1. The theory of coverings, presented in Section 1.2, applies to this setting. The universal cover of an $(r, q)_{gen}$-polycycle is, by definition, simply connected and so it is an (r, q)-polycycle.

Theorem 4.5.1 *For $r, q \leq 4$, the list of $(r, q)_{gen}$-polycycles that are not (r, q)-polycycles, consists of the following infinite series:*

1 $Prism_m$, $m \geq 2$ (on $\mathbb{S}^2$, with two m-gons seen as holes) and their non-orientable quotients, for $m \geq 2$ even (on projective plane, with one hole),

2 $APrism_m$, $m \geq 2$ (on $\mathbb{S}^2$, with two m-gons seen as holes) and their non-orientable quotients, for $m \geq 2$ odd (on projective plane, with one hole).

Proof. The universal cover of such a polycycle is an (r, q)-polycycle with a non-trivial group of fixed-point-free automorphisms. The list of (r, q)-polycycles for $r, q \leq 4$ is known (see Section 4.2). Inspection of this list gives only the infinite polycycles $Prism_\infty = P_2 \times P_\mathbb{Z}$ and $APrism_\infty = Tr_\mathbb{Z}$. Their orientable quotients are: the infinite series of prisms $Prism_m$ ($m \geq 2$), the infinite series of antiprisms $APrism_m$

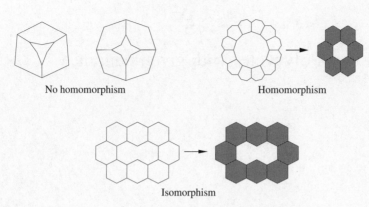

No homomorphism   Homomorphism

Isomorphism

Figure 4.4 Some $(6, 3)_{gen}$-polycycles and their homomorphisms to $\{6, 3\}$ when they exist

$(m \geq 2)$, and the non-orientable quotients (with respect to central symmetry) of $Prism_m$, for m even, and of $APrism_m$, for m odd. □

The two dualities, inner and outer, extend to this setting. For example, we have equalities:

$$Inn^*(\text{snub } Prism_m) = APrism_m \quad \text{and} \quad Inn^*(\text{snub } APrism_m) = \text{snub } Prism_m.$$

In general, an $(r, q)_{gen}$-polycycle does not admit homomorphism to $\{r, q\}$, since $(r, q)_{gen}$-polycycles are not simply connected. But, sometimes, such a homomorphism exists, see Figure 4.4. An (r, q)-*map* is a particular case of $(r, q)_{gen}$-polycycle, such that every vertex has degree q.

Finally, we mention two other relatives of finite (r, q)-polycycles. See [ArPe90] and references there (mainly authored by Perkel) for the study of *strict polygonal graphs*, i.e. graphs of girth $r \geq 3$ and vertex-degree q, such that any path P_3 (with two edges) belongs to a unique r-circuit of the graph. See [BrWi93, pages 546–547] for information on *equivelar* polyhedra, i.e. polyhedral embeddings with convex faces, of (r, q)-*map* into $\mathbb{R}^3$. So, in both these cases graphs are q-valent, have girth r, Euler–Poincaré characteristic $v - e + f = \frac{v(6-r)}{2r}$, and coincide with Platonic polyhedra in the case of genus 0. Recall that, for an (r, q)-polycycle P, any non-boundary path P_3 belongs to a unique r-circuit.

5

Polycycles with given boundary

The (r, q)-*boundary sequence* of a finite (r, q)-polycycle P is the sequence $b(P)$ of numbers enumerating, up to a cyclic shift or reversal, the consecutive degrees of vertices incident to the exterior face. For earlier applications of this (and other) codes, see [HeBr87, BCC92, HLZ96, DeGr99, CaHa98, DFG01].

Given an (r, q)-boundary sequence b, a plane graph P is called a (r, q)-*filling* of b if P is an (r, q)-polycycle such that $b = b(P)$.

In this chapter we consider the unicity of those (r, q)-fillings and algorithms used for their computations.

5.1 The problem of uniqueness of (r, q)-fillings

By inspecting the list of (r, q)-polycycles for $(r, q) = (3, 3), (3, 4)$, or $(4, 3)$ in Section 4.2, we find that the (r, q)-boundary sequence of an (r, q)-polycycle determines it uniquely. We expect that for any other pair (r, q) this is not so.

We show that the value $r = 3, 4$ are the only ones, such that the $(r, 3)$-boundary sequence always defines its $(r, 3)$-filling uniquely. Note that an (r, q)-polycycle, which is not an unique filling of its boundary, is, necessarily, a helicene. Some examples of non-uniqueness of $(r, 3)$-fillings are cases of boundaries b, admitting an $(r, 3)$-filling P with the symmetry group of b being larger than the symmetry group of P, implying the existence of several different $(r, 3)$-fillings. However, as the length of the boundary increases, the number of possibilities grows, see [DeSt06].

Theorem 5.1.1 *For any $r \geq 5$, there is an $(r, 3)$-boundary sequence b, such that there exist $(r, 3)$-fillings P and P' with $P \neq P'$ and $b(P) = b(P') = b$. For instance:*

(i) If $r = 5$, then such an example is given by $b_5 = u3u3^4u3u3^4$ with $u = 3232323$ of length 38 (see Figure 5.1).

(ii) If $r \geq 6$, then such an example is given by:

$$b_r = u3^{r-1}u2^{r-6}u3^{r-1}u2^{r-6} \text{ with } u = (32^{r-4})^{r-1}3$$

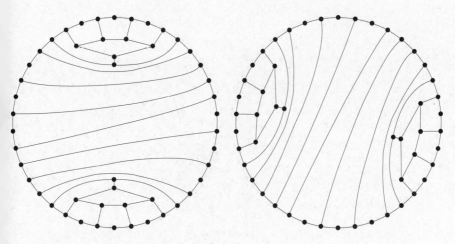

Figure 5.1 Two different (but isomorphic as maps) (5, 3)-fillings of the same (5, 3)-boundary

of length $4r^2 - 12r + 2$. Amongst the vertices of this boundary, $6r - 2$ vertices are of degree 3 and the remaining ones are of degree 2 (see Figure 5.2).

The $(r, 3)$-boundary sequence b_r can be filled by $4r$ r-gons in two ways obtained one from the other by reflection.

Proof. For any of the boundary sequences indicated in the theorem, the symmetry group is C_{2v} of size 4. Furthermore, the fillings on Figures 5.1 and 5.2, for $p = 5, 6, 7$, have symmetry C_2, which is a group of order 2. This is, in fact, true for any b_r with $r \geq 5$. So, the $(r, 3)$-boundary b_r has at least two fillings, which are isomorphic as maps but different. It proves that the $(r, 3)$-boundary sequence b_r does not define uniquely its $(r, 3)$-filling. □

The main difficulty of above theorem consists in finding the $(r, 3)$-boundary sequence b_r. It seems likely that our examples are minimal with respect to the number of r-gons of their corresponding fillings.

Remark 5.1.2 *The $(3, 5)$-boundary sequence $(43445544345)^2$ admits two different $(3, 5)$-fillings (by 36 3-gons and 30 vertices) shown on Figure 5.3. Those fillings are isomorphic and have only symmetry C_2, while the boundary has symmetry C_{2v}, as in Theorem 5.1.1.*

The $(3, 5)$-boundary sequence $(34345)^2 5^2 (34345)^2 5^2$ admits two different $(3, 5)$-fillings (by 34 3-gons and 30 vertices) shown on Figure 5.4. Those fillings are non-isomorphic and have the same symmetry as the boundary, i.e. C_2. This $(3, 5)$-boundary sequence might be minimal for the number of 3-gons.

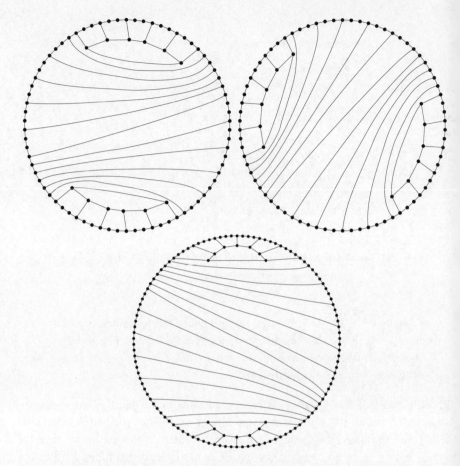

Figure 5.2 Two different (but isomorphic as maps) $(6, 3)$-fillings of the same $(6, 3)$-boundary; the left one and the lower one are the first cases $r = 6, 7$ of a series of $(r, 3)$-fillings, which are not defined by their boundaries

In the case of $(5, 3)$-polycycles, for every given number we have an example of a $(5, 3)$-boundary sequence, which admits exactly that number of fillings. The statement and the proof of this theorem used the elementary polycycles presented in Chapter 7 (especially, E_1, C_1, and C_3 from Figure 7.2).

Theorem 5.1.3 *The $(5, 3)$-boundary $b(n) = 223^{5n+1}223^{5n+3}223^{5n+1}223^{5n+3}$ admits exactly $n+1$ different $(5, 3)$-fillings. Each such filling corresponds to a number k, $0 \le k \le n$, where the kth filling is obtained by taking two elementary $(5, 3)$-polycycles E_1, gluing them and adding to them four open edges of $E_1 + E_1$, respectively, chains of k, $n - k$, k, and $n - k$ elementary polycycles C_1.*

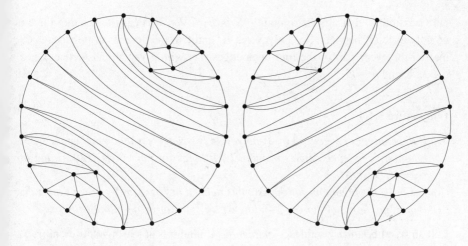

Figure 5.3 Two different (but isomorphic as maps) (3, 5)-fillings of the same boundary

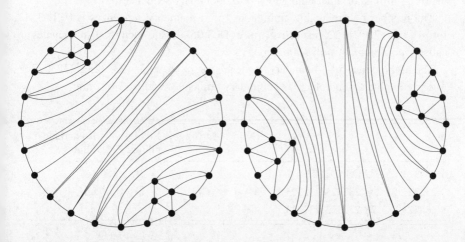

Figure 5.4 Two different non-isomorphic (3, 5)-fillings (with symmetry C_2) of the same boundary with symmetry C_2

Proof. We need to prove only that there are no more than these $n + 1$ fillings. First, since all runs of two are 22, the only possible elementary (5, 3)-polycycles are E_1, C_1, or C_3. Secondly, since the (5, 3)-boundary has exactly 4 runs of 2, the total number of (5, 3)-polycycles E_1 and C_3 is 2.

The addition of the (5, 3)-polycycle C_1 to an existing (5, 3)-polycycle, adds the symbol 3^5 at two emplacements of the (5, 3)-boundary sequence. The three possible cases $E_1 + E_1$, $E_1 + C_3$, and $C_3 + C_3$ correspond, respectively, to (5, 3)-boundary sequences $223^1223^3223^1223^3$, $223^2223^4223^1223^4$, and $223^2223^5223^2223^5$. So, the only possibility, that agrees to modulo 5, is $E_1 + E_1$. It is easy to see that there is no polycycle C_1 between two polycycles E_1 and that the number of polycycles C_1, inserted on every part of the four open edges (i.e. with 2-valent end vertices), is k, $n - k, k$, and $n - k$. □

Conjecture 5.1.4

(i) *If the number of r-gons of $(r, 3)$-polycycle is strictly less than $4r$, then the $(r, 3)$-boundary sequence defines it uniquely. It holds for $r = 6$ ([GHZ02]) and for $r = 5$ ([DeSt06]).*

(ii) *Let b_r be the boundary for the example defined in Theorem 5.1.1; we expect that it does not admit either $(r', 3)$-filling with $r' > r$, or (r, q)-filling with $q > 3$.*

Call an (r, q)-boundary sequence *ambiguous* if it admits at least two different (r, q)-fillings. Call it *irreducible* if its (r, q)-filling does not contain, as induced polycycle, the (r, q)-fillings of other ambiguous (r, q)-boundary sequences.

The number of irreducible ambiguous $(5, 3)$-boundary sequences is 0, 1, 3, 17, for $p_5 < 20, 20, 21, 22$, respectively (see [DeSt06], where such $(5, 3)$-polycycles are called *equi-boundary polypentagons*).

The ambiguous $(5, 3)$-boundary sequence for $p_5 = 20$, is $(2323233^5 2323233^2)^2$. The irreducible ambiguous $(5, 3)$-boundary sequences with $p_5 = 21$ are:

b	$|b|$	$Aut(b)$	Nr. of fillings	Isomorphic fillings
$3^2 232323^6 2323^4 2323^6 232323$	35	C_s	2	yes
$3^2 232323^6 23^2 2^2 3^4 232323^6 232323$	39	C_1	2	no
$3^2 232323^7 2^2 3^2 23^3 232323^6 232323$	39	C_1	2	no

Here $|b|$ is the length of the sequence, $Aut(b)$ is its automorphism group, and we indicate if the obtained fillings are isomorphic. In Figure 5.5, a $(6, 3)$-polycycle is presented, whose $(6, 3)$-boundary sequence admits eight different $(6, 3)$-fillings. More generally, by aggregating the examples, found in Theorem 5.1.1, we can obtain an $(r, 3)$-boundary, admitting an arbitrary large number of fillings. Furthermore, by adding one r-gon to an example, found in Theorem 5.1.1, we can obtain $(r, 3)$-boundaries, admitting two non-isomorphic $(r, 3)$-fillings.

Note that if two (r, q)-polycycles have the same (r, q)-boundary, then their image under the cell-homomorphism in $\{r, q\}$ is the same. Moreover, in [Gra03] it is proved that if the image of a $(6, 3)$-polycycle does not triply cover some point, then the $(6, 3)$-boundary sequence determines this $(6, 3)$-polycycle uniquely.

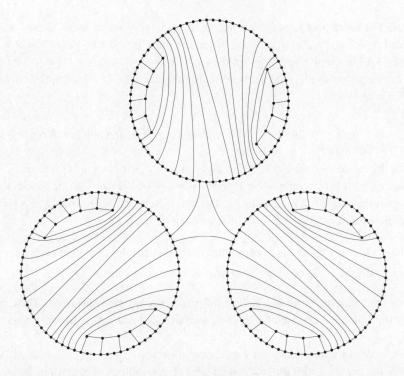

Figure 5.5 Several non-isomorphic $(6, 3)$-fillings of the same $(6, 3)$-boundary are obtained by different $(6, 3)$-fillings of above 3 components

5.2 $(r, 3)$-filling algorithms

The algorithmic problem treated here is: given an $(r, 3)$-boundary, find all possible $(r, 3)$-fillings of it.

Theorem 5.2.1 *Let P be an $(r, 3)$-polycycle. Denote by x the number of interior vertices of P and by p_r the number of r-gonal faces in P. Denote by v_2, v_3 the number of vertices of degree 2, 3, respectively, on the $(r, 3)$-boundary; then it holds that:*

$$\begin{cases} p_r - \frac{x}{2} = 1 + \frac{v_3}{2} \\ rp_r - 3x = v_2 + 2v_3 \end{cases} \tag{5.1}$$

Moreover, it holds that:

(i) if $r \neq 6$, then $x = \frac{2(v_2 - r) - (r-4)v_3}{r-6}$ and $p_r = \frac{v_2 - v_3 + 5}{r-6}$.
(ii) if $r = 6$, then $v_2 = v_3 + 6$.

Proof. The number of edges e satisfies to $2e = v_2 + v_3 + rp_r = 2v_2 + 3v_3 + 3x$, which implies $rp_r - 3x = v_2 + 2v_3$. Then the Euler formula $v - e + f = 2$ with $v = v_2 + v_3 + x$ and $f = 1 + p_r$ implies $p_r - \frac{x}{2} = 1 + \frac{v_3}{2}$.

The linear system (5.1) has a unique solution if and only if $r \neq 6$, thereby proving (i). (ii) is obvious. □

Actually, for $r = 6$ also the $(r, 3)$-boundary determines the number of r-gons (see [GHZ02, BDvN06]).

The basic idea of the $(r, 3)$-filling algorithm is the following. Given an $(r, 3)$-boundary b and two consecutive vertices x and y of degree 3 on it, consider all possible ways to add an r-gonal face to this pair of vertices. When adding an r-gonal face, there are several possibilities (illustrated in Figure 5.6):

- either the new r-gon does not split the $(r, 3)$-boundary into different components,
- or it splits the $(r, 3)$-boundary into two, or more, components.

Given one pair of vertices x and y, all cases should be considered. Then, the algorithm should be reapplied to the remaining boundaries, until we obtain an $(r, 3)$-polycycle.

The order in which cases are considered affects the speed of computation. We choose the pair x, y with the smallest number of possibilities of extension.

In particular, if the $(r, 3)$-boundary sequence contains the pattern 2^{r-1} or 2^{r-2}, then there is a unique way of adding an edge; so we obtains a unique smaller problem (see Figure 5.7).

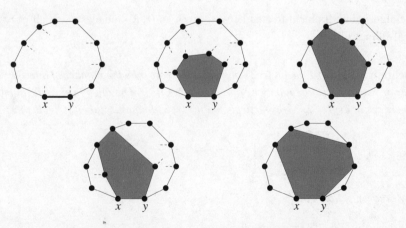

Figure 5.6 Some examples of possible ways to add a 6-gon between two vertices x and y of valency 3 on the $(6, 3)$-boundary

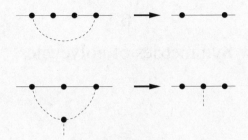

Figure 5.7 The unique completion cases for $r = 4$

Another speedup consists in showing that an $(r, 3)$-boundary b does not admit any $(r, 3)$-fillings. Two criteria are possible:

- Use Theorem 5.2.1 to compute the number of interior vertices and faces. If they are negative or non-integer, then the $(r, 3)$-boundary is non-extensible. This is a global criterion.
- If a pair of consecutive vertices of valency 3 on the $(r, 3)$-boundary does not admit any extension by an r-gon, i.e. its distance is lower than $r - 1$, then the $(r, 3)$-boundary is non-extensible. This is a local criterion.

By using these methods, we can find all $(r, 3)$-fillings of a given $(r, 3)$-boundary. The running time is, most likely, non-polynomial; if the length of the $(r, 3)$-boundary increases, the number of cases to consider becomes very large and the efficiency of the above criteria is limited. Nevertheless, if $r \neq 6$, the algorithm is guaranteed to terminate, since then we know how many r-gons will occur. If we want to check whether an (r, q)-boundary corresponds to a proper (r, q)-polycycle, then this is easier. The (r, q)-boundary sequence specifies a path in $\{r, q\}$; this path has to be closed and this is an easily checked condition. The second and sufficient condition is that the path has to be non- self-intersecting. Both conditions can be checked in polynomial time.

6

Symmetries of polycycles

Recall that the *automorphism group $Aut(P)$* of an (r, q)-polycycle P is the group of automorphisms of the plane graph preserving the set of interior faces (see Section 4.1). Call a polycycle P *isotoxal*, *isogonal*, or *isohedral* if $Aut(P)$ is transitive on edges, vertices, or interior faces, respectively. In this chapter we first consider the possible automorphism groups of an (r, q)-polycycle, then we list all isogonal or isohedral polycycles for elliptic (r, q) and present a general algorithm for their enumeration. We also present the problem of determining all isogonal and isohedral $(r, q)_{gen}$-polycycles.

6.1 Automorphism group of (r, q)-polycycles

If an (r, q)-polycycle P is finite, then it has a single boundary and $Aut(P)$ is a dihedral group consisting only of rotations and mirrors around this boundary. So its order divides $2r$, 4, or $2q$, depending on what $Aut(P)$ fixes: the center of an r-gon, the center of an edge, or a vertex.

None of $(3, 3)$-, $(3, 4)$-, $(4, 3)$-polycycles has, but almost all (r, q)-polycycles for any other (r, q) have, trivial $Aut(P)$.

The number of *chiral* (i.e. with $Aut(P)$ containing only rotations and translations) proper $(5, 3)$-, $(3, 5)$-polycycles is 12, 208 (amongst, respectively, all 39, 263.)

Given an (r, q)-polycycle P, consider the cell-homomorphism ϕ, presented in Section 4.3, from P to $\{r, q\}$. It maps the group $Aut(P)$ into $Aut(\{r, q\}) = T^*(l, m, n)$. The image $\phi(Aut(P))$ consists of automorphisms of $\phi(P)$. If P is a proper polycycle, then $Aut(P)$ coincides with $Aut(\phi(P))$. Otherwise, $Aut(P)$ is an extension of $Aut(\phi(P))$ by the kernel of this homomorphism.

Only r-gons and non-Platonic plane tilings $\{r, q\}$ are isotoxal; their respective automorphism groups are C_{rv} and $T^*(2, r, q)$. The group $Aut(\{r, q\} - f)$ is C_{rv} in five Platonic cases; none is isotoxal, isogonal, or isohedral polycycle, except of isohedral $\{3, 3\} - f = (3, 3)$-star.

We call the set of q r-gons with a common vertex, the (r, q)-*star* of q-gons.

6.2 Isohedral and isogonal (r, q)-polycycles

Theorem 6.2.1 *Let P be an isohedral (r, q)-polycycle; then it holds that:*

(i) *Every r-gon has the same number t of non-boundary edges.*
(ii) *For $t = 0$, r or 1, the polycycle P is, respectively, r-gon, non-elliptic $\{r, q\}$, or a pair of adjacent r-gons (with $Aut(P) = C_{2v}$).*
(iii) *If $t = 2$, then P is either an (r, q)-star, or an infinite outerplanar polycycle.*
(iv) *If $t \geq 3$, then P is infinite.*

Theorem 6.2.2 *(i) For a given pair (r, q), the number of isohedral (r, q)-polycycles is finite. Moreover, it is bounded by a function depending only on r.*

(ii) If P is an isogonal (r, q)-polycycle, then either it is r-gon, or non-elliptic $\{r, q\}$, or an infinite outerplanar polycycle and its outer dual $Out^*(P)$ (considering P as an $(r, q + 1)$-polycycle) is an isohedral infinite $(q + 1, r)$-polycycle.

Proof. Consider an isohedral (r, q)-polycycle P and fix a face F in it. The automorphism group of an r-gon is the dihedral group C_{rv}; the stabilizer $Stab(F)$ of F in P is a subgroup of C_{rv}. Since C_{rv} is finite, we have a finite number of possibilities for $Stab(F)$. The face F is adjacent in P to t r-gonal faces $F_1, \ldots, F_t$. By isohedrality, we have, for every i, a transformation ϕ_i of P that maps F to F_i. This transformation is defined up to an element of $Stab(F)$. Clearly, there are at most $2r$ possible transformations ϕ_i. One way to see it is that ϕ_i is defined by the image of a flag (v, e, F) into a flag (v', e', F_i) and that there are $2r$ such flags. So, the finiteness is established. Moreover, the number of choices for $Stab(F)$ depends only on r; so, the total number of choices is bounded by a function depending only on r.

If P is an isogonal (r, q)-polycycle, then either no vertex belongs to the boundary and we have no boundary, or every vertex belongs to the boundary and P is outerplanar. If P is outerplanar, then we can consider it as an $(r, q + 1)$-polycycle and so the outer dual is well defined and isohedral. □

Given a pair (r, q), the above theorem gives that the set of isohedral (r, q)-polycycles is finite but not precisely how we can describe this set. Given an r-gon F of P, we specify a transformation that maps F to adjacent r-gons to F. After this transformation is prescribed we check if the coherency is satisfied on F. Simple connectedness assures us that those local conditions are actually global (see [Dre87], for the proofs). The actual enumeration is then done by computer, using exhaustive enumeration schemes (see [Du07]).

Theorem 6.2.3 *For any r, there exists isohedral $(r, 4)$-polycycle with exactly one boundary edge on each r-gon.*

Proof. Take an r-gon F and select $r - 1$ edges. On every one of those edges define the line symmetry that maps the r-gon to its mirror along the edge; see below, for $r = 4$, the line symmetries along edges 1, 2, and 3:

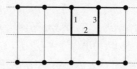

The r-gon has $r - 2$ interior vertices. All those transformations fit together around those vertices. So, they yield an isohedral $(r, 4)$-polycycle. $\qquad\square$

For non-elliptic (r, q), the practical representation of an isohedral (r, q)-polycycle is difficult, since the number of vertices at distance r from a given vertex grows very fast and we are led to draw smaller and smaller faces. The compressed presentation (see Theorems 6.2.5 and 6.2.6 below) will mimic the computer presentation of such polycycles: it presents one r-gon F and its adjacent r-gons F_i. Boundary edges and boundary vertices are boldfaced. The edges of F are marked by a number and the edges of the adjacent r-gons F_i have those numbers under a symmetry transformation (generally, non-unique) mapping F to F_i. The stabilizer $Stab(F)$ of F is a point group, whose type is indicated under picture. See below a representation of an infinite $(3, 5)$-polycycle and its corresponding code:

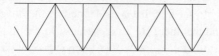 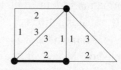

A similar representation can be done for isogonal (r, q)-polycycles.

Another way to get a representation, is to take a quotient $\widetilde{P}$ of an (r, q)-polycycle P by a subgroup of $Aut(P)$. $\widetilde{P}$ might be easier to represent and P is then obtained by taking the universal cover of the $(r, q)_{gen}$-polycycle $\widetilde{P}$ (see Figure 6.3 for two examples).

Using the algorithm of Theorem 6.2.2, in [DeSt00b], all elliptic isohedral polycycles were found (see Figure 6.1): 11 finite ones, 7 infinite ones with strip groups, and a $(5, 3)$-polycycle with $Aut(P) = T^*(2, 3, \infty)$.

The enumeration of elliptic isogonal (r, q)-polycycles yields nine polycycles: all three r-gons (as five polycycles), $Prism_\infty$, $APrism_\infty$ (as two polycycles), and the isogonal $(3, 5)$-polycycle represented in Figure 6.3.

We now list some existence and classification results for isohedral (r, q)-polycycles obtained by using the previous formalism.

Remark 6.2.4 *If an (r, q)-polycycle is proper, then we can realize its group of combinatorial transformations as a group of isometries of its image in $\{r, q\}$. If the*

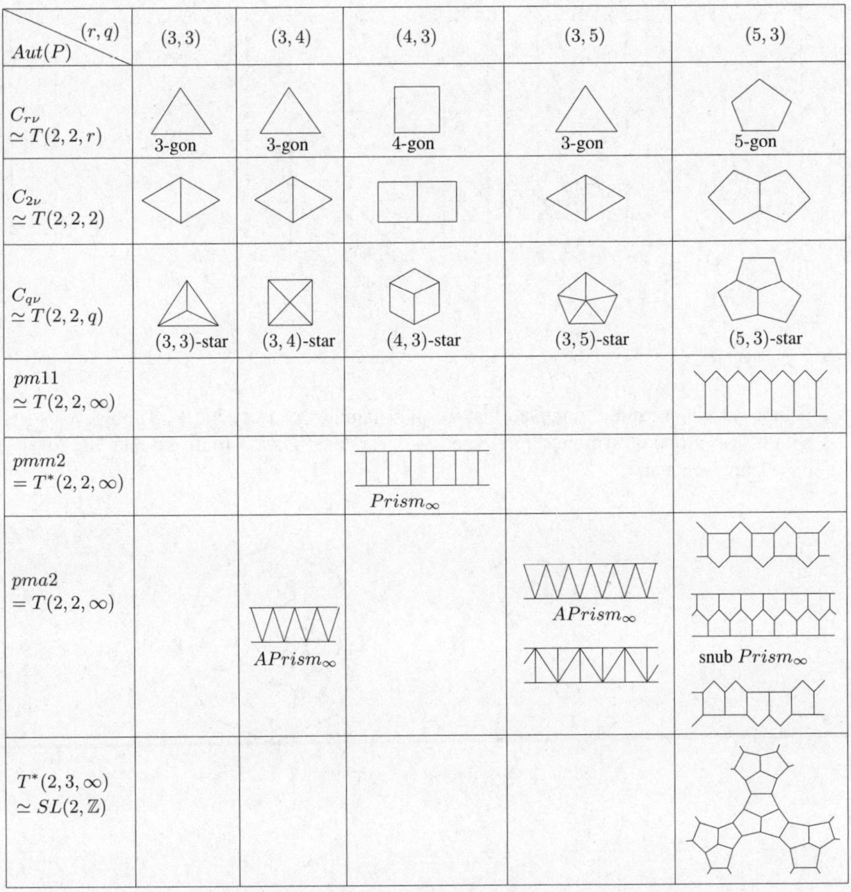

(r,q) $Aut(P)$	$(3,3)$	$(3,4)$	$(4,3)$	$(3,5)$	$(5,3)$
C_{rv} $\simeq T(2,2,r)$	3-gon	3-gon	4-gon	3-gon	5-gon
C_{2v} $\simeq T(2,2,2)$					
C_{qv} $\simeq T(2,2,q)$	$(3,3)$-star	$(3,4)$-star	$(4,3)$-star	$(3,5)$-star	$(5,3)$-star
$pm11$ $\simeq T(2,2,\infty)$					
$pmm2$ $= T^*(2,2,\infty)$			$Prism_\infty$		
$pma2$ $= T(2,2,\infty)$	$APrism_\infty$			$APrism_\infty$	snub $Prism_\infty$
$T^*(2,3,\infty)$ $\simeq SL(2,\mathbb{Z})$					

Figure 6.1 All 19 isohedral elliptic (r,q)-polycycles; only r-gons and $Prism_\infty$, $APrism_\infty$ are isogonal (the remaining 9th elliptic isogonal polycycle is given on the right-hand side of Figure 6.3)

(r,q)-polycycle is a helicene, then this is not possible; see, for example, the infinite elliptic isohedral (r,q)-polycycles in Figure 6.1. Unfortunately, we do not know any practical method for checking if a given infinite (r,q)-polycycle is proper or not.

Theorem 6.2.5 *All isohedral $(3,q)$-polycycles with $q \geq 3$ are:*

(i) 3-gon (isogonal), $\{3,q\}$ (isogonal), pair of adjacent 3-gons, $(3,q)$-star.

(ii) For $q \geq 4$, $APrism_\infty$ (isogonal) and, for $q \geq 5$, the infinite not isogonal $(3,5)$-polycycle from Figure 6.1.

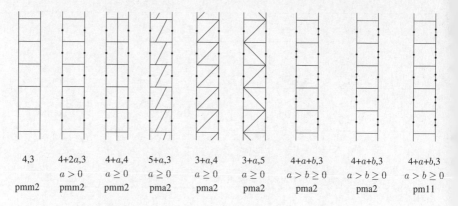

4,3	4+2a,3	4+a,4	5+a,3	3+a,4	3+a,5	4+a+b,3	4+a+b,3	4+a+b,3
	$a>0$	$a\geq0$	$a\geq0$	$a\geq0$	$a\geq0$	$a>b\geq0$	$a>b\geq0$	$a>b\geq0$
pmm2	pmm2	pmm2	pma2	pma2	pma2	pma2	pma2	pm11

Figure 6.2 Conjecturally complete list of eight families of isohedral (r,q)-polycycles with a strip group of symmetry presented as decorated $Prism_\infty$ (only 1st and 5th, for $a=0$, are isogonal)

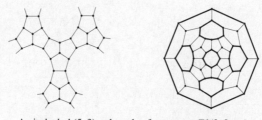

An isohedral $(5,3)$-polycycle of symmetry $T^*(2,3,\infty)$

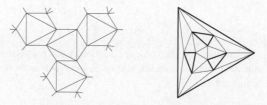

An isogonal $(3,5)$-polycycle of symmetry $T(2,3,\infty)$

Figure 6.3 Two examples of infinite (r,q)-polycycles with, for each of them, a quotient $(r,q)_{gen}$-polycycle (strictly face-regular Nr. 58 and Icosahedron with boldfaced six, respectively, four holes)

Theorem 6.2.6 *All isohedral $(4,q)$-polycycles with $q\geq3$ are:*

(i) 4-gon (isogonal), $\{4,q\}$ (isogonal), pair of adjacent 4-gons, $(4,q)$-star.

(ii) For $q\geq3$, respectively, $q\geq4$, the following outerplanar $(4,3)$-polycycle, respectively, $(4,4)$-polycycle:

C_{2v}, isogonal C_s

(iii) For $q \geq 4$, respectively, $q \geq 5$, the following $(4, q)$-polycycles:

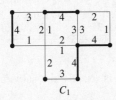

C_1

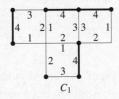

C_1

(iv) If $q \geq 4$ even, respectively, $q \geq 6$ even, the following $(4, q)$-polycycles:

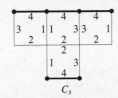

C_s

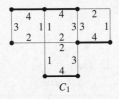

C_1

(v) For $q \geq 5$, the following three outerplanar $(4, 5)$-polycycles:

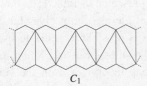

C_1

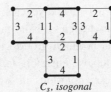

C_s, isogonal

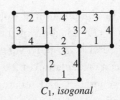

C_1, isogonal

(vi) For $q \geq 7$, the following two outerplanar $(4, 7)$-polycycles:

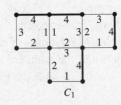

C_1 C_1

Table 6.1 *Triples (x, y, z) of numbers x of isohedral, y of isogonal, z of isohedral, and isogonal (r, q)-polycycles, different from r-gon and $\{r, q\}$, for $r, q \leq 8$*

$r \downarrow q \rightarrow$	3	4	5	6	7	8
3	2, 0, 0	3, 1, 1	4, 2, 1	4, 3, 1	4, 6, 1	4, 11, 1
4	3, 1, 1	6, 3, 1	9, 8, 3	11, 21, 3	11, 53, 3	13, 137, 3
5	7, 0, 0	17, 0, 0	24, 0, 0	38, 5, 5	37, 13, 5	51, 19, 5
6	12, 1, 1	45, 4, 3	67, 11, 3	130, 24, 3	123, 87, 20	196, 234, 20
7	28, 0, 0	157, 0, 0	257, 0, 0	518, 0, 0	452, 0, 0	896, 60, 60
8	58, 1, 1	486, 3, 1	894, 11, 6	2095, 35, 6	1781, 119, 6	3823, 367, 6

Proof. The proof is a case-by-case analysis. It is exactly the same as running the program; so, we refer to the program itself. □

Clearly, y and z, introduced in Table 6.1, are non-decreasing functions of q and $z \leq \min(x, y)$. From Theorem 6.2.5, it follows that if $r = 3$ and $q \geq 4$, then $z = 1$ and it is realized by $APrism_\infty$, seen as a $(3, q)$-polycycle.

Conjecture 6.2.7

(i) *If r is odd, then $y = z = 0$ for $3 \leq q \leq r$, $y = z > 0$ for $q = r + 1$ and $y > z > 0$, otherwise.*

(ii) *If r is even, then $y > z > 0$ for $q \geq 4$ and $y = z = 1$, for $q = 3$; it is realized by $(r, 3)$-cactus (infinite $(r, 3)$-polycycle obtained by growing from an r-gon, i.e. adding r-gon on $\frac{r}{2}$ disjoint edges, see two decorations of the $(6, 3)$-cactus on Figure 6.3).*

The computation of isohedral (r, q)-polycycles, presented in Table 6.1, shows that a full classification of them is hopeless. Note that the algorithm gives the list of isohedral (r, q)-polycycles but not their groups. It is possible to generate by computer a presentation of the group by generators and relations. But the groups, defined by generators and relations, are notorious in group theory for even the simplest questions to be *undecidable* (i.e. no algorithm can answer those questions in full generality; see, for example, [Coo04] for an introduction to this subject). It makes the identification with known groups somewhat of an art. See on Figure 6.4 an isohedral $(6, 3)$-polycycle (obtained in [DeSt01] with many other examples), whose automorphism group $\mathbb{Z}_3 \times T^*(\infty, \infty, \infty)$ is not a Coxeter group.

We can consider k-isotoxal, k-isohedral, and k-isogonal (r, q)-polycycles, i.e. (r, q)-polycycles with k orbits of edges, faces, and vertices, respectively. The finiteness still holds and the enumeration would still be possible by computer, but the number of possibilities would be much larger and the complexity of the computation is unknown.

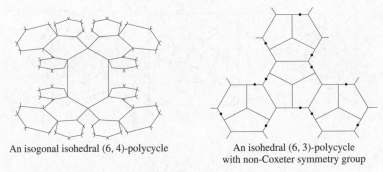

An isogonal isohedral $(6, 4)$-polycycle

An isohedral $(6, 3)$-polycycle
with non-Coxeter symmetry group

Figure 6.4 Two isohedral (r, q)-polycycles

Eight families of isohedral (r, q)-polycycles with $Aut(P)$ being a strip group are given in Figure 6.2; all have $q = 3, 4$. The points on edges indicate vertices of degree two of depicted polycycles producing those families. We think there is no other isohedral (r, q)-polycycles with a strip group.

For any $r \geq 5$ there exists a continuum of *quasi*-isohedral polycycles, i.e. not isohedral ones, but with all r-gons having the same corona, i.e. circuit of adjacent faces. In fact, let T be an infinite, in both directions, path of regular r-gons, such that for any of them the edges of adjacency to their neighbors are at distance $\lfloor \frac{r-3}{2} \rfloor$ and the sequence of (one of two possible) choices of joining each new r-gon, is aperiodic and different from its reversal. There is a continuum of such paths T for any $r \geq 5$. Any T is quasi-isohedral and its group of automorphisms is trivial. It is an $(r, 3)$-helicene if $r = 5, 6$ and isometric proper polycycle if $r \geq 7$.

6.3 Isohedral and isogonal $(r, q)_{gen}$-polycycles

The hypothesis of simple connectedness of (r, q)-polycycle radically simplifies the enumeration of the isohedral and isogonal ones.

Given an $(r, q)_{gen}$-polycycle, which is isohedral, isogonal, then its universal cover is also isohedral, isogonal, respectively. This gives, in principle, a method for enumerating the isohedral, respectively, isogonal $(r, q)_{gen}$-polycycles: enumerate such simple connected ones, i.e. such (r, q)-polycycles, then take their quotients by adequate groups.

As an illustration, consider the enumeration of isogonal $(3, 5)_{gen}$-polycycles. There are three isogonal $(3, 5)$-polycycles: 3-gon, $APrism_\infty$, and the isogonal $(3, 5)$-polycycle depicted on Figure 6.3, which we denote by $P_{3,5}$. The 3-gon has no possible quotients, the quotients of $APrism_\infty$ are enumerated in Section 4.5.

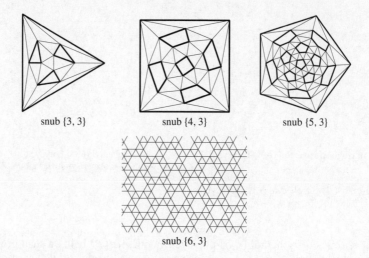

snub {3, 3}   snub {4, 3}   snub {5, 3}

snub {6, 3}

Figure 6.5 Twisted $\{b, 3\}$ for $b = 3, 4, 5, 6$

The automorphism group $Aut(P_{3,5})$ of $P_{3,5}$ is isomorphic to the modular group $PSL(2, \mathbb{Z}) \simeq T(2, 3, \infty)$ and the stabilizer of any vertex is trivial.

Let G be a group of fixed-point-free automorphisms of $P_{3,5}$ such that the quotient map $P_{3,5}/G$ is isogonal. Since the group of automorphisms of $P_{3,5}/G$ is isomorphic to the quotient $N_{Aut(P_{3,5})}(G)/G$, it holds that $N_{Aut(P_{3,5})}(G) = Aut(P_{3,5})$, since the stabilizer of any vertex is trivial. So, G has to be a normal subgroup of $Aut(P_{3,5})$, in order, for the quotient to be isogonal. There is a large variety of normal fixed-point-free subgroups of $PSL(2, \mathbb{Z})$ (see [New72]) and so a large variety of isogonal $(3, 5)_{gen}$-polycycles.

One class of them, *snub* $\{b, 3\}$, is obtained from the regular tiling $\{b, 3\}$ by replacing every vertex by a 3-gon and every edge by two 3-gons. The snub $\{b, 3\}$ is a $(\{3, b\}, 5)$-map (see Chapter 2). See snub $\{b, 3\}$ for $3 \leq b \leq 6$ on Figure 6.5, i.e. some $(3, 5)_{gen}$-polycycles with holes being b-gons. The corona of their vertices is $3^4.b$; so, their skeletons are Icosahedron, Snub Cube, Snub Dodecahedron, and Archimedean plane tiling $(3^4.6)$, respectively.

In practice, if we want normal subgroups of a specific finite index, then we use the computer algebra software MAGMA ([Mag07]), which uses the algorithms explained in [CoDo05]. This method has been used in [CoDo02], [CoDo01] to determine small symmetric trivalent maps and small regular maps on oriented surfaces.

7

Elementary polycycles

We have seen in Section 4.5 a full classification of $(3, 3)_{gen}$-, $(3, 4)_{gen}$-, and $(4, 3)_{gen}$-polycycles. We have also seen that, for all other (r, q), there is a continuum of (r, q)-polycycles. The purpose of this chapter is to introduce a decomposition of polycycles into elementary components in an analogous way to decompose the molecules into atoms. This method will prove to be very effective but only in the elliptic case, since, for all other cases, we will show that there is a continuum of such elementary components (see Theorem 7.2.1). The first occurrence of the method is in [DeSt02b], followed by [DDS05b] and [DDS05c].

7.1 Decomposition of polycycles

Given an integer $q \geq 3$ and a set $R \subset \mathbb{N} - \{1\}$ (so, 2-gons will be permitted in this chapter), a $(R, q)_{gen}$-*polycycle* is a non-empty 2-connected map on a surface S with faces partitioned in two non-empty sets F_1 and F_2, so it holds that:

(i) all elements of F_1 (called *proper faces*) are combinatorial i-gons with $i \in R$;
(ii) all elements of F_2 (called *holes*, the exterior face(s) are amongst them) are pairwisely disjoint, i.e. have no common vertices;
(iii) all vertices have degree within $\{2, \ldots, q\}$ and all *interior* (i.e. not on the boundary of a hole) vertices are q-valent.

The map can be finite or infinite and some holes can be i-gons with $i \in R$. If $R = \{r\}$, then the above definition corresponds to $(r, q)_{gen}$-polycycles. If an $(R, q)_{gen}$-polycycle is simply connected, then we call it an (R, q)-polycycle; those polycycles can be drawn on the plane with the holes being exterior faces. (R, q)-polycycles with $R = \{r\}$ are exactly the (r, q)-polycycles considered in Chapters 4–6. One motivation for allowing several possible sizes for the interior faces is that polycyclic hydrocarbons in Chemistry have a molecular formula, which can modeled on such polycycles, see Figure 7.1. The definition of (R, q)-polycycles given here is combinatorial; we no longer have the cell-homomorphism into $\{r, q\}$. We will define later on elliptic,

73

parabolic, hyperbolic (R, q)-polycycles, but this will no longer have a direct relation to the sign of the curvature of $\{r, q\}$.

A *boundary* of an $(R, q)_{gen}$-polycycle P is the boundary of any of its holes.

A *bridge* of an $(R, q)_{gen}$-polycycle is an edge, which is not on a boundary and goes from a hole to a hole (possibly, the same). An $(R, q)_{gen}$-polycycle is called *elementary* if it has no bridges. See below illustrations of these notions:

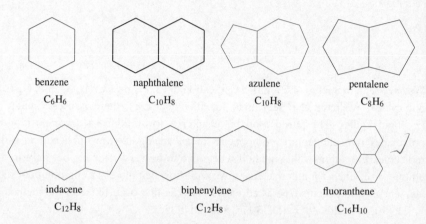

benzene	naphthalene	azulene	pentalene
C_6H_6	$C_{10}H_8$	$C_{10}H_8$	C_8H_6

indacene	biphenylene	fluoranthene
$C_{12}H_8$	$C_{12}H_8$	$C_{16}H_{10}$

Figure 7.1 Some small polycyclic hydrocarbon molecules; here C_nH_m means n vertices (carbon atoms) including m 2-valent ones, where a hydrogen atom can be attached (double bonds are omitted for simplicity):

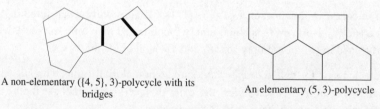

A non-elementary $(\{4, 5\}, 3)$-polycycle with its bridges

An elementary $(5, 3)$-polycycle

An *open edge* of an $(R, q)_{gen}$-polycycle is an edge on a boundary, such that each of its end-vertices have degree less than q. See below the open edges of some (R, q)-polycycles:

The open edges of a $(5, 3)$-polycycle

The open edge of a $(\{2, 3\}, 5)$-polycycle

Theorem 7.1.1 *Every $(R, q)_{gen}$-polycycle is uniquely formed by the agglomeration of elementary $(R, q)_{gen}$-polycycles along open edges or, in other words, it can be uniquely cut, along the bridges, into elementary $(R, q)_{gen}$-polycycles.*

See below for an example of such decomposition of a $(5, 3)_{gen}$-polycycle on the plane:

A $(5, 3)_{gen}$-polycycle with its bridges being boldfaced

The elementary components of this polycycle

Theorem 7.1.1, together with the determination of elliptic elementary $(r, q)_{gen}$-polycycles, is used in Chapters 5, 8, 12, 13, 14, and 18 for classification purposes.

Theorem 7.1.1 gives a simple way to describe an (r, q)-polycycle: give the names of its elementary components and use the symbol $+$. But, in many cases, this is ambiguous, i.e. the same elementary component can be used to form an (r, q)-polycycle in different ways, in the same way as the formula of a molecule, giving its number of atoms, does not define it in general. For example, with D denoting the $(5, 3)$-polycycle formed by a 5-gon, $D + D + D$ refers unambiguously to the following $(5, 3)$-polycycle:

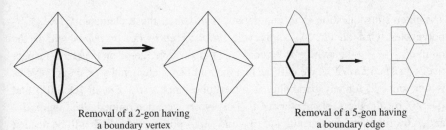

There is another $(5, 3)$-polycycle with three 5-gons sharing a vertex; it is elementary and named E_1 according to Figure 7.2. But $D + D + D + D$ is ambiguous, since there are two $(5, 3)$-polycycles having four elementary components D.

Given an $(R, q)_{gen}$-polycycle P, we can define another $(R, q)_{gen}$-polycycle P' by removing a face f from F_1, i.e. by considering it as a hole. If f has no common vertices with other faces from F_1, then removing it leaves unchanged the plane graph G and only changes the pair (F_1, F_2). If f has some edges in common with a hole, then we remove them. If f has a vertex v in common with a hole and if v does not belong to a common edge, then we split v into two vertices. See below two examples of this operation:

Removal of a 2-gon having
a boundary vertex

Removal of a 5-gon having
a boundary edge

The reverse operation is the *addition* of a face. An $(R, q)_{gen}$-polycycle P is called *extensible* if there exists another $(R, q)_{gen}$-polycycle P', such that the removal of a face of P' yields P, i.e. if it is possible to add a face to it.

In all pictures below, we put under an (R, q)-polycycle P, its symmetry group $Aut(P)$ and mark *nonext.* for non-extensible P (see Section 8.2 for details on this notion). Also, we put in parenthesis the group $Aut(G)$ of the plane graph G of P if $Aut(P) \neq Aut(G)$ and no other polycycle with the same plane graph exists. Our setting here is more general than in Chapter 4; the plane graph no longer determines the polycyclic realization. In fact, the same plane graph G can admit several realizations as a (R, q)-polycycle; see examples in Appendices 1 and 2.

A natural question to ask is if it is possible to further enlarge the class of polycycles.

There will be only some technical difficulties if we try to obtain the catalog of elementary (R, Q)-*polycycles*, i.e. the generalization of (R, q)-polycycles allowing the set Q for values of degree of interior vertices. Such polycycle is called *elliptic, parabolic,* or *hyperbolic* if $\frac{1}{q} + \frac{1}{r} - \frac{1}{2}$ (where $r = \max_{i \in R} i, q = \max_{i \in Q} i$) is positive, zero, or negative, respectively. The decomposition and other main notions could be applied directly.

We required 2-connectedness and that any two holes do not share a vertex. If we remove those two hypothesis, then many other graphs do appear.

The omitted cases $(R, q) = (2, q)$ are not interesting. In fact, consider infinite series of $(2, 6)$-polycycles, tripled m-gons, $m \geq 2$ (i.e. m-gon with each edge being tripled). The central edge is a bridge for those polycycles, for both 2-gons of the triple of edges. But if we remove those two 2-gons, then the resulting plane graph has two holes sharing a face, i.e. it violates the crucial point (ii) of the definition of (R, q)-polycycle. For even m, each even edge (for some order $1, \ldots, m$ of them) can be duplicated t times (for fixed t, $1 \leq t \leq 5$), and each odd edge duplicated $6 - t$ times, so the degrees of all vertices will still be 6. On the other hand, two holes (m-gons inside and outside of the tripled m-gon) have common vertices, so again it is not our polycycle.

7.2 Parabolic and hyperbolic elementary $(R, q)_{gen}$-polycycles

The interesting question is to enumerate, if possible, those elementary $(R, q)_{gen}$-polycycles. Call an $(R, q)_{gen}$-polycycle *elliptic, parabolic,* or *hyperbolic* if the number $\frac{1}{r} + \frac{1}{q} - \frac{1}{2}$ (where $r = \max_{i \in R} i$) is positive, zero, or negative, respectively. In Theorem 7.2.1, we will see that the number of elementary (r, q)-polycycles is uncountable for any parabolic or hyperbolic pairs (r, q). But in [DeSt01] and [DeSt02b], all elliptic elementary (r, q)-polycycles were determined. See Figures 7.2 and 7.3 for the list of elementary $(5, 3)$- and $(3, 5)$-polycycles, which will be needed

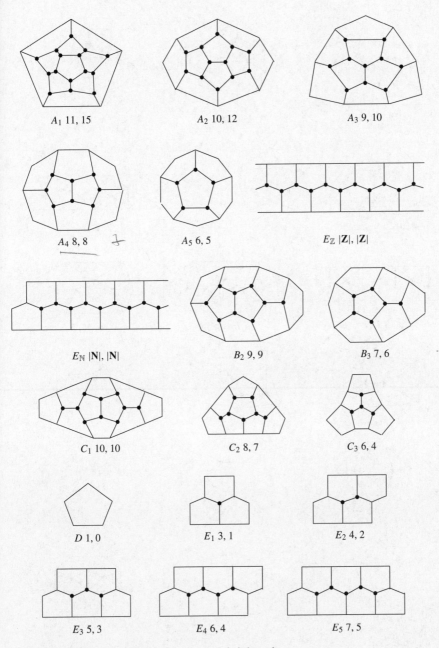

A_1 11, 15

A_2 10, 12

A_3 9, 10

A_4 8, 8

A_5 6, 5

$E_{\mathbb{Z}}$ |**Z**|, |**Z**|

$E_{\mathbb{N}}$ |**N**|, |**N**|

B_2 9, 9

B_3 7, 6

C_1 10, 10

C_2 8, 7

C_3 6, 4

D 1, 0

E_1 3, 1

E_2 4, 2

E_3 5, 3

E_4 6, 4

E_5 7, 5

Figure 7.2 Elementary (5, 3)-polycycles and their kernels

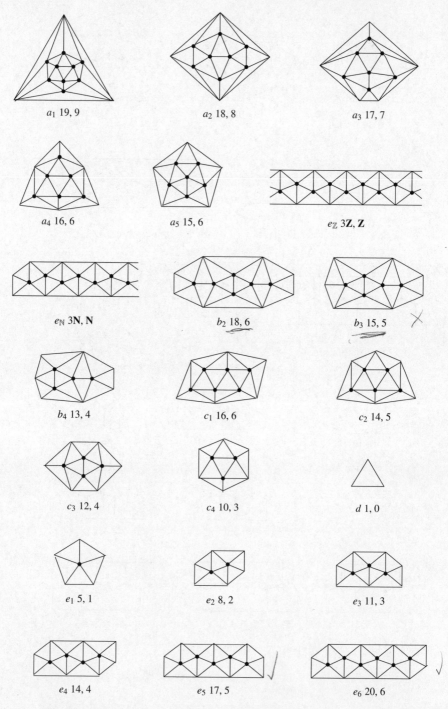

Figure 7.3 Elementary (3, 5)-polycycles and their kernels

later. For (3, 3)-, (3, 4)-, and (4, 3)-polycycles, we have full classification of them in Chapter 4. In fact, we will consider the case $R = \{i : 2 \leq i \leq r\}$ covering all elliptic possibilities: $(\{2, 3, 4, 5\}, 3)$-, $(\{2, 3\}, 4)$-, and $(\{2, 3\}, 5)$-polycycles in Sections 7.4, 7.5, and 7.6, respectively. Call the *kernel* of a polycycle, the cell-complex of its vertices, edges, and faces, which are not incident with its boundary.

Theorem 7.2.1 *For parabolic and hyperbolic parameters (r, q), there exists a continuum of non-isomorphic elementary (r, q)-polycycles.*

Proof. Consider a semi-infinite (to the right) chain of 4-gons that fill a strip between two parallel rays. Inside two horizontal sides of each 4-gon of the chain, we put $r - 5$ and one new vertex to obtain an r-gon instead of a 4-gon. There are two alternatives: either $r - 5$ new vertices are placed on the upper side and one on the lower side or, vice versa, one new vertex is placed on the upper side and $r - 5$ on the lower. Such a choice is made independently on each 4-gon when we move to the right along this chain. Therefore, there is a continuum of various (non-isomorphic) chains of this kind. All of them are chains in the tiling $\{r, q\}$ for $r \geq 7$ and $q \geq 3$, as well as for $r = 5$ and $q \geq 4$. It is also clear that this (r, q)-polycycle is the kernel of an elementary (r, q)-polycycle consisting of this polycycle supplemented with all r-gons that are incident to it in the tiling $\{r, q\}$.

Now, consider the case of parabolic parameters (r, q), i.e. $(r, q) = (6, 3), (4, 4)$, and $(3, 6)$. In the square lattice, i.e. in the regular tiling $\{4, 4\}$ of the Euclidean plane $\mathbb{R}^2$, we construct a chain of 4-gons semi-infinite in the upper right direction. On each step of this construction, there are two alternatives for choosing the next 4-gon: we can choose an adjacent 4-gon, either on the right on the same level, or one level higher. It is clear that there is a continuum of various (non-isomorphic) chains of this kind in the tiling $\{4, 4\}$ and each of these chains is the kernel of a certain elementary $(4, 4)$-polycycle. Infinite chains of hexagons in the tiling $\{6, 3\}$ are constructed analogously. As for the tiling $\{3, 6\}$, combining two adjacent 3-gons in it into a rhomb and transforming the entire tiling $\{3, 6\}$ into a rhombic lattice combinatorially equivalent to the tiling $\{4, 4\}$, we can apply the same line of reasoning as in the case of the square lattice. The kernels of those parabolic polycycles are outerplanar. They can be interpreted as the kernels of hyperbolic polycycles (by increasing the value of the parameter q). $\qquad\square$

7.3 Kernel-elementary polycycles

Call an (r, q)-polycycle *kernel-elementary* if it is an r-gon or if it has a non-empty connected kernel, such that the deletion of any face from the kernel will diminish it (i.e. any face of the polycycle is incident to its kernel).

Theorem 7.3.1 *(i) If an (r, q)-polycycle is kernel-elementary, then it is elementary.*

(ii) If (r, q) is elliptic, then any elementary (r, q)-polycycle is also kernel-elementary.

Proof. (i) Take a kernel-elementary (r, q)-polycycle P; we can assume it to be different from an r-gon. Let $P_1, \ldots, P_r$ be the elementary components of this polycycle. The connectedness condition on the kernel gives that all P_i but one are r-gons. But removing the P_i that are r-gons does not change the kernel; so, $r = 1$ and P is elementary.

(ii) Consider any two vertices of an r-gon of an elliptic (r, q)-polycycle that belongs to the kernel of this polycycle. The shortest edge path between these vertices lies inside the union of two stars of r-gons with the centers at these two vertices; this result can easily be verified in each particular case for any elliptic parameters $(r, q) = (3, 3), (3, 4), (3, 5), (4, 3),$ and $(5, 3)$. Hence, any r-gon of an elliptic (r, q)-polycycle is only incident with one simply connected component of its kernel. All r-gons that are incident with the same non-empty connected component of the kernel constitute a non-trivial elementary summand. Since the polycycle is elementary, this is its totality and the kernel is connected. $\square$

In [DeSt02b] the notion of kernel-elementary was called *elementary*. See below for an example of a $(6, 3)$-polycycle, which is elementary but not kernel-elementary (since its kernel is not connected):

The decomposition Theorem 7.1.1 (of (r, q)-polycycles into elementary polycycles) is the main reason why we prefer the property to be elementary to kernel-elementary. Another reason is that if an $(r, q)_{gen}$-polycycle is elementary, then its universal cover is also elementary. However, the notion of kernel elementariness will be useful in the classification of infinite elementary $(\{3, 4, 5\}, 3)$- and $(\{2, 3\}, 5)$-polycycles.

In Figures 7.2 and 7.3, each elementary polycycle is denoted by a certain letter with a subscript; two numbers indicate the values of the parameters p_r (the number of interior faces) and v_{int} (the number of interior vertices). The infinite series E_s, respectively e_s have $(p_r, v_{int}) = (s+2, s)$, respectively $(3s+2, s)$ and are represented for $s \leq 5$, respectively $s \leq 6$.

Theorem 7.3.2 *The list of elementary* $(5, 3)$-*polycycles (see Figure 7.2) consists of:*

  (i) 11 *sporadic finite* $(5, 3)$-*polycycles,*
  (ii) *an infinite series* E_n, $n \geq 1$,
  (iii) *two infinite polycycles* $E_\mathbb{N}$ *and* $E_\mathbb{Z}$ *(snub Prism$_\infty$).*

Proof. Take an elementary $(5, 3)$-polycycle P, which, by Theorem 7.3.1, is kernel-elementary. If its kernel is empty, then P is simply D. If the kernel is reduced to a vertex, then P is simply E_1. If each 5-gon of an elementary $(5, 3)$-polycycle has at most three vertices from the kernel, that are arranged in succession along the perimeter of the 5-gon, then the kernel does not contain 5-gons and has the form of a *geodesic* (see the elementary $(5, 3)$-polycycles E_i, $i \geq 1$, $E_\mathbb{N}$, and $E_\mathbb{Z}$) or a *propeller* (see the elementary $(5, 3)$-polycycle C_3). If at least one 5-gon of the elementary $(5, 3)$-polycycle contains three vertices of the kernel that are arranged along the perimeter not in succession, then the whole 5-gon belongs to the kernel. Only in the case of one or two 5-gons, the kernel can additionally contain one or two pendant edges (see $(5, 3)$-polycycles A_5, B_3, C_2 and A_4, B_2, C_1). If the kernel contains more than two 5-gons, then the total number of these 5-gons can only be 3, 4, or 6 (see A_3, A_2, and A_1). $\square$

Theorem 7.3.3 *The list of elementary* $(3, 5)$-*polycycles (see Figure 7.3) consists of:*

  (i) 13 *sporadic finite* $(3, 5)$-*polycycles,*
  (ii) *an infinite series* e_n, $n \geq 1$,
  (iii) *two infinite polycycles* $e_\mathbb{N}$ *and* $e_\mathbb{Z}$ *(snub APrism$_\infty$).*

Proof. Take an elementary $(3, 5)$-polycycle P which, by Proposition 7.3.1, is kernel-elementary. If its kernel is empty, then P is simply d. If the kernel is reduced to a vertex, then P is simply e_1. If there are at most two vertices from a kernel in each 3-gon, then the kernel does not contain 3-gons and has the form of a *geodesic* (see e_i, $i \geq 1$, as well as $e_\mathbb{N}$ and $e_\mathbb{Z}$). If there is one 3-gon in a kernel, the latter may additionally have one pendant edge (see c_4 and b_4). If there are two 3-gons in a kernel, the latter may additionally have one or two pendant edges (see c_3, b_3, and b_2). If there are more than two 3-gons in a kernel, then their total number may only be 3, 4, 5, 6, 8, or 10 (see a_1, a_2, a_3, a_4, a_5, c_1, and c_2). $\square$

Remark that the kernels of $(5, 3)$-polycycles A_i, $1 \leq i \leq 5$, of Figure 7.2 are all non-trivial isometric subgraphs $(5, 3)$-polycycles; they are also all *circumscribed* $(5, 3)$-polycycles, i.e. r-gons can be added around the perimeter, so that they will form a simple circuit. (They were found in [CCBBZGT93]; such $(5, 3)$-polycycles are useful in Organic Chemistry.) All circumscribed $(3, 5)$-polycycles are the kernels of polycycles a_i, $1 \leq i \leq 5$, and c_i, $1 \leq i \leq 4$ of Figure 7.3. It turns out that all polycycles P in either Figure 7.2 or 7.3, admitting the inner dual polycycle $Inn^*(P)$ in another one, are as follows:

$$Inn^*(A_1) = a_5 \quad Inn^*(A_2) = c_3 \quad Inn^*(A_3) = c_4 \quad Inn^*(A_4) = e_2$$
$$Inn^*(A_5) = e_1 \quad Inn^*(E_1) = d \quad Inn^*(a_1) = A_3 \quad Inn^*(a_2) = A_4$$
$$Inn^*(a_4) = C_3 \quad Inn^*(a_5) = A_5 \quad Inn^*(c_1) = E_4 \quad Inn^*(c_2) = E_3$$
$$Inn^*(c_3) = E_2 \quad Inn^*(c_4) = E_1 \quad Inn^*(e_1) = D$$

$E_{\mathbb{Z}} = $ snub $Prism_\infty$ occurs also in Figure 6.1; its inner dual is infinite $(3, 4)$-polycycle $APrism_\infty$.

Theorem 7.3.4 *All elementary* $(5, 3)_{gen}$*-polycycles that are not* $(5, 3)$*-polycycles, are:*

(i) *the plane graphs snub* $Prism_m$ *with two holes (both m-gonal faces removed), for any* $m \geq 2$,

(ii) *the (non-orientable, on Möbius surface) quotient of snub* $Prism_m$, *for any odd* m, *with respect to central symmetry.*

Proof. Take such a polycycle. Its universal cover is an elementary $(5, 3)$-polycycle, whose automorphism group contains some fixed-point-free transformation. Inspection of the list of elementary $(5, 3)$-polycycles in Figure 7.2 yields only $E_{\mathbb{Z}} = $ snub $Prism_\infty$ as a possibility. Snub $Prism_m$ is obtained from the group of translations by m faces and the non-orientable quotients if the group contains also some translation followed by reflection.   □

Theorem 7.3.5 *All elementary* $(3, 5)_{gen}$*-polycycles that are not* $(3, 5)$*-polycycles, are:*

(i) *the plane graphs snub* $APrism_m$ *with two holes (both m-gonal faces removed) for any* $m \geq 2$,

(ii) *the (non-orientable, on Möbius surface) quotient of snub* $APrism_m$ *for any odd* m, *with respect to central symmetry.*

Proof. Take such a polycycle. Its universal cover is an elementary $(3, 5)$-polycycle, whose automorphism group contains some fixed-point-free transformations. Inspection of the list of elementary $(3, 5)$-polycycles in Figure 7.3 yields only $e_{\mathbb{Z}} = $ snub $APrism_\infty$ as a possibility. The arguments follow as before.   □

We can now classify all kernel-elementary elliptic $(r, q)_{gen}$-polycycles obtained in [DDS05b]. Such a polycycle is either elementary or it is obtained by self-gluing of an elementary (r, q)-polycycle. Hence, the list is as follows:

1 All kernel-elementary $(5, 3)_{gen}$-polycycles, obtained by self-gluing of elementary $(5, 3)$-polycycles, are $E_{n,or}^*$, $E_{n,nor}^*$, $C_{1,or}^*$, $C_{1,nor}^*$, $C_{2,or}^*$, $C_{2,nor}^*$, $C_{3,or}^*$, $C_{3,nor}^*$. The

symbol $*$ refers to the self-gluing and subscripts *or*, *nor* are used if the self-gluing is orientable or not, respectively. See below $E^*_{13,or}$ and $E^*_{14,or}$:

2 All kernel-elementary $(5, 3)_{gen}$-polycycles, obtained by self-gluing of elementary $(5, 3)$-polycycles are $e^*_{n,or}, e^*_{n,nor}, b^*_{2,or}, b^*_{2,nor}$. The symbols $*$, *or*, *nor* are used as above. See below $e^*_{15,or}$ and $e^*_{16,or}$:

7.4 Classification of elementary $(\{2, 3, 4, 5\}, 3)_{gen}$-polycycles

Theorem 7.4.1 *The list of elementary $(\{2, 3, 4, 5\}, 3)_{gen}$-polycycles consists of:*

(i) 204 *sporadic $(\{2, 3, 4, 5\}, 3)$-polycycles given in Appendix 1.*

(ii) *Six $(\{3, 4, 5\}, 3)$-polycycles, infinite in one direction:*

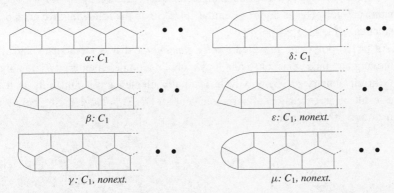

(iii) $21 = \binom{6+1}{2}$ *infinite series obtained by taking two endings of the infinite polycycles of (ii) above and concatenating them.*

For example, merging of α with itself produces the infinite series of elementary $(5, 3)$-polycycles, denoted on Figure 7.2 by E_n. See Figure 7.4 for the first three members (starting with 6 faces) of two such series: $\alpha\alpha$ and $\beta\varepsilon$.

(iv) *The infinite series of snub Prism$_m$, $2 \le m \le \infty$, and its non-orientable quotient for m odd.*

Infinite series $\alpha\alpha$ of elementary ($\{2, 3, 4, 5\}$, 3)-polycycles (E_4-E_6):

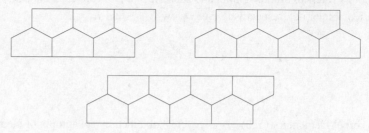

Infinite series $\beta\varepsilon$ of elementary ($\{2, 3, 4, 5\}$, 3)-polycycles:

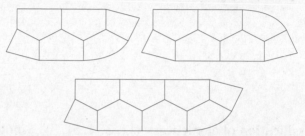

Figure 7.4 The first 3 members (starting with 6 faces) of two infinite series, amongst 21 series of ($\{2, 3, 4, 5\}$, 3)-polycycles in Theorem 7.4.1 (v)

Proof. The first step is to determine all elementary ($\{2, 3, 4, 5\}$, 3)-polycycles, which contain a 2-gon. This is done in Lemma 7.4.2. So, in the following, we consider only elementary ($\{3, 4, 5\}$, 3)-polycycles.

An (R, 3)-polycycle is called *totally elementary* if it is elementary and if, after removing any face adjacent to a hole, we obtain a non-elementary (R, 3)-polycycle. So, an elementary (R, 3)-polycycle is totally elementary if and only if it is not the result of an extension of some elementary (R, 3)-polycycle. See below for an illustration of this notion:

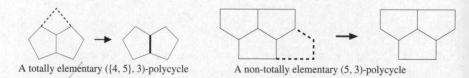

A totally elementary ($\{4, 5\}$, 3)-polycycle   A non-totally elementary (5, 3)-polycycle

All totally elementary ($\{3, 4, 5\}$, 3)-polycycles are enumerated in Lemma 7.4.3. We will now classify all finite elementary ($\{3, 4, 5\}$, 3)-polycycles with one hole. Such a polycycle with N interior faces is either totally elementary, or it is obtained from another such polycycle with $N - 1$ interior faces by addition of a face.

There is no elementary $(\{3, 4, 5\}, 3)$-polycycles with 2 interior faces; so, all elementary $(\{3, 4, 5\}, 3)$-polycycles with 3 interior faces are totally elementary and, by Lemma 7.4.3, we know them. Also, by Lemma 7.4.3, there are no finite totally elementary $(\{3, 4, 5\}, 3)$-polycycles with more than 3 interior faces. We iterate the following procedure starting at $N = 3$:

1 Take the list of finite elementary $(\{3, 4, 5\}, 3)$-polycycles with N interior faces and add a face in all possible ways, while preserving the property to be an elementary $(\{3, 4, 5\}, 3)$-polycycle.
2 Reduce by isomorphism and obtain the list of elementary $(\{3, 4, 5\}, 3)$-polycycles with $N + 1$ interior faces.

So, for any fixed N, we obtain the list of elementary $(\{3, 4, 5\}, 3)$-polycycles with N interior faces. The enumeration is done in the following way: run the computation up to $N = 13$ and obtain the sporadic elementary $(\{3, 4, 5\}, 3)$-polycycles and the members of the infinite series. Then undertake, by hand, the operation of addition of a face and reduction by isomorphism; we obtain only the 21 infinite series for all $N \geq 13$. This completes the enumeration of finite elementary $(\{3, 4, 5\}, 3)$-polycycles.

Take now an elementary infinite $(\{3, 4, 5\}, 3)$-polycycle P. Remove all 3- or 4-gonal faces. The resulting graph P' is not necessarily connected, but its connected components are $(5, 3)_{gen}$-polycycles, though not necessarily elementary ones. We will now use the classification of elementary $(5, 3)_{gen}$-polycycles (possibly, infinite) in Theorem 7.3.2. If the infinite $(5, 3)$-polycycle snub $Prism_\infty$ appears in the decomposition, then, clearly, P is reduced to it. If the infinite polycycle $\alpha = E_\mathbb{N}$ appears in the decomposition, then there are two possibilities for extending it, as indicated below:

If a 3- or 4-gonal face is adjacent on the dotted line, then there should be another face on the boldfaced edges. So, in any case, there is a face, adjacent on the boldfaced edges, and we can assume that it is a 3- or 4-gonal face. Then, consideration of all possibilities to extend it yields $\beta, \ldots, \mu$. Suppose now, that P' does not contain any infinite $(5, 3)_{gen}$-polycycles. Then we can find an infinite path $f_0, \ldots, f_i, \ldots$ of distinct faces of P in F_1, the set of proper faces, such that f_i is adjacent to f_{i+1} and f_{i-1} is not adjacent to f_{i+1}. The condition on P implies that an infinite number of faces are 3- or 4-gons, but the condition of non-adjacency of f_{i-1} with f_{i+1} forbids 3-gons. Take now a 4-gon f_i and assume that f_{i-1} and f_{i+1} are 5-gons. The consideration of all possibilities of extension around that face, leads us to an impossibility. If some of f_{i-1} or f_{i+1} are 4-gons, then we have a path of 4-gons and the case is even simpler.

Let us now determine all elementary $(\{2, 3, 4, 5\}, 3)_{gen}$-polycycles. The universal cover $\widetilde{P}$ of such a polycycle P is an elementary $(\{2, 3, 4, 5\}, 3)$-polycycle, which has a non-trivial fixed-point-free automorphism group in $Aut(\widetilde{P})$. Consideration of the above list of polycycles yields snub $Prism_\infty$ as the only possibility. The polycycles snub $Prism_m$ and its non-orientable quotients arise in this process.   □

Lemma 7.4.2 *All elementary* $(\{2, 3, 4, 5\}, 3)$-*polycycles, containing a 2-gon, are the following eight ones:*

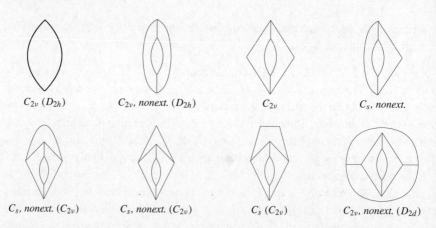

C_{2v} (D_{2h}) $\qquad$ C_{2v}, *nonext.* (D_{2h}) $\qquad$ C_{2v} $\qquad$ C_s, *nonext.*

C_s, *nonext.* (C_{2v}) $\qquad$ C_s, *nonext.* (C_{2v}) $\qquad$ C_s (C_{2v}) $\qquad$ C_{2v}, *nonext.* (D_{2d})

Proof. Let P be such a polycycle. Clearly, the 2-gon is the only possibility if the number of proper $|F_1|$ is 1. If $|F_1| = 2$, then it is not elementary. If $|F_1| \geq 3$, then the 2-gon should be inside of the structure. So, P contains, as a subgraph, one of three following graphs:

Therefore, the only possibilities for P are those given in above lemma.   □

Lemma 7.4.3 *The list of totally elementary* $(\{3, 4, 5\}, 3)$-*polycycles consists of:*

(i) Three isolated i-gons, $i \in \{3, 4, 5\}$:

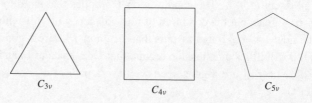

C_{3v} $\qquad\qquad$ C_{4v} $\qquad\qquad$ C_{5v}

(ii) Ten triples of i-gons, $i \in \{3, 4, 5\}$:

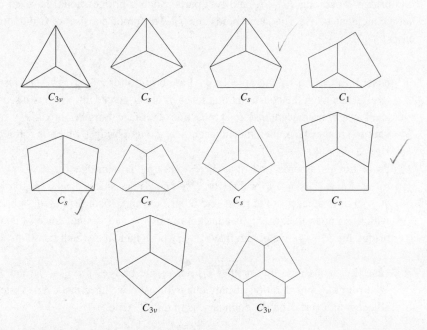

C_{3v} C_s C_s C_1

C_s C_s C_s C_s

C_{3v} C_{3v}

(iii) snub Prism$_\infty$.

Proof. Take a totally elementary ($\{2, 3, 4, 5\}$, 3)-polycycle P. If $|F_1| = 1$, then P is, clearly, totally elementary; so, let us assume that $|F_1| \geq 2$. If $|F_1| = 2$, then it is, clearly, not elementary; so, assume $|F_1| \geq 3$. Of course, P has at least one interior vertex; let v be such a vertex. Furthermore, we can assume that v is adjacent to a vertex v', which is incident to a hole.

The vertex v is incident to three faces f_1, f_2, f_3. Let us denote by v_{ij} unique vertex incident to f_i, f_j, and adjacent to v. Without loss of generality, we can suppose that v' is incident to the faces f_1 and f_2, i.e. that $v' = v_{12}$:

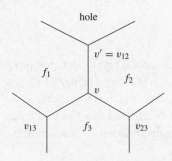

The removal of the face f_1 yields a non-elementary polycycle, so there is at least one bridge separating $P - f_1$ into two parts. Such a bridge should have an end-vertex incident to f_1. The same holds for f_2. The proof consists of a number of cases.

1st **case**: If $e_1 = \{v, v_{23}\}$ and $e_2 = \{v, v_{13}\}$ are bridges for $P - f_1$, $P - f_2$, respectively, then from the constraint that faces f_i are p-gons with $p \leq 5$, we see that each face f_i is adjacent in P to at most one other face. Furthermore, if f_i is adjacent to another face, then this adjacency is along a bridge, which is forbidden. Hence, $F_1 = \{f_1, f_2, f_3\}$.

2nd **case**: Let us assume now that $e_1 = \{v, v_{23}\}$ is a bridge for $P - f_1$, but $e_2 = \{v, v_{13}\}$ is not a bridge for $P - f_2$. Then, since f_2 is a p-gon with $p \leq 5$, it is adjacent to at most one other face and, if so, then along a bridge, which is impossible. So, f_2 is adjacent to only f_1 and f_3 and, since e_2 is not a bridge for $P - f_2$, we obtain that $P - f_2$ is elementary, which contradicts the hypothesis.

3rd **case**: Let us assume that neither e_1, nor e_2 are bridges for $P - f_1$ and $P - f_2$. From the consideration of previous two cases, we have that every vertex v, adjacent to a vertex on the boundary, is in this 3rd case.

The first subcase, which can happen only if f_1 is 5-gon, happens when the boldfaced edge e', in the drawing below, is a bridge:

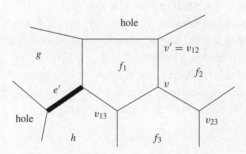

The face g is adjacent to the faces h and f_1 and, possibly, to another face g'. But if g is adjacent to such a face g', it is along a bridge of P; hence, g is adjacent only to h and f_1. So $P - g$ is elementary, which is impossible.

So, the edge e' is not a bridge and this forces the face h to be 5-gonal. Hence, the vertex v_{13} is in the same situation as the vertex v, described in the diagram below:

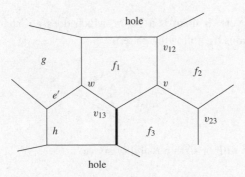

So, we can repeat the construction. If, at some point, e_1 is a bridge, then the construction stops; otherwise, we continue indefinitely and obtain snub $Prism_\infty$. $\quad\square$

7.5 Classification of elementary $(\{2, 3\}, 4)_{gen}$-polycycles

Theorem 7.5.1 *Any elementary* $(\{2, 3\}, 4)_{gen}$-*polycycle is one of the following eight:*

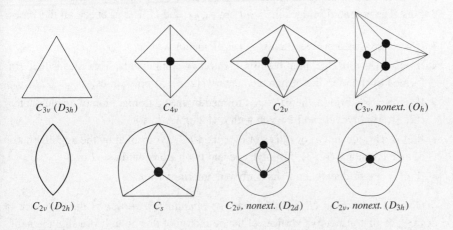

Proof. The list of elementary $(3, 4)$-polycycles is determined by inspecting the list in Section 4.2 and consists of the first four graphs of this theorem. Let P be a $(\{2, 3\}, 4)$-polycycle, containing a 2-gon. If $|F_1| = 1$, then it is the 2-gon. Clearly, the case, in which two 2-gons share one edge, is impossible. Assume that P contains two 2-gons, which share a vertex. Then we should add 3-gons on both sides and so, obtain the eighth above polycycle. If there is a 2-gon, which does not share a vertex with a 2-gon, then P contains the following pattern:

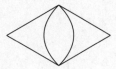

So, clearly, P is the fifth or sixth possibility above. $\square$

Note that seventh and fourth polycycles in Theorem 7.5.1 are, respectively, $APrism_2$ and $APrism_3$; here the exterior face is the unique hole.

7.6 Classification of elementary $(\{2, 3\}, 5)_{gen}$-polycycles

Let us consider an elementary $(\{2, 3\}, 5)$-polycycle P. Assume that P is not an i-gon and has a 2-gonal face f. If f is adjacent to a hole, then the polycycle is not elementary. So, holes are adjacent only to 3-gons. If we remove such a 3-gon t, then the third vertex v of t, which is necessarily interior in P, becomes non-interior in $P - t$. The polycycle $P - t$ is not necessarily elementary. Let us denote by $e_1, \dots, e_5$ the edges incident to v and assume that e_1, e_2 are edges of t. The boundary is adjacent only to 3-gons. The potential bridges in $P - t$ are e_3, e_4, and e_5. Let us check all five cases:

- If no edge e_k is a bridge, then $P - t$ is elementary.
- If only e_4 is a bridge, then it splits P into two components. This means that P is formed by the merging of two elementary $(\{2, 3\}, 5)$-polycycles.
- If e_3 or e_5 is a bridge, then $P - t$ is formed by the agglomeration of an elementary $(\{2, 3\}, 5)$-polycycle and a i-gon with $i = 2$ or 3.
- If e_4 is a bridge and e_3 or e_5 is a bridge, then $P - t$ is formed by the agglomeration of an elementary $(\{2, 3\}, 5)$-polycycle and two i-gons with $i = 2$ or 3.
- If all e_k are bridges, then P has only one interior vertex.

Given a hole of an (R, q)-polycycle, its *boundary sequence* is the sequence of degrees of all consecutive vertices of the boundary of this hole. It is a slight generalization of Chapter 5, where the considered polycycles have only one exterior face.

Theorem 7.6.1 *The list of elementary* $(\{2, 3\}, 5)_{gen}$*-polycycles consists of:*

(i) 57 sporadic $(\{2, 3\}, 5)$-polycycles given in Appendix 2,
(ii) three following infinite $(\{2, 3\}, 5)$-polycycles:

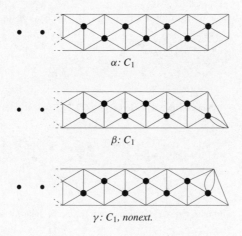

$\alpha: C_1$

$\beta: C_1$

$\gamma: C_1,$ *nonext.*

(iii) *Six infinite series of* $(\{2, 3\}, 5)$*-polycycles with one hole (they are obtained by concatenating endings of a pair of polycycles, given in (ii); see Figure 7.5 for the first graphs),*

(iv) *the infinite series of snub* $APrism_m$, *for* $2 \leq m \leq \infty$, *and its non-orientable quotients for m odd.*

Proof. Let us take an elementary $(\{2, 3\}, 5)$-polycycle, which is finite. Then, by removing a 3-gon, which is adjacent to a boundary, we are led to the situation described above. Hence, the algorithm for enumerating finite elementary $(\{2, 3\}, 5)$-polycycles is the following:

1 Begin with isolated i-gons with $i = 2$ or 3.
2 For every vertex v of an elementary polycycle with n interior vertices, consider all possibilities of adding 2- and 3-gons incident to v, such that the obtained polycycle is elementary and v has become an interior vertex.
3 Reduce by isomorphism.

The above algorithm first finds some sporadic elementary $(\{2, 3\}, 5)$-polycycles and the first elements of the infinite series and then find only the elements of the infinite series. In order to prove that this is the complete list of all finite elementary $(\{2, 3\}, 5)$-polycycles, we need to consider the case, in which only e_4 is a bridge going from a hole to the same hole. So, we need to consider all possibilities where the addition of two elementary $(\{2, 3\}, 5)$-polycycles and one 3-gon makes a larger elementary $(\{2, 3\}, 5)$-polycycle. Given a sequence $a_1, \ldots, a_n$, say that a sequence

Infinite series $\alpha\alpha$ of elementary ($\{2, 3\}$, 5)-polycycles (e_1-e_6):

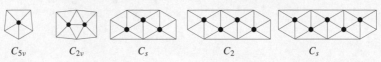

C_{5v} $\qquad$ C_{2v} $\qquad$ C_s $\qquad$ C_2 $\qquad$ C_s

Infinite series $\alpha\beta$ of elementary ($\{2, 3\}$, 5)-polycycles:

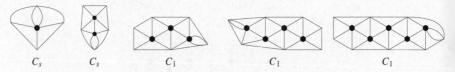

C_s $\qquad$ C_s $\qquad$ C_1 $\qquad$ C_1 $\qquad$ C_1

Infinite series $\alpha\gamma$ of elementary ($\{2, 3\}$, 5)-polycycles:

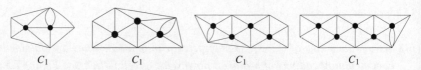

C_1 $\qquad$ C_1 $\qquad$ C_1 $\qquad$ C_1

Infinite series $\beta\beta$ of elementary ($\{2, 3\}$, 5)-polycycles:

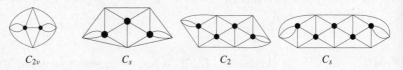

C_{2v} $\qquad$ C_s $\qquad$ C_2 $\qquad$ C_s

Infinite series $\beta\gamma$ of elementary ($\{2, 3\}$, 5)-polycycles:

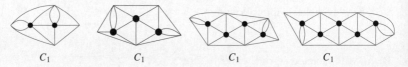

C_1 $\qquad$ C_1 $\qquad$ C_1 $\qquad$ C_1

Infinite series $\gamma\gamma$ of elementary ($\{2, 3\}$, 5)-polycycles:

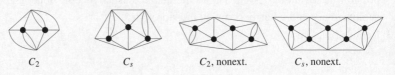

C_2 $\qquad$ C_s $\qquad$ C_2, nonext. $\qquad$ C_s, nonext.

Figure 7.5 The first 5 members of the six infinite series of ($\{2, 3\}$, 5)-polycycles

$b_1, \ldots, b_p$ with $p < n$ is a *pattern* of that sequence if, for some n_0, we have $a_{n_0+j-1} = b_j$ or $a_{n_0+1-j} = b_j$ with the addition being modulo n. The ($\{2, 3\}$, 5)-polycycles, used in that construction, should have the pattern $3, 3, x$ with $x \leq 4$ in their boundary sequence. Only the polycycles, which belong to the six infinite series, satisfy this and it is easy to see that the result of the operation is still one of the six infinite series. So, the list of finite elementary $(\{2, 3\}, 5)$-polycycles is given in the theorem.

Consider now an elementary infinite $(\{2, 3\}, 5)$-polycycle P. Eliminate all 2-gonal faces of P and obtain another $(3, 5)$-polycycle P', which is not necessarily elementary. We do a decomposition of P' along its elementary components, which are enumerated in Section 7.3. If snub $APrism_\infty$ is one of the components, then we are finished and $P = P'$ is snub $APrism_\infty$. If α is one of the components, then we have two edges along which we can extend the polycycle; they are depicted below:

Clearly, if we extend the polycycle along only one of those edges, then the result is not an elementary polycycle. The consideration of all possibilities yields β and γ. Suppose now that P' has no infinite components. Then P has at least one infinite path $f_0, \ldots, f_i, \ldots$, such that f_i is adjacent to f_{i+1}, but f_{i-1} is not adjacent to f_{i+1}. The considerations, analogous to the 3-valent case, yield the result for $(\{2, 3\}, 5)$-polycycles.

If P is an elementary $(\{2, 3\}, 5)_{gen}$-polycycle, which is not a $(\{2, 3\}, 5)$-polycycle, then its universal cover $\widetilde{P}$ is an elementary $(\{2, 3\}, 5)$-polycycle, which has a fixed-point-free automorphism group included in $Aut(\widetilde{P})$. Clearly, only snub $APrism_\infty$ is such and it yields the infinite series of snub $APrism_m$ and its non-orientable quotients. $\qquad \square$

7.7 Appendix 1: 204 sporadic elementary $(\{2, 3, 4, 5\}, 3)$-polycycles

Below (11 cases), when several elementary sporadic $(\{2, 3, 4, 5\}, 3)$-polycycles correspond to the same plane graph, we always add the sign **x** with $1 \leq x \leq 11$.

List of 4 sporadic elementary $(\{2, 3, 4, 5\}, 3)$-polycycles with 1 proper face:

C_{2v} (D_{2h}) C_{3v} (D_{3h}) C_{4v} (D_{4h}) C_{5v} (D_{5h})

List of 13 sporadic elementary ({2, 3, 4, 5}, 3)-polycycles with 3 proper faces:

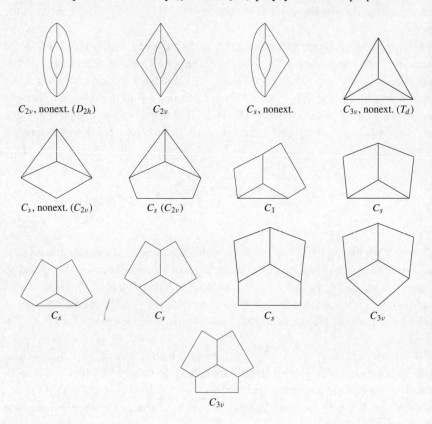

C_{2v}, nonext. (D_{2h}) C_{2v} C_s, nonext. C_{3v}, nonext. (T_d)

C_s, nonext. (C_{2v}) C_s (C_{2v}) C_1 C_s

C_s C_s C_s C_{3v}

C_{3v}

List of 26 sporadic elementary ({2, 3, 4, 5}, 3)-polycycles with 4 proper faces:

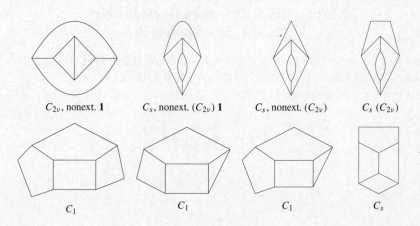

C_{2v}, nonext. **1** C_s, nonext. (C_{2v}) **1** C_s, nonext. (C_{2v}) C_s (C_{2v})

C_1 C_1 C_1 C_s

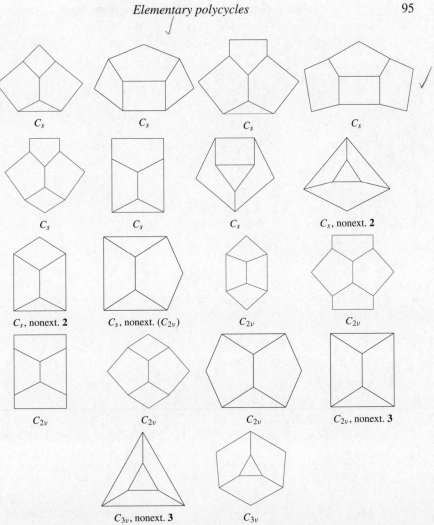

C_s C_s C_s C_s

C_s C_s C_s C_s, nonext. **2**

C_s, nonext. **2** C_s, nonext. (C_{2v}) C_{2v} C_{2v}

C_{2v} C_{2v} C_{2v} C_{2v}, nonext. **3**

C_{3v}, nonext. **3** C_{3v}

List of 36 sporadic elementary ($\{2, 3, 4, 5\}$, 3)-polycycles with 5 proper faces:

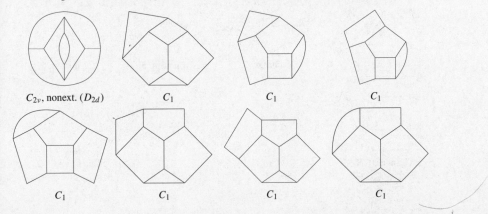

C_{2v}, nonext. (D_{2d}) C_1 C_1 C_1

C_1 C_1 C_1 C_1

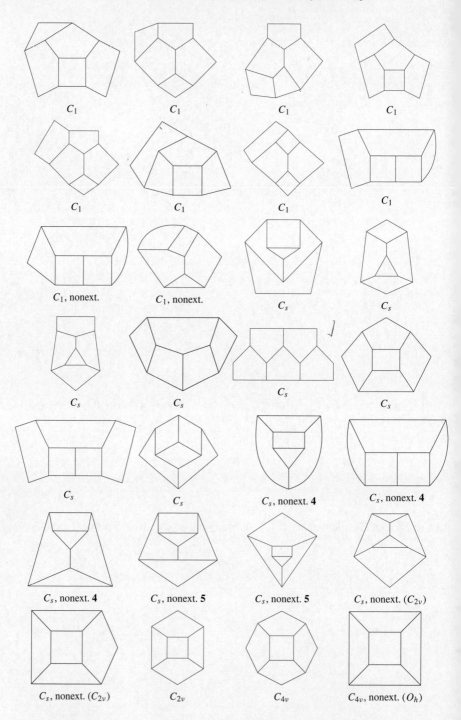

C_1 C_1 C_1 C_1

C_1 C_1 C_1 C_1

C_1, nonext. C_1, nonext. C_s C_s

C_s C_s C_s C_s

C_s C_s C_s, nonext. **4** C_s, nonext. **4**

C_s, nonext. **4** C_s, nonext. **5** C_s, nonext. **5** C_s, nonext. (C_{2v})

C_s, nonext. (C_{2v}) C_{2v} C_{4v} C_{4v}, nonext. (O_h)

List of 34 sporadic elementary ({2, 3, 4, 5}, 3)-polycycles with 6 proper faces:

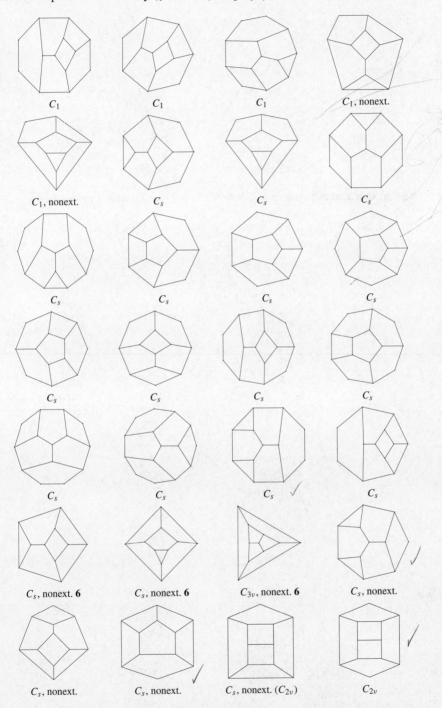

C_1 C_1 C_1 C_1, nonext.

C_1, nonext. C_s C_s C_s

C_s C_s C_s C_s

C_s C_s C_s C_s

C_s C_s C_s C_s

C_s, nonext. **6** C_s, nonext. **6** C_{3v}, nonext. **6** C_s, nonext.

C_s, nonext. C_s, nonext. C_s, nonext. (C_{2v}) C_{2v}

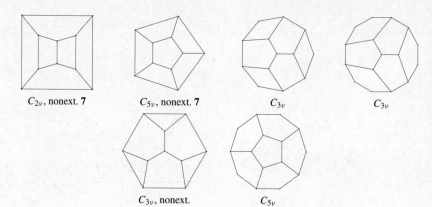

List of 36 sporadic elementary ({2, 3, 4, 5}, 3)-polycycles with 7 proper faces:

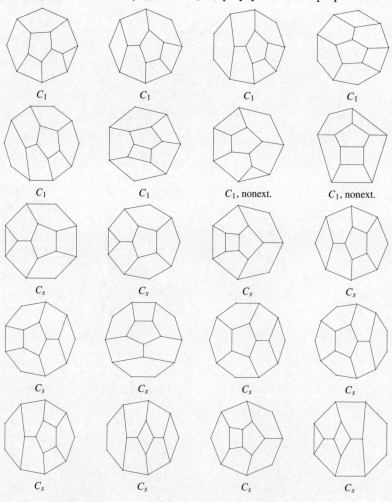

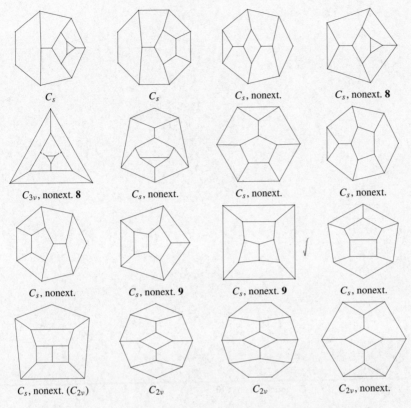

C_s C_s C_s, nonext. C_s, nonext. **8**

C_{3v}, nonext. **8** C_s, nonext. C_s, nonext. C_s, nonext.

C_s, nonext. C_s, nonext. **9** C_s, nonext. **9** C_s, nonext.

C_s, nonext. (C_{2v}) C_{2v} C_{2v} C_{2v}, nonext.

List of 29 sporadic elementary ($\{2, 3, 4, 5\}$, 3)-polycycles with 8 proper faces:

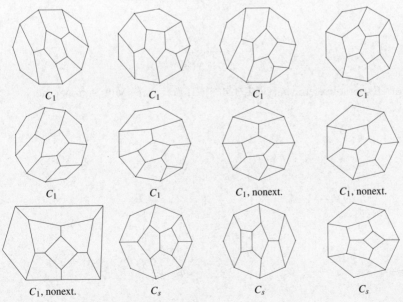

C_1 C_1 C_1 C_1

C_1 C_1 C_1, nonext. C_1, nonext.

C_1, nonext. C_s C_s C_s

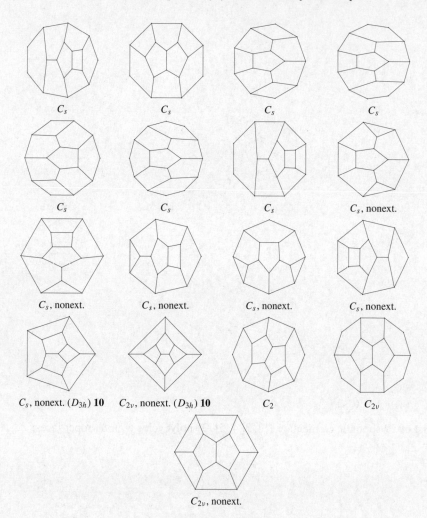

List of 16 sporadic elementary ({2, 3, 4, 5}, 3)-polycycles with 9 proper faces:

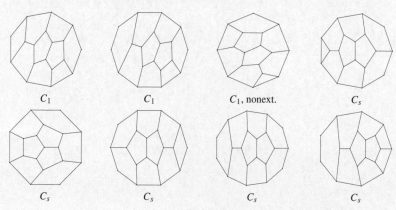

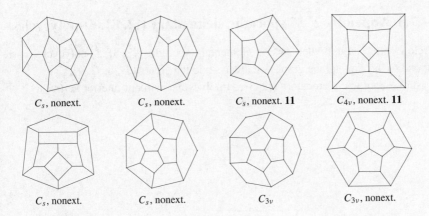

C_s, nonext. C_s, nonext. C_s, nonext. **11** C_{4v}, nonext. **11**

C_s, nonext. C_s, nonext. C_{3v} C_{3v}, nonext.

List of 9 sporadic elementary ({2, 3, 4, 5}, 3)-polycycles with 10 proper faces:

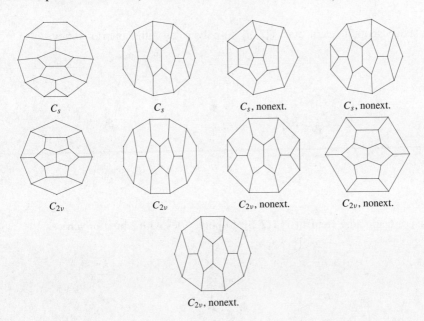

C_s C_s C_s, nonext. C_s, nonext.

C_{2v} C_{2v} C_{2v}, nonext. C_{2v}, nonext.

C_{2v}, nonext.

Unique sporadic elementary ({2, 3, 4, 5}, 3)-polycycle with at least 11 proper faces:

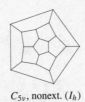

C_{5v}, nonext. (I_h)

7.8 Appendix 2: 57 sporadic elementary ({2, 3}, 5)-polycycles

Below (three cases) when several elementary sporadic ({2, 3}, 5)-polycycles correspond to the same plane graph, we always add the sign **A**, **B**, or **C**.
List of 2 sporadic elementary ({2, 3}, 5)-polycycles without interior vertices:

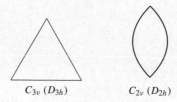

C_{3v} (D_{3h}) C_{2v} (D_{2h})

List of 3 sporadic elementary ({2, 3}, 5)-polycycles with 1 interior vertex:

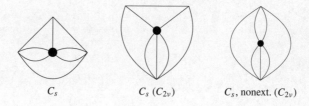

C_s C_s (C_{2v}) C_s, nonext. (C_{2v})

List of 6 sporadic elementary ({2, 3}, 5)-polycycles with 2 interior vertices:

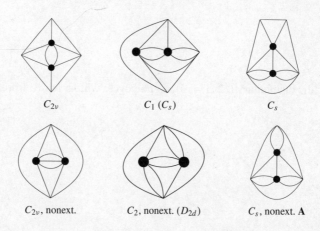

C_{2v} C_1 (C_s) C_s

C_{2v}, nonext. C_2, nonext. (D_{2d}) C_s, nonext. **A**

List of 10 sporadic elementary ({2, 3}, 5)-polycycles with 3 interior vertices:

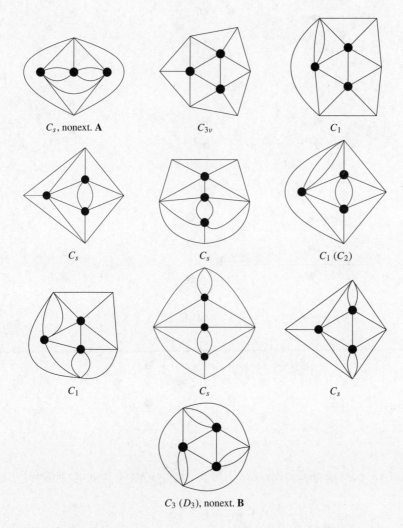

C_s, nonext. **A** C_{3v} C_1

C_s C_s C_1 (C_2)

C_1 C_s C_s

C_3 (D_3), nonext. **B**

List of 14 sporadic elementary ({2, 3}, 5)-polycycles with 4 interior vertices:

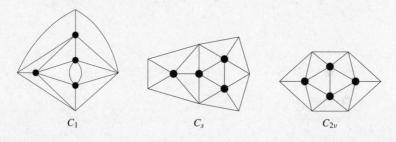

C_1 C_s C_{2v}

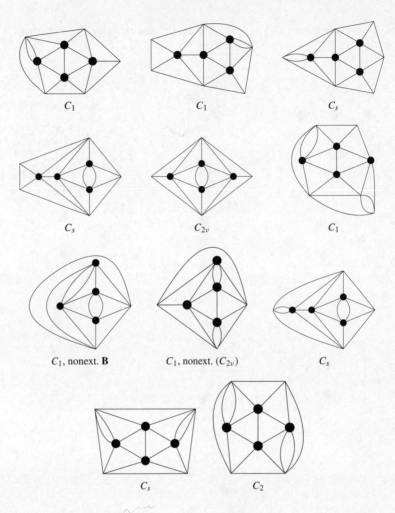

List of 10 sporadic elementary ({2, 3}, 5)-polycycles with 5 interior vertices:

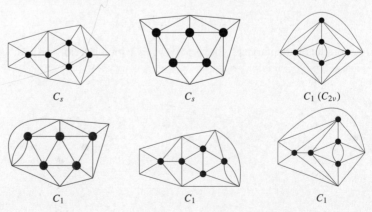

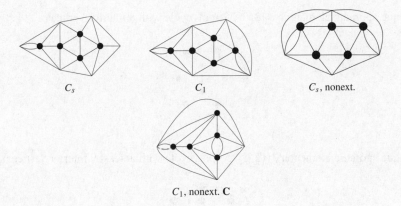

C_s C_1 C_s, nonext.

C_1, nonext. **C**

List of 9 sporadic elementary ({2, 3}, 5)-polycycles with 6 interior vertices:

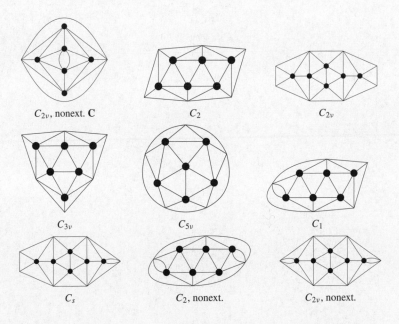

C_{2v}, nonext. **C** C_2 C_{2v}

C_{3v} C_{5v} C_1

C_s C_2, nonext. C_{2v}, nonext.

Unique sporadic elementary ({2, 3}, 5)-polycycle with 7 interior vertices:

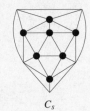

C_s

Unique sporadic elementary ({2, 3}, 5)-polycycle with 8 interior vertices:

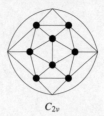

C_{2v}

Unique sporadic elementary ({2, 3}, 5)-polycycle with at least 9 interior vertices:

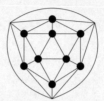

C_{3v}, nonext. (I_h)

8

Applications of elementary decompositions to (r, q)-polycycles

We present here applications of elementary polycycle decomposition (in particular, of lists in Figures 7.2 and 7.3) to three problems:

1 The determination of (r, q)-polycycles having the maximal number of interior vertices for a fixed number of interior faces. Complete solution for the elliptic case is presented.
2 The determination of all *non-extensible* (r, q)-polycycles, i.e. ones that cannot be extended by adding an r-gon. In particular, besides 5 platonic cases and 2 exceptional elliptic polycycles, they are infinite.
3 The determination of all 2-*embeddable* (r, q)-polycycles, i.e. ones whose skeleton can be embedded into a hypercube with scale 1 or 2. All parabolic and hyperbolic (r, q)-polycycles are 2-embeddable and a characterization, by induced subgraphs, of such elliptic (r, q)-polycycles is presented.

None of those applications is considered for other classes of polycycles, such as (R, q)-polycycles and $(r, q)_{gen}$-polycycles. A fourth main application, to face-regular maps, will be considered in Chapters 12, 13, 14, and 18.

Given an $(r, q)_{gen}$-polycycle P, its *major skeleton* $Maj(P)$ is the plane graph formed by its elementary components with two components being adjacent if they share an open edge. A *tree* is a connected graph with no cycles.

Theorem 8.0.1 *An $(r, q)_{gen}$-polycycle P is simply connected, i.e. is an (r, q)-polycycle, if and only if it holds:*

(i) its elementary components are simply connected and
(ii) its major skeleton $Maj(P)$ is a tree.

Proof. Assume (i) and (ii) hold, and take a closed cycle c in P. The set of elementary polycycles, passed by c, is a finite connected subgraph $Maj_c(P)$ of $Maj(P)$, so also a tree. If c pass though only one elementary component, then, by (i), we are done. Suppose that c pass thought more than one elementary component, then one vertex of $Maj_c(P)$ is of degree 1. Denote by e the open edge connecting the elementary component to the rest of $Maj_c(P)$. By (i), we can continuously transform c into a path c' that passes only through e, i.e. eliminate one vertex of $Maj_c(P)$. Iterating, we are led to $Maj_c(P)$ being a single vertex and we conclude again by (i).

Condition (i) is, clearly, necessary. If (ii) is not satisfied, then $Maj(P)$ contains a cycle. This cycle corresponds to a cycle in P, which, clearly, is not homotopic to 0.   □

8.1 Extremal polycycles

Denote by $p_r(P)$, $v_{int}(P)$ (or, simply, p_r, v_{int}) the number of interior faces and interior vertices of given finite (r, q)-polycycle P. Denote by $dens(P)$ and call *density* of a finite (r, q)-polycycle P the quotient $dens(P) = \frac{v_{int}(P)}{p_r(P)}$. Denote by $N_{r,q}(x)$ the maximum of $v_{int}(P)$ over all (r, q)-polycycles P with $p_r(P) = x$; call *extremal* any (r, q)-polycycle P with $v_{int}(P) = N_{r,q}(x)$ and $p_r(P) = x$. So, extremal polycycles represent the opposite case to outerplanar ones, in the class of all (r, q)-polycycles with the same p_r.

Remark 8.1.1 *Amongst all (r, q)-polycycles with the same p_r, the following statements are equivalent:*

1 v_{int} *is maximal,*
2 e_{int} *(the number of non-boundary edges) is maximal,*
3 *the* perimeter Per *(the number of boundary edges) is minimal,*
4 *the number v of vertices is minimal,*
5 *the number e of edges is minimal.*

This follows easily from Euler formula $(v_{int} + Per) - (e_{int} + Per) + (p_r + 1) = 2$ and equality $rp_r = 2e_{int} + Per$.

For $(5, 3)$-polycycles with $x \leq 11$, $N_{5,3}(x)$ was found in [CCBBZGT93]; all such extremal $(5, 3)$-polycycles turn out to be proper and unique. Moreover, the $(5, 3)$-polycycles, which are reciprocal (see Section 4.1) to any such extremal ones, turn out to be also extremal. [CCBBZGT93] asked about $N_{5,3}(x)$ for $x \geq 12$; this section answers this question for any x and for all elliptic (p, q).

In spite of the negative result of Theorem 7.2.1, we can easily obtain the following general density estimate.

Theorem 8.1.2 *(i) For any finite (r, q)-polycycle P, it holds that:*

$$dens(P) < \frac{r}{q}.$$

(ii) For parabolic (r, q), there exists a sequence of finite (r, q)-polycycles P_R with:

$$\lim_{R \to \infty} dens(P_R) = \frac{r}{q}.$$

Proof. (i) Take an arbitrary (r, q)-polycycle P. We tile each r-gon into 4-gons by connecting its center with the midpoints of the sides. Then the number of 4-gons in each r-gon is equal to r and the number of 4-gons, incident to any internal vertex, is equal to q. Then the number of 4-gons, incident only to internal vertices of the polycycle P, is equal to $v_{int}q$, while the total number of 4-gons is equal to rp_r. Hence, $v_{int}q < rp_r$.

(ii) On the Euclidean plane $\{r, q\}$, take a disk $C(0, R)$ with center 0 and radius R. Define the (r, q)-polycycle P_R to be formed by all r-gons contained in $C(0, R)$. It is easy to see that $\lim_{R \to \infty} dens(P_R) = \frac{r}{q}$. $\qquad\square$

All $(3, 3)$-, $(4, 3)$-, and $(3, 4)$-polycycles were obtained in [Har90] (proper) and [DeSt98] (improper ones). In the case of $(r, q) = (3, 3)$, the pairs (p_r, v_{int}) are $(1, 0)$, $(2, 0)$, and $(3, 1)$; in the case of $(r, q) = (4, 3)$, the pairs (p_r, v_{int}) are $(m, 0)$ for any $m \geq 1$, $(|\mathbb{N}|, 0)$ and[1] $(|\mathbb{Z}|, 0)$, $(3, 1)$, $(4, 2)$, and $(5, 4)$; in the case of $(r, q) = (3, 4)$, the pairs (p_r, v_{int}) are $(m, 0)$ for any $m \geq 1$, $(|\mathbb{N}|, 0)$ and $(|\mathbb{Z}|, 0)$, $(4, 1)$, $(5, 1)$, $(6, 1)$, $(6, 2)$, and $(7, 3)$. Amongst these pairs, those with $v_{int} \geq 1$, except for the pair $(p_r, v_{int}) = (6, 1)$, are realized only by proper polycycles; all improper polycycles, except for the case $(p_r, v_{int}) = (6, 1)$, have $v_{int} = 0$; i.e. they are outerplanar.

Theorem 8.1.3 *If an (r, q)-polycycle P is decomposed into elementary (r, q)-polycycles $(EP_i)_{i \in I}$ appearing x_i times, then we have:*

$$\begin{cases} v_{int}(P) = \sum_{i \in I} x_i v_{int}(EP_i) \\ p_r(P) = \sum_{i \in I} x_i p_r(EP_i). \end{cases}$$

If we solve the linear programming problem:

maximize $\sum_{i \in I} x_i v_{int}(EP_i)$
subject to $x = \sum_{i \in I} x_i p_r(EP_i)$ and $x_i \in \mathbb{N}$

---

[1] Here we distinguish two cases that are formally denoted by $(|\mathbb{N}|, 0)$ and $(|\mathbb{Z}|, 0)$, depending on whether a polycycle, considered as a chain, is infinite only in one direction or in two opposite directions.

and if $(x_i)_{i \in I}$, realizing the maximum, can be realized as an (r, q)-polycycle, then $N_{r,q}(x)$ is equal to the value of the objective function.

If this is not the case, then we have more possibilities to consider.

8.1.1 Extremal (5, 3)-polycycles

Theorem 8.1.4 *(i) If $x \leq 12$, then $N_{5,3}(x)$ is as given in Figure 8.1 with all the extremal (5, 3)-polycycles realizing the extremum.*

(ii) For any $x \geq 12$, we have:
$N_{5,3}(x) = x$ *if $x \equiv 0, 8, 9$ (mod 10) with the maximum realized by the following unique extremal (5, 3)-polycycle:*

- *If $x \equiv 0$ (mod 10), it is $\frac{x}{10}C_1$:*

- *If $x \equiv 9$ (mod 10), it is $\frac{x-9}{10}C_1 + B_2$:*

- *If $x \equiv 8$ (mod 10), it is $B_2 + \frac{x-18}{10}C_1 + B_2$:*

$N_{5,3}(x) = x - 1$ *if $x \equiv 6, 7$ (mod 10) with the maximum realized by the following (non-unique) extremal (5, 3)-polycycle:*

- *If $x \equiv 7$ (mod 10), it is $\frac{x-7}{10}C_1 + B_3$:*

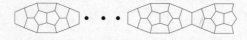

- *If $x \equiv 6$ (mod 10), it is $B_2 + \frac{x-16}{10}C_1 + B_3$:*

$N_{5,3}(x) = x - 2$ *if $x \equiv 1, 2, 3, 4, 5$ (mod 10) and the extremum is realized by (non-unique) E_{x+2}.*

x	$N_{5,3}(x)$	extremal polycycles	elem. components
1	0	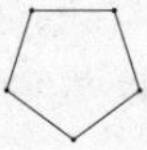	D
2	0	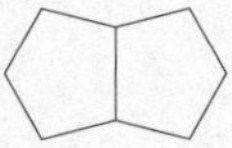	$2D$
3	1	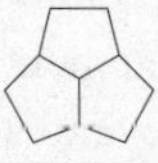	E_1
4	2	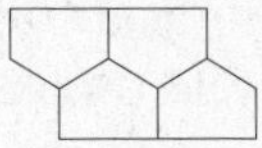	E_2
5	3	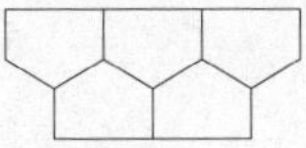	E_3
6	5	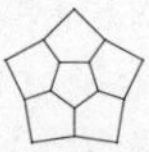	A_5
7	6		B_3
8	8	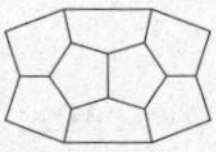	A_4
9	10		A_3
10	12		A_2
11	15		A_1
12	10	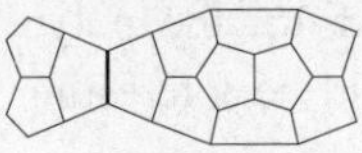	$E_1 + B_2$
		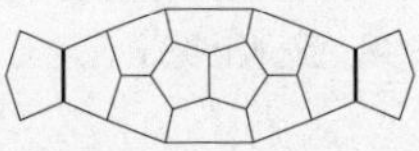	$D + C_1 + D$
		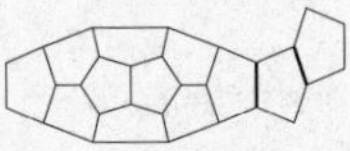	$C_1 + D + D$
		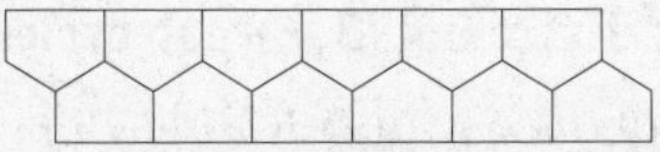	E_{10}

Figure 8.1 Extremal $(5, 3)$-polycycles with at most 12 faces

(iii) All possible densities of finite $(5, 3)$-polycycles, except for three cases $p_5 = 9, 10, 11$ are all rational numbers of the segment $[0, 1]$. All possible densities of polycycles, any of whose faces contain an internal vertex, are all rational numbers of the segment $[\frac{1}{3}, 1]$.

Proof. The proof of (i) and (ii) uses decomposition of $(5, 3)$-polycycles into elementary ones and the classification of elementary $(5, 3)$-polycycles. For example, for $p_5 \equiv 0 \pmod{10}$, an extremal polycycle is obtained by gluing only the copies of the polycycle C_1, while, for $p_5 \equiv 9 \pmod{10}$ or $p_5 \equiv 8 \pmod{10}$, we should glue together the copies of the polycycle C_1 and one or two copies of the polycycle B_2 (always at a deadlock). An elementary polycycle E_{x-2} is extremal for $n(x) = x - 2 \geq 10$; however, even for $x = 12$, there are three other extremal $(5, 3)$-polycycles.

The $(5, 3)$-polycycles E_1, C_1 have densities $\frac{1}{3}$, 1, respectively. Furthermore, if P is a $(5, 3)$-polycycle of the form $mE_1 + nC_1$, then its density is:

$$dens(P) = \frac{m + 10n}{3m + 10n} .$$

It is easy to see that we can find $n, m \in \mathbb{N}$ realizing all rational densities in $[\frac{1}{3}, 1]$. If every 5-gons is incident to an interior vertex, then E_1 does not occur as an elementary component of P. If $p_5 \geq 12$, then A_i cannot occur as an elementary component. All remaining elementary components have densities between $\frac{1}{3}$ and 1, thereby proving that for $p_5 \notin \{9, 10, 11\}$ the densities belong to $[\frac{1}{3}, 1]$. If we allow for the polycycle D to occur, then all rational densities in $[0, 1]$ can be realized. $\qquad\square$

As x grows, the number of extremal polycycles in the non-unique case grows. For example, for $x = 13, 14$, and 15, the polycycles $C_1 + E_1$, $C_1 + E_2$, and $C_1 + E_3$ are also extremal. It would be interesting to extend the notion of densities to infinite (r, q)-polycycles, but it is sometimes impossible to define densities for infinite settings (see [FeKuKu98, FeKu93] for a possible methodology in the hyperbolic case).

8.1.2 Extremal $(3, 5)$-polycycles

Theorem 8.1.5 *(i) If $x \leq 19$, then $N_{3,5}(x)$ is as given in Figure 8.2 with all the extremal $(5, 3)$-polycycles realizing the extremum.*

x	$N_{3,5}(x)$	extremal	comp.
1	0		d
2	0	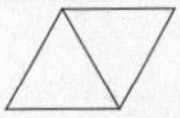	$2d$
3	0	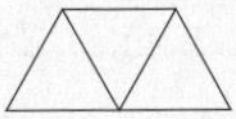	$3d$
4	0	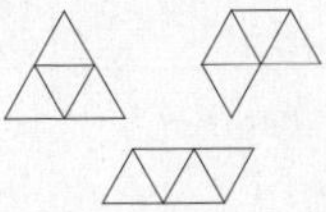	$4d$
5	1	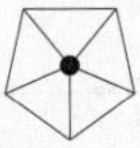	e_1
6	1	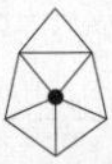	$e_1 + d$
7	1	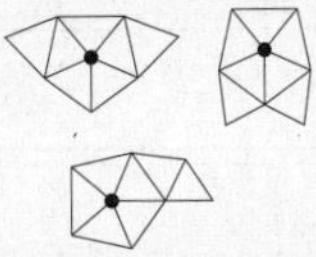	$d + e_1 + d$
8	2	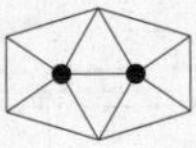	e_2
9	2	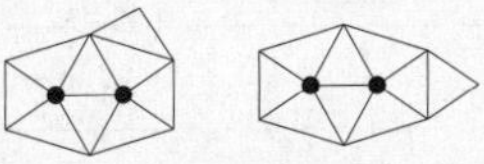	$e_2 + d$
10	3	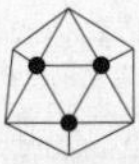	c_4
11	3	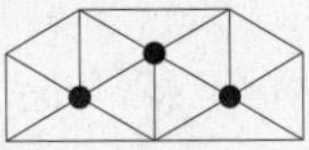	e_3
		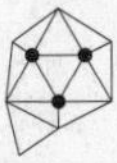	$c_4 + d$

x	$N_{3,5}(x)$	extremal	comp.
12	4	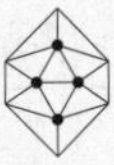	c_3
13	4	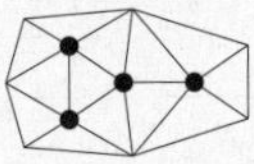	b_4
		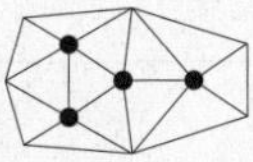	b_4
14	5	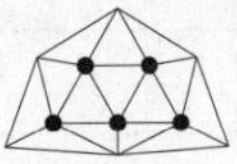	c_2
15	6	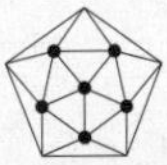	a_5
16	6	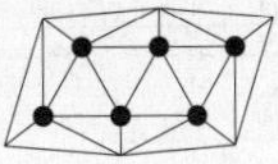	c_1
		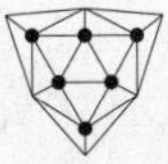	a_4
		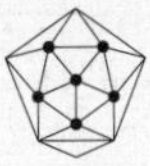	$a_5 + d$
17	7	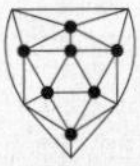	a_3
18	8	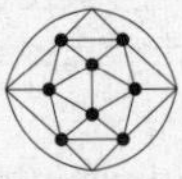	a_2
19	9	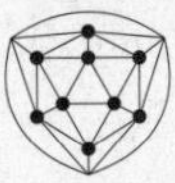	a_1

Figure 8.2 Extremal (3, 5)-polycycles with less than 20 faces

(ii) For any $x \geq 20$, we have:

$x \equiv a \pmod{18}$	$N_{3,5}(x)$	example of extremal polycycle	unicity
$a = 0$	$\frac{x}{3}$	$\frac{x}{18}b_2$	*no*
$a = 1$	$\frac{x-1}{3}$	$d + \frac{x-1}{18}b_2$	*no*
$a = 10$	$\frac{x-1}{3}$	$b_3 + \frac{x-28}{18}b_2 + b_4$ $\frac{x-10}{18}(c_1 + 2d) + c_4$	*no*
$a = 12$	$\frac{x}{3}$	$b_3 + \frac{x-30}{18}b_2 + b_3$ $\frac{x-12}{18}(c_1 + 2d) + c_3$	*no*
$a = 13$	$\frac{x-1}{3}$	$b_4 + \frac{x-13}{18}b_2$	*yes*
$a = 14$	$\frac{x+1}{3}$	$\frac{x-14}{18}(c_1 + 2d) + c_2$	*no*
$a = 15$	$\frac{x}{3}$	$b_3 + \frac{x-15}{18}b_2$	*yes*
$a = 16$	$\frac{x+2}{3}$	$\frac{x-16}{18}(c_1 + 2d) + c_1$	*yes*
$a = 17$	$\frac{x+1}{3}$	$\frac{x-17}{18}(c_1 + 2d) + c_1 + d$	*yes*

and, if x is not listed above, then the following applies:

$x \equiv a \pmod{3}$	$N_{3,5}(x)$	example of extremal polycycle	unicity
$a = 2$	$\frac{x-2}{3}$	$e_{N_{3,5}(x)}$	*no*
$a = 0$	$\frac{x-3}{3}$	$d + e_{N_{3,5}(x)}$	*no*
$a = 1$	$\frac{x-4}{3}$	$2d + e_{N_{3,5}(x)}$	*no*

(iii) All possible densities of finite $(3, 5)$-polycycles, except for $x \in \{14, \ldots, 19\}$ are all rational numbers of the segment $[0, \frac{1}{3}]$.

Proof. To prove Theorem 8.1.5, we apply the same strategy as for Theorem 8.1.4 except that now the number of cases to consider is larger (see Figure 7.3 of elementary $(3, 5)$-polycycles and their kernels).

For $x \leq 19$, the enumeration was done by hand since it contains some sporadic cases. For other values of x, we first remark that the a_i and b_1 cannot be elementary components of extremal polycycles. Afterward, we undertake exhaustive enumeration of all possibilities. □

8.1.3 Parabolic and hyperbolic extremal (r, q)-polycycles

The results presented here are very partial, but we do not expect to have fundamental difficulty in obtaining general results.

For parabolic or hyperbolic parameters (r, q), the *spiral $Sp_{r,q}(n)$* is defined as the proper (r, q)-polycycle with n r-gons obtained by taking an r-gon and adding r-gons in sequence, always rotating in the same direction. A formal definition is difficult to write (see [HaHa76, Gre01]), so we show some examples below:

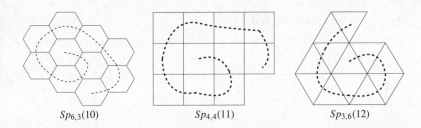

$Sp_{6,3}(10)$ $Sp_{4,4}(11)$ $Sp_{3,6}(12)$

It is proved in [HaHa76] that, for parabolic (r, q), the $Sp_{r,q}(n)$ (called there *extremal animal*) is an (r, q)-polycycle with the shortest perimeter amongst all proper (r, q)-polycycles. It is proved in [BBG03, Gre01] that, moreover, $Sp_{6,3}(n)$ has the shortest perimeter amongst all $(6, 3)$-polycycles. We recall, that minimizing the perimeter, when the number of r-gons is fixed, is equivalent to maximizing the number of interior vertices (see previous section).

It is conjectured that, for any parabolic or hyperbolic (r, q) and any $n \geq 1$, $Sp_{r,q}(n)$ is an extremal (r, q)-polycycle. A related conjecture is that all extremal polycycles are proper in the non-elliptic case. Moreover, for hyperbolic (r, q), it is likely that we have $\frac{N_{r,q}(x)}{x} < C_{r,q}$ for some constant $C_{r,q} < \frac{r}{q}$; see Theorem 8.1.2 for the parabolic case. The reason would be that in the hyperbolic space, the boundary is not negligible compared to the faces, which do not contain boundary edges; this point is well illustrated in [Mor97]. Also, the phenomenon, occurring in cases $(r, q) = (5, 3)$ or $(3, 5)$, of having small (r, q)-polycycle with higher density $\frac{N_{r,q}(x)}{x}$ than any other (r, q)-polycycle (for example, $\frac{N_{5,3}(9)}{9} > \frac{N_{5,3}(x)}{x}$ for any $x \geq 12$) will not occur for parabolic or hyperbolic (r, q). The reason is that, for parabolic or hyperbolic (r, q), any two (r, q)-polycycles can be joined to form a bigger (r, q)-polycycle.

Furthermore, the above extends to some (R, q)-polycycles. For example, one such problem is to determine the $(\{3, 4, 5, 6\}, 3)$-polycycles (see Chapter 7) with the shortest perimeter for a fixed p-vector. If $3p_3 + 2p_4 + p_5 \leq 6$, then the solution is shown in [Egg05] to be a kind of generalized spiral; previous work of Greinus [BBG03, Gre01] solved the $(\{5, 6\}, 3)$-polycycle case.

8.2 Non-extensible polycycles

Now we consider another natural notion of maximality for polycycles. An (r, q)-polycycle is called *non-extensible* if it is not a partial subgraph of any other (r, q)-polycycle, i.e. if an addition of any new r-gon removes it from the class of (r, q)-polycycles. It is clear that any tiling $\{r, q\}$ ($\{r, q\} - f$ in elliptic case) is non-extensible, while all other non-extensible polycycles are helicenes. It is also clear that any 3-connected $(r, 3)$-polycycle is non-extensible.

Four exceptional non-extensible polycycles, depicted on Figure 8.3 are: vertex-split Octahedron, vertex-split Icosahedron, and two infinite ones: $P_2 \times P_\mathbb{Z} = Prism_\infty$ and $Tr_\mathbb{Z} = APrism_\infty$ (see Section 4.2).

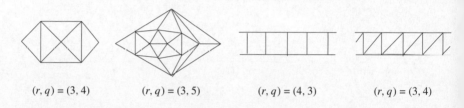

$(r, q) = (3, 4)$ $(r, q) = (3, 5)$ $(r, q) = (4, 3)$ $(r, q) = (3, 4)$

Figure 8.3 Exceptional non-extensible elliptic (r, q)-polycycles

Theorem 8.2.1 *([DeSt02b]) All non-extensible (r, q)-polycycles are:*

- *regular tilings $\{r, q\}$, ($\{r, q\} - f$ in elliptic case),*
- *4 exceptional elliptic polycycles from Figure 8.3,*
- *a continuum of infinite ones for any $(r, q) \neq (3, 3), (3, 4), (4, 3)$.*

Proof. The case $(r, q) = (3, 3), (3, 4), (4, 3)$ follows immediately from the list of these polycycles given in Chapter 4. It is clear that doubly infinite and non-periodic (at least in one direction) sequences of glued copies of the elementary polycycles b_2 and e_6 (from Figure 7.3) yield a continuum of infinite non-extensible $(3, 5)$-polycycles. By gluing the elementary $(5, 3)$-polycycles C_2 (from Figure 7.2) and C_2' (obtained from C_2 by rotation through π), we obtain infinite non-extensible $(5, 3)$-polycycles. Clearly, there is a continuum of such. In Lemma 8.2.4, we will construct a continuum of non-extensible (r, q)-polycycles for non-elliptic (r, q).

For the parameters $(r, q) = (3, 5), (5, 3)$, in Lemma 8.2.3, we will prove that the vertex-split Icosahedron is such a unique finite polycycle.   □

Lemma 8.2.2 *Any finite non-extensible (r, q)-polycycle is elliptic.*

Proof. The proof is based on curvature estimates. In fact, it is the counting of vertex-face incidences plus using the Euler formula. We choose to use the curvature viewpoint (see Section 4.4) since it express nicely ellipticity.

Since the angle of a regular r-gon is equal to $\frac{r-2}{r}\pi$ and the number of regular r-gons that meet at an internal vertex of the polycycle P is equal to q, the curvature of any internal vertex of the polycycle P is equal to:

$$\omega = 2\pi - \frac{r-2}{r}q\pi.$$

Hence, the total curvature of the polycycle P is equal to:

$$\Omega = v_{int}\frac{2(r+q)-rq}{r}\pi.$$

If $v_{int} = 0$, i.e. an (r, q)-polycycle P is outerplanar, then the curvature Ω is equal to zero for any parameters (r, q). If $v_{int} > 0$, then the curvature Ω is positive, zero, or negative, depending on whether the parameters (r, q) are elliptic, parabolic, or hyperbolic, respectively.

Any internal edge of a polycycle P belongs to exactly two r-gons, while any boundary edge belongs to only one r-gon. Therefore, the following equality holds:

$$rp_r = 2e_{int} + k,$$

where, again, e_{int} is the number of non-boundary edges of the polycycle P, and k is the number of boundary edges. On the other hand, since the number of boundary vertices and the number of boundary edges of P are equal to the same number k (the perimeter of the polycycle), these two numbers in the Euler formula cancel out, and a condensed version of this formula reads as:

$$v_{int} - e_{int} + p_r = 1.$$

From the last two formulas, we obtain:

$$(r-2)p_r = 2v_{int} + (k-2).$$

Now, let us calculate the sum of plane angles of the polycycle P. We do this in two different ways: (i) we first calculate the sum of angles in separate r-gons and then sum up over all r-gons and (ii) we first calculate the sum of angles at separate vertices and then sum up over all vertices, both boundary and internal. As a result, we obtain the equality:

$$p_r(r-2)\pi = \sum_{i=1}^{k} \varphi_i + v_{int}\frac{r-2}{r}q\pi,$$

where φ_i denotes the total angle at the ith boundary vertex of P. Combining this formula with the preceding one yields:

$$v_{int}\left(2\pi - \frac{r-2}{r}q\pi\right) = \sum_{i=1}^{k} \varphi_i - (k-2)\pi. \qquad (8.1)$$

This formula is called Gauss–Bonnet formula (see [Ale50]). Expressed differently, Euler formula $v - e + f = 2$ is for plane graphs with no boundaries, but it can be extended to plane graphs with boundaries.

Let k_j be the number of vertices of degree j (where $j = 2, 3, \ldots, q - 1, q$) on the boundary of P, and k be the total number of vertices of the boundary circuit of P, i.e. its perimeter. Then it holds that:

$$k = k_2 + k_3 + \ldots + k_{q-1} + k_q. \tag{8.2}$$

Let us calculate the sum $\sum_{i=1}^{k} \varphi_i - (k-2)\pi$ on the right-hand side of equality (8.1) for a finite polycycle P considered as a geodesic k-gon. Since it holds that:

$$\sum_{i=1}^{k} \varphi_i = (1k_2 + 2k_3 + \ldots + (q-2)k_{q-1} + (q-1)k_q)\frac{r-2}{r}\pi,$$

by formula (8.2), we obtain the equality:

$$\sum_{i=1}^{k} \varphi_i - (k-2)\pi = ((1\tfrac{r-2}{r} - 1)k_2 + (2\tfrac{r-2}{r} - 1)k_3 + \ldots$$
$$+ \left((q-2)\tfrac{r-2}{r} - 1\right) k_{q-1} + \left((q-1)\tfrac{r-2}{r} - 1\right) k_q)\pi + 2\pi. \tag{8.3}$$

Consider a particular case of a finite polycycle P in which each vertex has degree q. In this case, $2e = qn$, where n is the total number of vertices and e is the total number of edges of P; in view of the equality $2e = rp_r + k$, we can rewrite the Euler formula $n - e + p_r = 1$ as follows:

$$n\frac{2(q+r) - qr}{2r} = 1 + \frac{k}{r}.$$

Hence, $2(q+r) - qr > 0$, i.e. the parameters (r, q) are elliptic. For any of the five elliptic pairs $(r, q) = (3, 3), (3, 4), (3, 5), (4, 3), (5, 3)$, we directly verify that the equality $k = r$ holds and that P is, in fact, the surface of a Platonic body without one face.

Assume now that there exists a vertex of degree different from q. By Theorem 4.3.3, there are at least two such vertices.

Suppose that the parameters (r, q) are parabolic or hyperbolic, i.e. the inequality $qr - 2(q + r) \geq 0$ holds. Then, the following estimate is valid for the coefficient of k_q in Equation (8.3):

$$(q-1)\frac{r-2}{r} - 1 = \frac{qr - 2(q+r)}{r} + \frac{2}{r} \geq \frac{2}{r}.$$

By Corollary 4.3.5, the total number of vertices on the boundary must be greater than r Any two vertices of degree less than q should be separated by at least $r - 1$ vertices of degree q; otherwise, the polycycle P would be extensible.

From the aforesaid and condition $r \geq 3$, it follows that:

$$k_q \geq (r-1) \sum_{j=2}^{q-1} k_j \geq 2 \sum_{j=2}^{q-1} k_j.$$

Therefore, by (3), the quantity $\sum_{i=1}^{k} \varphi_i - (k-2)\pi$, calculated for the polycycle P, follows the bound:

$$\sum_{i=1}^{k} \varphi_i - (k-2)\pi \geq ((1\tfrac{r-2}{r} - 1 + 2\tfrac{2}{r})k_2 + (2\tfrac{r-2}{r} - 1 + 2\tfrac{2}{r})k_3 + \cdots \\ + ((q-2)\tfrac{r-2}{r} - 1 + 2\tfrac{2}{r})k_{q-1})\pi + 2\pi. \quad (8.4)$$

The coefficient of k_j on the right-hand side of Inequality (8.4) increases as the index j increases. Since the least coefficient (that of k_2) is positive, all the other coefficients of k_j are also positive. The values of k_j themselves are non-negative (and there even exists one positive k_j, because there are vertices on the boundary of P, whose degree is less than q). Hence, it holds:

$$\sum_{i=1}^{k} \varphi_i - (k-2)\pi \geq 2\pi.$$

The positivity of the right-hand side of Inequality (8.4) implies the positivity of the left-hand side. Thus, in view of Equation (8.1), the curvature Ω of the geodesic k-gon is positive. The resulting inequality $2(r+q) - qr > 0$ contradicts the assumption made. Hence, a finite non-extensible polycycle cannot have parabolic or hyperbolic parameters (r, q); these parameters are elliptic. □

Lemma 8.2.3 *([DSS06]) All finite elliptic non-extensible (r, q)-polycycles are two vertex-splittings (of Octahedron and Icosahedron; see first two in Figure 8.3) and five Platonic $\{r, q\}$ (with a face deleted).*

Proof. The case of $(3, 3)$-, $(3, 4)$-, and $(4, 3)$-polycycles is resolved by using the classification of such polycycles in Section 4.2. Consider now an (r, q)-polycycle P with $(r, q) = (3, 5)$ or $(5, 3)$. Then we can use the classification of the elementary (r, q)-polycycles given in Figure 7.2 and 7.3. Consider now its elementary components and the major skeleton $Maj(P)$ formed by them with two components being adjacent if and only if they are adjacent on an open edge. Clearly, $Maj(P)$ is a plane graph. But, by Theorem 8.0.1, it is also a tree. So, either $Maj(P)$ is reduced to a point, or $Maj(P)$ has a vertex of degree 1.

Consider now the case $(r, q) = (5, 3)$. It is easy to see that the only finite elementary $(5, 3)$-polycycle, which is non-extensible, is $A_1 = \{5, 3\} - f$. Assume now that $Maj(P)$ has a vertex of degree 1. Then, the elementary polycycle corresponding to this vertex is different from A_i. It is easy to see that for all other finite elementary $(5, 3)$-polycycles, we can extend, i.e. add one more face.

Consider now the case $(r, q) = (3, 5)$ and take a non-extensible $(3, 5)$-polycycle P. The only finite elementary non-extensible $(3, 5)$-polycycle is $a_1 = \{3, 5\} - f$. Assume now that P is different from a_1, then it has more than one elementary component. So, the major skeleton $Maj(P)$ has vertices of degree 1.

It is easy to see that all e_i, b_i, and c_i cannot be a vertex of degree 1 in $Maj(P)$ since otherwise there will be an open edge on which we can add at least a 3-gon. So, d and a_3 are the only possibilities for vertices of degree 1 in $Maj(P)$. If a_3 occurs, then P is reduced to $d + a_3$ as expected. So, let us assume that d occurs as vertices of degree 1. If d is adjacent to the elementary polycycle P_{el}, then the open edge to which d is incident, should have both vertices of degree 4 since, otherwise, we can add another 3-gon to d. The only elementary $(3, 5)$-polycycles having two vertices of degree 4 in succession are a_3, c_2, and c_3. If a_3 occurs, then we are done. The following diagram shows, up to symmetry, why c_2 is impossible:

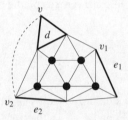

It is conceivable that we can add a polycycle on the open edge e_1 to forbid the extension by an edge (v, v_1). But it is not possible to add a polycycle on e_2; so, we are always able to add the edge (v, v_2) and P is extensible.

If c_3 occurs, then we have the following diagram:

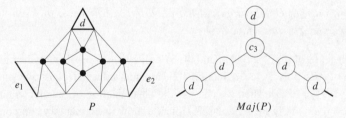

$$P \qquad\qquad\qquad Maj(P)$$

Since the polycycle P is non-extensible, there are some polycycles incident to the edges e_1 and e_2. So, we have two paths starting from d. Since $Maj(P)$ is a finite tree, those two paths will eventually terminate on a vertex of degree 1, which, by the above analysis, has to be another d. Furthermore, the elementary $(3, 5)$-polycycle preceding it has to be c_3. So, again we have two paths, one of them new. This argument does not terminate. We do not find a cycle in $Maj(P)$ since it is a tree and so we proved that P is infinite. This is impossible by the hypothesis and so the only possibility is $d + a_3$. $\qquad\qquad\qquad\qquad\qquad\qquad\qquad\qquad\qquad\qquad\qquad\qquad\square$

Lemma 8.2.4 *For non-elliptic (r, q), there is a continuum of non-extensible (r, q)-polycycles.*

Proof. We consider infinite non-extensible polycycles obtained from $\{r, q\}$ by deleting certain non-adjacent r-gons followed by taking the universal cover. If we delete a countable number of r-gons using non-periodic sequences of deleted r-gons, then (due to an ambiguous choice of the deleted r-gons at each step) we obtain a continuum of different polycycles. Consider two non-congruent sequences S_1 and S_2 of r-gons in $\{r, q\}$, the $(r, q)_{gen}$-polycycles $P_1 = \{r, q\} - S_1$ and $P_2 = \{r, q\} - S_2$ are not isomorphic, their universal covers $\widetilde{P}_1$ and $\widetilde{P}_2$ are non-extensible (r, q)-polycycles, whose image in $\{r, q\}$ by the cell-homomorphism (see Theorem 4.3.1) is P_1 and P_2. Therefore, $\widetilde{P}_1$ and $\widetilde{P}_2$ are not isomorphic. $\qquad\square$

Finally, consider (r, q)-polycycles, such that any interior point of any interior face has degree 1 by the cell-homomorphism onto $\{r, q\}$. The number of such polycycles, which are not extensible without losing this property, is equal to 0 for $(p, q) = (3, 3), (3, 4)$ and equal to 1 for $(p, q) = (4, 3)$ (it is $P_2 \times P_5$). This number is finite for $(p, q) = (5, 3), (3, 5)$ and infinite, otherwise. The finiteness of this number for the parameters $(r, q) = (5, 3)$ and $(3, 5)$ follows from the fact that the number of 5-gons and 3-gons must be no greater than 12 and 20, respectively.

8.3 2-embeddable polycycles

We will shortly present 2-embedding of polycycles just as an application of elementary polycycles. 2-embedding of graphs is the main subject of the book [DGS04]. Let us only mention recent enumeration of 2-embeddable $(\{a, 6\}, 3)$-spheres $(a = 3, 4, 5)$. There are 1, 5, 5, respectively, such graphs for $a = 3, 4, 5$ (see [DDS05, MaSh07]).

Given a set S, the *Hamming distance* on $|S|$-hypercube $\{0, 1\}^{|S|}$ is defined by $d(x, y) = |\{i \in S : x_i \neq y_i\}|$. Given two vertices u, v of a graph G, the *path-distance* $d_G(u, v)$ is the minimal number of edges needed to connect u to v.

A graph G is said to be *2-embeddable* if there exist a set S and a function:

$$\psi : V(G) \to \{0, 1\}^{|S|}$$

$$v \mapsto \psi(v)$$

such that for all vertices v, v' of G, we have $d(\psi(v), \psi(v')) = 2d_G(v, v')$ with d_G being the path-distance on G. In fact, a graph is 2-embeddable if and only if it is an isometric subgraph of an *half-cube*.

For more information on this subject, see [DeLa97, DGS04] and references therein. For finite graphs, an efficient polynomial time algorithm for recognizing 2-embeddable plane graphs is given in [DeSh96] (see an implementation in [Dut03]).

Given a plane graph G, an *alternating zone* is a sequence (e_i) of edges such that e_i and e_{i+1} belong to the same face F_i. If F_i has an even number of vertices, then we require e_i and e_{i+1} to be on opposite side in F_i. If F_i has an odd number of faces, then e_i and e_{i+1} are again in opposition but then we have two choices denoted, up to rotation of the plane, $+$ and $-$. We require that the choices $+$ and $-$ are alternating. The final edges of the zone (incident to exterior faces) are called the *ends* of the zone. If the zone is not self-intersecting, then, after removal of the edges of the zone, we obtain two graphs G_1 and G_2. If each shortest path in G between two vertices of G_i for a fixed $i = 1, 2$ consists only of edges in this subgraph G_i, then we say that the zone realizes a *convex cut* of the given (r, q)-polycycle. If every alternating zone realizes a convex cut, then the graph is 2-embeddable (see [CDG97] for a proof).

A (r, q)-*graph* is a plane graph such that all interior faces have at least r edges and all interior vertices have degree at least q. In [PSC90] it is proved that $(4, 4)$-graphs are 2-embeddable and it is proved in [CDV06] that $(6, 3)$- and $(3, 6)$-graphs are 2-embeddable. This implies that all (r, q)-polycycles with parabolic or hyperbolic (r, q) are 2-embeddable.

This and a check for elliptic (r, q) in Section 4.2 gives the following:

Theorem 8.3.1 *For $(r, q) \neq (5, 3), (3, 5)$, only three finite (r, q)-polycycles are not 2-embeddable:*

So, it remains to solve the 2-embedding problem only for $(r, q) = (3, 5)$ and $(5, 3)$. If an elementary polycycle P appears in the decomposition of a polycycle P', then P is an isometric subgraph of P'.

Theorem 8.3.2 *([DeSt00b]) A $(5, 3)$-polycycle different from Dodecahedron $\{5, 3\}$ is 2-embeddable if and only if it does not contain any of $(5, 3)$-polycycles E_4 and $D + E_2 + D$ (see Figure 8.4) as an induced subgraph.*

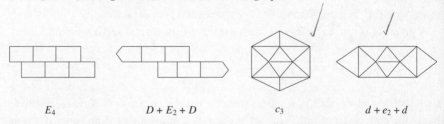

$\qquad\quad E_4 \qquad\qquad\qquad D + E_2 + D \qquad\qquad\qquad c_3 \qquad\qquad\qquad d + e_2 + d$

Figure 8.4 Forbidden subgraphs of non-embeddable $(5, 3)$- and $(3, 5)$-polycycles

Proof. The $(5, 3)$-polycycles E_4 and $D + E_2 + D$ are non-2-embeddable. So, some of their alternating zones do not define convex cuts:

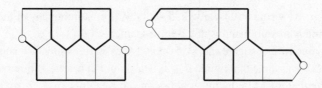

The $(5, 3)$-polycycle A_1 is 2-embeddable. The elementary $(5, 3)$-polycycles A_2, A_3, A_4, A_5, $E_\mathbb{Z}$, $E_\mathbb{N}$, B_2, B_3, C_1, C_2 contain E_4 as an induced subgraph and so they are not 2-embeddable.

Consider now a $(5, 3)$-polycycle, different from A_1, which does not contain E_4 and $D + E_2 + D$ as partial subgraph. From the above we know that its possible elementary components are D, E_2, E_3, C_3. Take an alternating zone Z passing through the elementary components $\ldots, E P_i, \ldots$. We know that if E_2 appears in this list, then it is the end of the sequence, since $D + E_2 + D$ is forbidden. Now, looking at the zone itself, we see that E_3 and C_3 also have to be at the end. Therefore, we have only D in the middle of the sequence $E P_i$. But D can be glued to only two other elementary (r, q)-polycycles. Then we check that the alternating zones are convex (this is long and cumbersome). $\qquad \square$

Theorem 8.3.3 *([DeSt02a]) A $(3, 5)$-polycycle different from Icosahedron $\{3, 5\}$ and $\{3, 5\} - v$ (Icosahedron with one vertex removed) is 2-embeddable if and only if it does not contain, as an induced subgraph, any of $(3, 5)$-polycycles c_3 and $d + e_2 + d$ shown in Figure 8.4.*

Proof. The $(3, 5)$-polycycles c_3 and $d + e_2 + d$ are not 2-embeddable. So, some of their alternating zones do not define convex cuts:

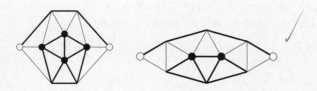

The edge of induced graphs G_i defined at the beginning of this section are boldfaced. Between the vertices in white there are the two shortest paths, one outside of the subgraph. The two polycycles $a_1 = \{3, 5\} - f$ and $a_5 = \{3, 5\} - v$ are 2-embeddable. The two polycycles $a_5 + d$ and $d + a_5 + d$, containing $d + e_2 + d$, are, on the other hand, not 2-embeddable. There are no other $(3, 5)$-polycycles containing a_5 as an elementary summand. The polycycles a_2, a_3, a_4, b_2, b_3, c_1, c_2, c_3 have at least one non-2-embeddable polycycle c_3 amongst their isometric sub-polycycles. The polycycles e_i, $i \geq 4$, b_1 and a_6 contain an isometric non-2-embeddable sub-polycycle

$d + e_2 + d$. By virtue of this, any $(3, 5)$-polycycle containing any of these polycycles as an elementary summand is not 2-embeddable.

It remains to investigate the $(3, 5)$-polycycles with elementary summands c_4, d, e_1, e_2, e_3. We assume that c_3 and $d + e_2 + d$ do not appear as sub-polycycles and we will prove that their alternating zones define convex cuts.

Let us take an alternating zone. The lateral sides of a 3-gon lie interior to the zone; only the extreme left and extreme right of them are the end of the zone; the base of a 3-gon lies on the boundary of the zone. Suppose that the base of a 3-gon belongs to the kernel. Then a polycycle e_2, that has this base as the kernel, is a sub-polycycle of the given $(3, 5)$-polycycle. There is only one edge of e_2, both ends of which have degree 3, to which a 3-gon d can be added (the polycycle $d + e_2 + d$ is forbidden); another such edge of e_2 is always an end of the zone. If the boundary of the zone does not contain two bases in succession belonging to the kernel, then another such base can only occur on the boundary of the zone at its other end. Of all the vertices on the boundaries of the zone not belonging to these bases in the kernel, at least every other one lies on the boundary of the surrounding $(3, 5)$-polycycle. By this, the zone defines a convex cut. But if two bases in succession on the boundary of the zone belong to the kernel, then the zone has 8 edges in all; both its ends are edges of an elementary polycycle e_3 with ends of degree 3. They define convex cuts.   ·   □

What we considered above was 2-embedding into $\{0, 1\}^{|S|}$. If the graph G is finite, then S is finite. If G is infinite, we can consider 2-embedding of G into $\mathbb{Z}^{|S|}$ with the distance $d(x, y) = \sum_{i \in S} |x_i - y_i|$. Any 2-embedding into $\mathbb{Z}^{|S|}$ gives a 2-embedding into $\{0, 1\}^{|T|}$ for some T. But we might be able to embed into $\mathbb{Z}^{|S|}$ with finite S even when G is infinite. This actually happens for the parabolic tilings $\{4, 4\} = \mathbb{Z}^2$, $\{3, 6\}$, $\{6, 3\}$ (both are 2-embeddable into $\mathbb{Z}^3$) and for infinite $(4, 4)$-, $(3, 6)$-, $(6, 3)$-graphs, but not for the hyperbolic tilings. Also there are infinite parabolic (r, q)-polycycles, which are 2-embeddable into $\mathbb{Z}^{|S|}$ only for infinite S.

9

Strictly face-regular spheres and tori

We say that a $(\{a, b\}, k)$-map is aR_i if every a-gonal face is adjacent exactly i times to a-gonal faces. It is said to be bR_j if every b-gonal face is adjacent j times to b-gonal faces. If the map is a cell-complex, then above i and j are just the numbers of a- and b-gonal neighbors.

An $(\{a, b\}, k)$-map is said to be *strictly face-regular map* if it is aR_i and bR_j for some i and j. It is *weakly face-regular* if it is aR_i and/or bR_j. In this chapter we will enumerate all strictly face-regular maps on sphere or plane. The classification on surfaces of higher genus is very difficult because there is an infinity of possibilities.

Denote by p_a, p_b the number of a-, b-gonal faces of a face-regular map having a finite number v of vertices. Denote by e_{a-b} the number of $a - b$ edges, i.e. those separating a- and b-gons. If the map is k-valent and orientable, then the Euler formula $v - e + f = 2(1 - g)$ with g being the genus can be rewritten (see Theorem 1.2.3) as:

$$p_a(2k - a(k - 2)) + p_b(2k - b(k - 2)) = 4k(1 - g).$$

But we have, clearly, $e_{a-b} = (a - i)p_a = (b - j)p_b$. Therefore, it follows:

$$e_{a-b}\left(\frac{2k - a(k - 2)}{a - i} + \frac{2k - b(k - 2)}{b - j}\right) = 4k(1 - g). \tag{9.1}$$

Write $\alpha(k, a, b, i, j) = \dfrac{2k - a(k - 2)}{a - i} + \dfrac{2k - b(k - 2)}{b - j}$ and find:

1 If $\alpha(k, a, b, i, j) > 0$, then $g = 0$, the map exists only on the sphere and the number of vertices depends only on $\alpha(k, a, b, i, j)$.

2 If $\alpha(k, a, b, i, j) = 0$, then $g = 1$, the map exists only on the torus.

3 If $\alpha(k, a, b, i, j) < 0$, then $g > 1$, the map exists only on surfaces of higher genus g and the number of vertices is determined by the genus and $\alpha(k, a, b, i, j)$.

We classify only the strictly face-regular spheres and strictly face-regular normal balanced planes. The plane case contains the torus case as a subcase. For the plane, the Euler formula does not hold, but the condition of normality, discussed thereafter,

125

allows us to use a version of it and obtain meaningful results. Again, the absence of a classification in the case of $g > 1$ is understandable, e.g. we cannot classify regular tilings of Riemann surfaces of genus $g > 1$.

Our classification is analogous to the following classification problems:

1. Classify maps, whose group acts transitively on vertices, i.e. Archimedean maps. The answer is known for the sphere and the plane.
2. Classify maps, whose symmetry group has two orbits on faces. Again, the answer is known for the sphere and the plane only.
3. Classify the tilings by regular r-gons. This is done for the sphere by Johnson (see [Joh66, Zal69]) and for the plane in [Cha89].
4. Classify maps with two types of edges (see [Jen90, GrSh87b]).

The list of all strictly face-regular two-faced polyhedra (completing the list in [BrDe99]) was found in [Dez02]. The list of all strictly face-regular planes (completing the list in [Dez02]) is given here and compared with the list of all 39 2-homeohedral types of such tilings of ([GLST85]); there are 33 realizable sets of parameters and a continuum of face-regular tilings for 21 of them. The results are presented in the form of Table 9.1 (illustrated by drawings of polyhedra) for 71 polyhedra and, for plane tilings, Table 9.3.

9.1 Strictly face-regular spheres

All strictly face-regular two-faced polyhedra are organized in 68 sporadic ones and 3 infinite families: prisms $Prism_b$ ($b \geq 5$), antiprisms $APrism_b$ ($b \geq 4$), and snub $Prism_b$ ($b \geq 6$).

A sphere is called 2-isohedral if its symmetry group has two orbits of faces. All 41 2-isohedral two-faced polyhedra (so, amongst those 71) are identified. They are given with symmetry groups and constructions.

The three infinite families are represented as Nrs. 15, 44, and 61 in Table 9.1, by their smallest members. Nrs. 2, 18 of Table 9.1 can be seen as snub $Prism_3$, snub $Prism_4$. Now, $Prism_4$, $APrism_3$, snub $Prism_5$ are Platonic polyhedra. $Prism_3$ is given separately under Nr. 1 and not as a case in Nr. 15 of $Prism_b$, because it has $a \neq 4$.

Theorem 9.1.1 *The list of strictly face-regular 3-connected ($\{a, b\}, k$)-spheres $a R_i$, $b R_j$ is the one of Table 9.1.*

Proof. We have the relation:

$$(2k - a(k - 2))(b - j) + (2k - b(k - 2))(a - i) > 0. \tag{9.2}$$

All elliptic ($\{a, b\}, k$)-spheres, i.e. those with $2k > b(k - 2)$ are listed in Chapter 2, we can just pick them here. All five there, having no 2-gons, are strictly face-regular.

Table 9.1 *All strictly face-regular* $(\{a, b\}, k)$*-polyhedra* aR_i *and* bR_j

Nr.	k	a, b	v	i, j	Aut	2-isohedral	Description
1	3	3,4	6	0,2	D_{3h}	+	*Prism*$_3$
2	3	3,5	12	0,4	D_{3d}	+	snub *Prism*$_3$
3	3	3,6	12	0,3	T_d	+	Truncated Tetrahedron $= GC_{1,1}(Tetrahedron)$
4	3	3,6	16	0,4	T_d	+	4−truncated Cube $= GC_{2,0}(Tetrahedron)$
5	3	3,6	16	0,4	D_{2h}	−	twisted Nr. 4
6	3	3,6	28	0,5	T	+	4-truncated Dodec. $= G_{2,1}(Tetrahedron)$
7	3	3,7	20	0,4	D_{3d}	+	6-truncated Cube
8	3	3,7	36	0,5	T_h	+	8-truncated Dodecahedron
9	3	3,7	36	0,5	D_3	−	twisted Nr. 8
10	3	3,8	24	0,4	O_h	+	Truncated Cube
11	3	3,8	44	0,5	T_h	−	12-truncated Dodecahedron
12	3	3,8	44	0,5	D_3	−	twisted Nr. 11
13	3	3,9	52	0,5	T	−	16-truncated Dodecahedron
14	3	3,10	60	0,5	I_h	+	Truncated Dodecahedron
15	3	4,**b**	2**b**	2,0	$D_{\mathbf{b}d}$	+	series *Prism*$_\mathbf{b}$, $\mathbf{b} \geq 5$
16	3	4,5	12	1,2	D_{2d}	+	decorated Cube
17	3	4,5	14	0,3	D_{3h}	+	$(b$-cap *Prism*$_3)^*$
18	3	4,5	16	0,4	D_{4d}	+	$(b$-cap *APrism*$_4)^* =$ snub *Prism*$_4$
19	3	4,6	14	2,2	D_{3h}	+	4-triakon Nr. 1
20	3	4,6	20	2,4	D_{3d}	+	4-triakon Nr. 2
21	3	4,6	20	1,3	D_3	+	4-halved Nr. 17
22	3	4,6	24	0,3	O_h	+	Truncated Octahedron $= GC_{1,1}(Cube)$
23	3	4,6	26	1,4	D_{3h}	−	decorated Nr. 17
24	3	4,6	32	0,4	O_h	+	$GC_{2,0}(Cube)$
25	3	4,6	32	0,4	D_{3h}	−	twisted $GC_{2,0}(Cube)$
26	3	4,6	56	0,5	O	+	$GC_{2,1}(Cube)$
27	3	4,7	44	1,4	T_h	+	4-halved Nr. 24
28	3	4,7	44	1,4	D_3	−	4-halved Nr. 25
29	3	4,7	44	2,5	T	+	4-triakon Nr. 6
30	3	4,7	80	0,4	O_h	−	$(b$-cap Rhombicuboctahedron$)^*$
31	3	4,7	80	0,4	D_{4d}	−	$(b$-cap tw.Rhombicuboctahedron$)^*$
32	3	4,8	32	2,4	T_d	+	4-triakon Nr. 4
33	3	4,8	32	2,4	D_{2h}	−	4-triakon Nr. 5
34	3	4,8	80	1,4	D_3	−	decorated Nr. 20
35	3	4,9	28	2,3	T_d	+	4-triakon Nr. 3
36	3	4,9	68	2,5	T_h	−	4-triakon Nr. 8
37	3	4,9	68	2,5	D_3	−	4-triakon Nr. 9
38	3	4,10	44	2,4	D_{3d}	−	4-triakon Nr. 7
39	3	4,11	92	2,5	T_h	−	4-triakon Nr. 11
40	3	4,11	92	2,5	D_3	−	4-triakon Nr. 12
41	3	4,12	56	2,4	O_h	+	4-triakon Nr. 10
42	3	4,13	116	2,5	T	−	4-triakon Nr. 13
43	3	4,15	140	2,5	I_h	−	4-triakon Nr. 14

Table 9.1 *(cont.)*

Nr.	k	a,b	v	i,j	Aut	2-isohedral	Description
44	3	5,**b**	**4b**	4,0	$D_{\mathbf{b}d}$	+	series snub $Prism_{\mathbf{b}}$, $\mathbf{b} \geq 6$
45	3	5,6	28	3,0	T_d	+	(b-cap truncated Tetrahedron)*
46	3	5,6	32	3,2	D_{3h}	−	decorated Nr. 23
47	3	5,6	38	2,2	C_{3v}	−	decorated snub $Prism_6$
48	3	5,6	44	2,3	T	+	decorated Nr. 45
49	3	5,6	52	1,3	T	−	decorated Nr. 48
50	3	5,6	56	2,4	T_d	−	decorated Nr. 22
51	3	5,6	60	0,3	I_h	+	Truncated Icosahedron $= GC_{1,1}(Dodecahedron)$
52	3	5,6	68	1,4	T_d	−	decorated Nr. 50
53	3	5,6	80	0,4	I_h	+	(b-cap Icosidodecahedron)*
54	3	5,6	80	0,4	D_{5h}	−	$GC_{2,0}(Dodecahedron)$
55	3	5,6	140	0,5	I	+	$GC_{2,1}(Dodecahedron)$
56	3	5,7	44	3,1	D_{3h}	−	6-halved Nr. 46
57	3	5,7	92	2,2	C_{3v}	−	decorated Nr. 47
58	3	5,8	56	3,0	O_h	+	decorated Nr. 22
59	3	5,8	92	3,2	T_d	−	decorated Truncated Octahedron
60	3	5,10	140	3,0	I_h	+	(b-cap Truncated Dodecahedron)*
61	4	3,**b**	**2b**	2,0	$D_{\mathbf{b}d}$	+	series $APrism_{\mathbf{b}}$, $\mathbf{b} \geq 4$
62	4	3,4	10	2,2	D_{4h}	+	capp. 1-Trunc. Octahedron
63	4	3,4	12	0,0	O_h	+	Cuboctahedron
64	4	3,4	14	1,2	D_{4h}	+	decorated (*Cuboctahedron*)*
65	4	3,4	14	1,2	D_{2d}	−	decorated (cuboct.)*
66	4	3,4	14	2,3	D_{4h}	+	capp. 2-Trunc. Octahedron
67	4	3,4	22	1,3	D_{2d}	−	decorated trunc. Tetrahedron
68	4	3,4	30	0,3	O	+	$GC_{2,1}(Octahedron)$
69	4	3,5	22	2,3	D_{4h}	+	4-cap 4-Trunc. Octahedron
70	4	3,5	30	0,0	I_h	+	Icosidodecahedron
71	4	3,6	30	2,3	O_h	+	4-cap Trunc. Octahedron

All parabolic ($(\{a,b\},k)$-spheres, i.e. those with $2k = b(k-2)$ that are bR_j, are listed in Chapter 10; so, assume in the following $(\{a,b\},k)$ to be hyperbolic, i.e. $2k < b(k-2)$.

Consider first the case $k = 3$. If $a = 3$, then $i = 0$. If we change those 3-gons into vertices, then we obtain a $(j,3)$-sphere G, which has to be *Bundle*$_3$ (3-connectedness is not necessarily preserved), Tetrahedron, Cube, or Dodecahedron. Those 3-gons determine a set of vertices Y of G such that every face of G is incident to $b - j$ vertices in Y. If G is *Bundle*$_3$, then Y consists of two vertices and G is *Prism*$_3$. If G is Tetrahedron, then Y consists of four vertices and G is Truncated Tetrahedron. For G being a Cube, Y has 2, 4, 6, and 8 elements, respectively. The corresponding spheres are listed in Table 9.1. For G a Dodecahedron, Y has 4, 8, 12, 16, or 20 elements, and the corresponding spheres are listed in Table 9.1. In the following, assume $a = 4$ or 5.

If $i = 0$, then b-gons can be adjacent only to at most $\frac{b}{2}$ a-gons and so we have $2(b - j) \leq b$. The only case of (aR_0, bR_j) satisfying those relations is $(4R_0, 7R_4)$. There are two such spheres (Nrs. 30 and 31 were found by computer enumeration).

Now assume $i = 1$. Every b-gon is adjacent to $b - j$ a-gons. For every b-gon F, denote by $n_1(F)$ the number of times the pattern bab occurs and by $n_2(F)$ the number of times $baab$ occurs. We have:

$$n_1(F) + 2n_2(F) = b - j \text{ and } 2n_1(F) + 3n_2(F) \leq b.$$

If $a = 4$, then summing over all b-gons gives $\sum n_1(F) = \sum n_2(F)$ implying the inequality $5\frac{b-j}{3} \leq b$, i.e. $2b \leq 5j$. The only cases of $(4R_1, bR_j)$ satisfying those relations are $(4R_1, 7R_3)$, $(4R_1, 7R_4)$, $(4R_1, 7R_5)$, $(4R_1, 8R_4)$, $(4R_1, 9R_4)$. The formula (9.1) rules out $(4R_1, 7R_3)$, since the number of vertices of such a map would be non-integral. The cases $(4R_1, 7R_5)$, $(4R_1, 9R_4)$, and $(4R_1, 8R_4)$ are ruled out by computer. If $a = 5$, then summing over all b-gons gives $\sum n_1(F) = 2\sum n_2(F)$ implying the inequality $7\frac{b-j}{4} \leq b$, i.e. $3b \leq 7j$ and this leaves no case to consider.

If $i = 2$ and $a = 4$, then either 4-gons are organized in a ring and we have *Prism*$_b$, or they are organized in triples. If they are in triples, then we can reduce them to 3-gons and obtain strictly face-regular $(\{3, b'\}, 3)$-spheres; this establishes a one-to-one correspondence, which is apparent in Table 9.1. If $i = 2$ and $a = 5$, then the possible cases are $b = 7$ with $0 \leq j \leq 3$ and $b = 8$ with $j = 0$ or 1. The case $b = 7$ is dealt with in Chapter 15 and we found only one sphere. If $b = 8$, then we have either a 8-gon, or two adjacent 8-gons. Those 8-gons contain 55 in their corona. Denote by F the unique other 8-gon adjacent to those two 5-gons; we see that F is adjacent to at least two other 8-gons, which is impossible.

If $i = 3$, then $a = 5$. The possibilities are $7 \leq b \leq 11$ with $12 - j - b > 0$. Computer enumeration determines what is possible (see also Lemma 12.5.5, Theorem 12.5.11, Theorem 12.5.12, and Lemma 12.5.4).

If $k = 4$, then $a = 3$ and recall that we already treated the $(\{3, 4\}, 4)$-spheres. If $i = 0$, then $b = 5$ and $j = 0$ or 1. Recall that the *vertex corona* of a vertex is the sequence of gonalities of faces to which the vertex belongs. We see easily that the vertex corona can only be 3535, which implies $j = 0$ and the sphere is Icosidodecahedron (Nr. 70).

If $i = 1$, then a vertex contained in the common edge of two 3-gons has vertex corona $33bb$. This implies that $j = 0$ is impossible. Combined with $8 - j - b > 0$, this yields $(b, j) = (5, 1)$, $(5, 2)$ or $(6, 1)$. If $j = 1$, then the vertices of 3-gons that are not contained in an edge between two 3-gons have vertex corona $3b3b$. We see that the b-gons around those vertices are adjacent to at least two other b-gons and we reach a contradiction. If $(b, j) = (5, 2)$, then the 5-gons are organized in rings that are non-intersecting in the sphere. At least one of those rings contains only 3-gons on one of its sides. We then obtain the following structure:

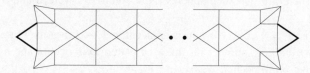

the boldfaced edges cannot be extended by adding an 3-gon without breaking the property $3R_1$. This proves that this case does not occur.

If $i = 2$, then either 3-gons are organized in a cycle and we obtain $APrism_b$, or they are in quadruples. If they are in quadruples, then we can reduce them to simple vertices and obtain a $(b', 4)$-sphere G. The only possibility is G being Octahedron. The quadruples correspond to a set of vertices Y of G. The possible sets Y are easy to obtain and there are two spheres (Nr. 67 and 69).

If $k = 5$, then $a = 3$. Equation (9.2) reads $(b - j) + (10 - 3b)(3 - i) > 0$. Using the condition $b \geq 4$, we see that $i = 0$ and $i = 1$ are impossible. If $i = 2$, then $b = 4$ and $j = 0$ or 1. The coronas of a 4-gon, respectively, pair of 4-gons are:

and so, we are forced to add a ring of 4-gons, which is impossible.   □

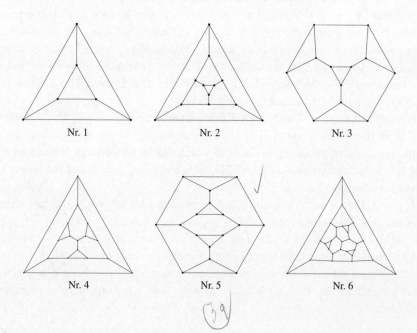

Nr. 1   Nr. 2   Nr. 3

Nr. 4   Nr. 5   Nr. 6

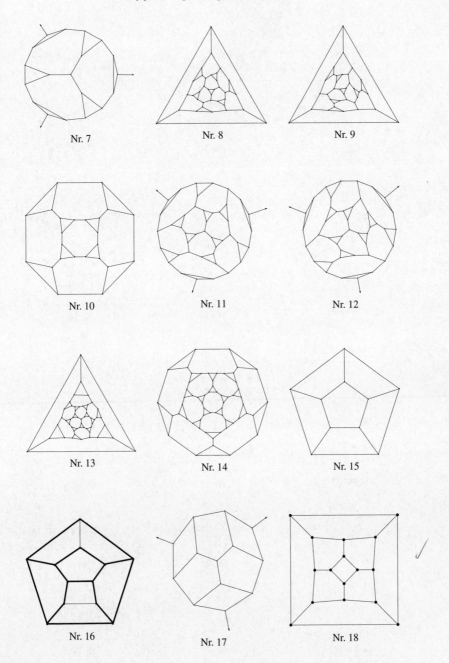

Nr. 7

Nr. 8

Nr. 9

Nr. 10

Nr. 11

Nr. 12

Nr. 13

Nr. 14

Nr. 15

Nr. 16

Nr. 17

Nr. 18

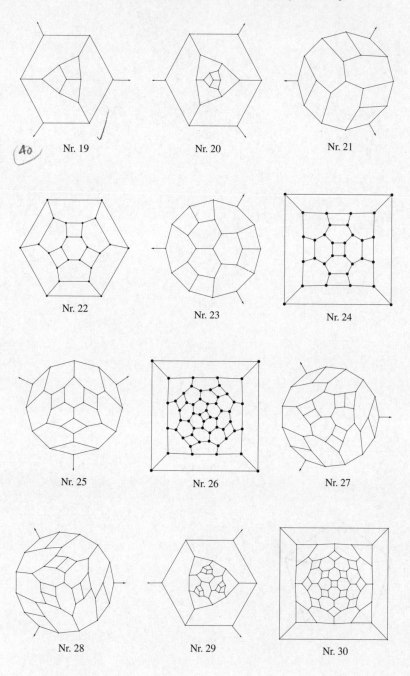

Nr. 19

Nr. 20

Nr. 21

Nr. 22

Nr. 23

Nr. 24

Nr. 25

Nr. 26

Nr. 27

Nr. 28

Nr. 29

Nr. 30

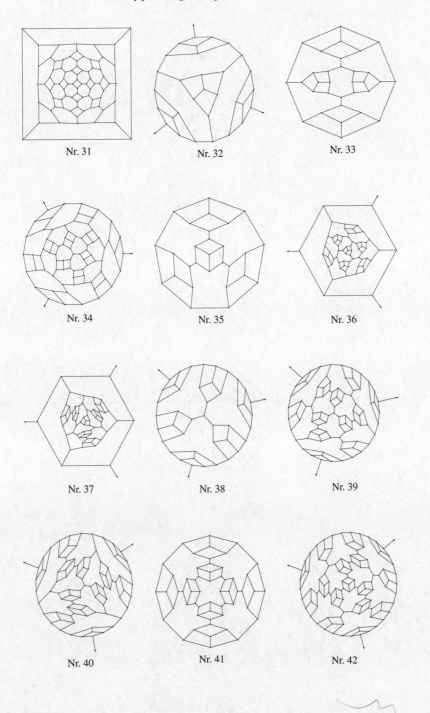

Nr. 31   Nr. 32   Nr. 33

Nr. 34   Nr. 35   Nr. 36

Nr. 37   Nr. 38   Nr. 39

Nr. 40   Nr. 41   Nr. 42

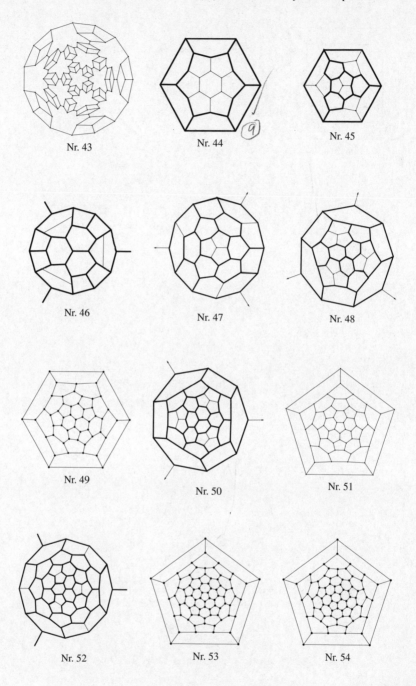

Nr. 43

Nr. 44

Nr. 45

Nr. 46

Nr. 47

Nr. 48

Nr. 49

Nr. 50

Nr. 51

Nr. 52

Nr. 53

Nr. 54

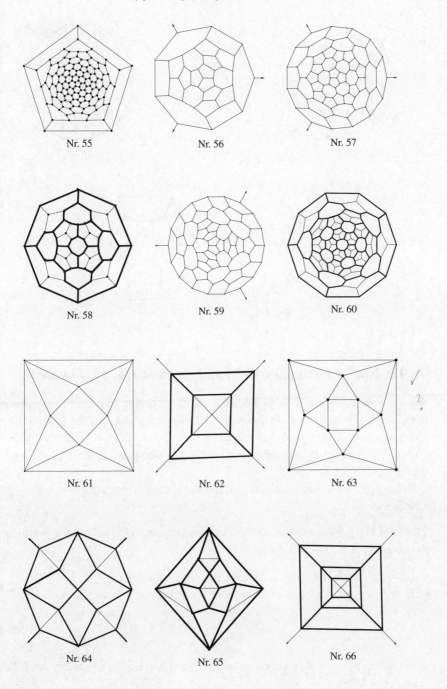

Nr. 55

Nr. 56

Nr. 57

Nr. 58

Nr. 59

Nr. 60

Nr. 61

Nr. 62

Nr. 63

Nr. 64

Nr. 65

Nr. 66

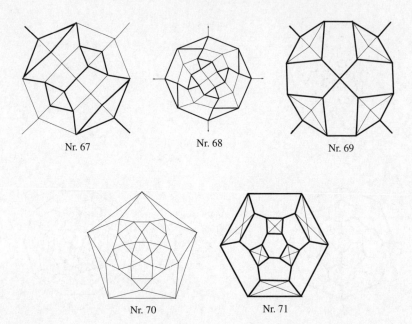

Nr. 67   Nr. 68   Nr. 69

Nr. 70   Nr. 71

9.2 Non-polyhedral strictly face-regular ({a, b}, k)-spheres

We list now the ({a, b}, k)-spheres that are 2-connected but are not polyhedra. Looking back at the proof of Theorem 9.1.1, we can notice that the hypothesis of 3-connectedness is used only for excluding $a = 2$ and, for $k = 3$, the case $a = 3$, $i = 1$. We consider those cases here, which, in particular, implies that the vertex degree k can be higher than 5.

Call *single ring* the map formed by a single edge, two faces and zero vertices (just a circle).

Theorem 9.2.1 *All* ({a, b}, 3)-*spheres that are* $a R_i$ *and* $b R_j$ *and are not polyhedra, are obtained in the following way:*

- *Let G be an* (a, 3)-*sphere or a single ring; fix an integer C.*
- *To every edge e of G, associate an integer parameter* x_e, *so that, for any face F of G, we have:*

$$\sum_{e \in F} x_e = C. \tag{9.3}$$

- *By putting* x_e *2-gons on edges e, we obtain a* ({2, $a + 2C$}, 3)-*sphere that is* $2R_0$ *and* $(a + 2C)R_{a+C}$.
- *By putting* x_e (3, 3)-*polycycles* {3, 3} $- e$ *on each edge e, instead of 2-gons, we obtain a* ({3, $a + 3C$}, 3)-*sphere that is* $3R_1$ *and* $(a + 3C)R_{a+C}$.

Proof. From the above discussion, a strictly face-regular ($\{a, b\}$, 3)-sphere that is 2- but not 3-connected, has $(a, i) = (2, 0)$ or $(3, 1)$. If G' is a ($\{3, b\}$, 3)-sphere that is $3R_1$ and bR_j, then we replace each pair of 3-gons by a 2-gon and obtain a ($\{2, b_1 = \frac{b+j}{2}\}$, 3)-sphere $2R_0$ and $b_1 R_j$.

Furthermore, by removal of each 2-gon (and both its vertices), we obtain the regular tiling $\{a, 3\}$ with $a = b_1 - (b - j) = \frac{3j-b}{2}$ and $a \leq 5$ or the simple ring. Reversing those operations means to put x_f 2-gons or $(3, 3)$-polycycles $\{3, 3\} - e$ on every edge f of $\{a, 3\}$. □

For example, the single ring yields the sphere formed by C 2-gons and two $2C$-gons and the sphere formed by $2C$ 3-gons and two $3C$-gons. Many other such spheres arise from $Bundle_3$, Tetrahedron, Cube, Dodecahedron by enumerating the integer solutions of equation (9.3). There is an extensive theory on this kind of problems; see, for example, [BeSi07].

Theorem 9.2.2 *All ($\{a, b\}$, 4)-spheres that are not polyhedra, are the following:*

(i) *The doubled b-gons ($2R_0$ and bR_0), i.e. those with edges split in two.*

(ii) *The ($\{2, b\}$, 4)-sphere $2R_0$ and bR_2, obtained by putting $b - 2$ 2-gons on the boldfaced edges of the diagram below:*

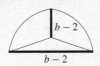

(iii) *For any integer $0 \leq h \leq b - 3$, the ($\{2, b\}$, 4)-sphere $2R_0$ and bR_3, obtained by putting h, $b - 3$ or $b - 3 - h$ 2-gons on the boldfaced edges of the diagram below:*

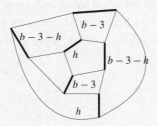

(iv) *If b is even, all half-tripled b-gons ($2R_1$ and $bR_{\frac{b}{2}}$), i.e. those with half edges tripled.*

(v) *If b is even, all spheres obtained by putting h Bundle₃ on one half of the edges and $\frac{b-2}{2} - h$ edges on the other half in the following way:*

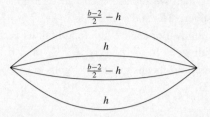

for some $0 \leq h \leq \frac{b-2}{2}$.

(vi) *If $b \equiv 3$ (mod 2), all spheres obtained by putting x_e Bundle₃ on every edge e of Octahedron, so that, for any face F of Octahedron, we have:*

$$\sum_{e \in F} x_e = \frac{b-3}{2}.$$

Proof. From the discussion at the beginning of this section, we can assume $a = 2$. If G is a $(\{2, b\}, 4)$-sphere that is $2R_0$ and bR_j, then we collapse every 2-gon of G to a vertex and obtain a 4-valent sphere G', whose faces are j-gons.

If $j = 0$, then G' is a single point and G is the sphere obtained from a b-gon by doubling every edge. If $j = 2$, then G' is $Bundle_4$ and, up to isomorphism, we have two ways to add the 2-gons on the boldfaced edges:

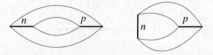

Only the second one is possible and we should have $n = p$. If $j = 3$, then G' is Octahedron. At every vertex of Octahedron, we have two ways to put 2-gons. Those possibilities are indexed by a pair of 3-gons sharing only an edge. Since we have 6 vertices, this makes 64 possibilities and, up to isomorphism, 7 possibilities. Clearly, every 3-gon should belong to at least one pair. This restricts us to three possibilities. The first possibility is shown below with the number associated to each vertex:

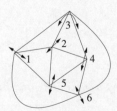

Denote by n_i the number of 2-gons put on the vertex i. After putting those 2-gons, the gonalities of the faces are:

face	gonality	face	gonality	face	gonality	face	gonality
123	$3 + n_1 + n_2$	234	$3 + n_3$	245	$3 + n_2 + n_4$	125	$3 + n_5$
456	$3 + n_5$	156	$3 + n_1 + n_6$	346	$3 + n_4 + n_6$	136	$3 + n_3$

All faces should have gonality b; so, $n_4 = n_1$, $n_6 = n_2$, $n_3 = n_5 = n_1 + n_2$. The second and third configurations are shown below:

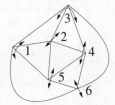 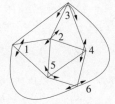

We can check that the second configuration has $n_1 = n_4 = 0$ and reduce to the first one, while the third configuration has $n_i = 0$.

If G is a $(\{2, 3\}, 4)$-sphere that is $2R_1$ and bR_j, then we can collapse the 2-gon and obtain a 4-valent sphere G' with faces of gonality j. The proof is then similar to the one of Theorem 9.2.1. $\qquad\square$

Theorem 9.2.3 *The list of 2-connected non-polyhedral strictly face-regular $(\{a, b\}, k)$-spheres with $k \geq 5$ is the one given on Table 9.2 and Figures 9.1 and 9.2.*

Proof. The non-polyhedrality implies $a = 2$. If $i = 0, 1$, then we get from Formula 9.2, $k \leq 12$, $k \leq 18$, respectively.

If $k = 5$ and $i = 1$, then we collapse the pair of adjacent 2-gons to a point and get a vertex of degree 4. Doing this for all vertices, we obtain a sphere G', whose vertices are 4- or 5-valent and whose faces are j-gons, i.e. the dual of a $(\{4, 5\}, j)$-sphere. Using Euler Formula (1.1), it is easy to see that $j = 2$ or 3 are the only possibilities. If $j = 2$, then G' is a 4-gon and we get sphere 1 in Figure 9.1. If $j = 3$, then we take the dual of the list of $(\{4, 5\}, 3)$-spheres from Chapter 2 and Figure 2.1. After considering all possibilities, we get only sphere 4 in Figure 9.1.

All the remaining enumeration is done with Formula 9.2 and exhaustive analysis of local configurations. $\qquad\square$

Table 9.2 *All strictly face-regular* $(\{a, b\}, k)$*-spheres* aR_i *and* bR_j *with* $k \geq 5$ *that are not polyhedra*

Nr.	k	a, b	v	i, j	Aut	2-isohedral	Description
1	5	2,3	4	1,1	D_{2d}	+	decorated Tetrahedron
2	5	2,3	4	0,1	D_{2d}	+	decorated Tetrahedron
3	5	2,4	12	0,2	D_{5d}	+	decorated $(APrism_5)^*$
4	5	2,4	12	1,3	D_5	+	decorated $(APrism_5)^*$
5	5	2,5	32	0,2	D_{5d}	−	decorated $(\text{snub } APrism_5)^*$
6	6	2,3	4	0,0	T_d	+	doubled Tetrahedron
7	6	2,3	5	0,1	D_{3h}	+	decorated $(Prism_3)^*$
8	6	2,3	8	0,2	D_{6h}	+	decorated $(Prism_6)^*$
9	6	2,4	8	0,0	O_h	+	doubled Cube
10	6	2,5	20	0,0	I_h	+	doubled Dodecahedron
11	6	2,**b**	**b**	1,0	$D_{\mathbf{b}h}$	+	tripled **b**-gon
12	7	2,3	4	1,1	D_{2d}	+	decorated Tetrahedron
13	7	2,3	16	0,2	D_{7d}	−	decorated $(\text{snub } Prism_7)^*$
14	7	2,4	16	1,2	D_{7d}	+	decorated $(APrism_7)^*$
15	7	2,5	44	1,2	D_{7d}	−	decorated $(\text{snub } APrism_7)^*$
16	8	2,3	6	0,0	O_h	+	doubled Octahedron
17	8	2,3	10	1,2	D_{8h}	+	decorated $(Prism_8)^*$
18	9	2,3	4	1,0	T_d	+	tripled Tetrahedron
19	9	2,3	20	1,2	D_{9d}	−	decorated $(\text{snub } Prism_9)^*$
20	9	2,4	8	1,0	O_h	+	tripled Cube
21	9	2,5	20	1,0	I_h	+	tripled Dodecahedron
22	10	2,3	12	0,0	I_h	+	doubled Icosahedron
23	12	2,3	6	1,0	O_h	+	tripled Octahedron
24	12	2,3	14	1,1	O_h	+	decorated (Truncated $Octahedron)^*$
25	15	2,3	12	1,0	I_h	+	tripled Icosahedron

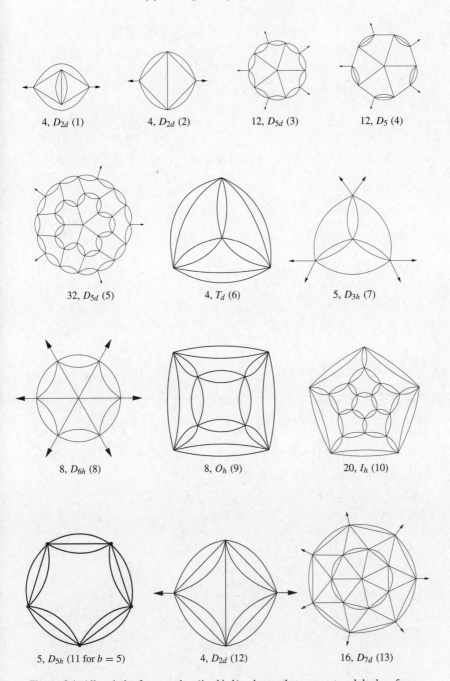

4, D_{2d} (1)   4, D_{2d} (2)   12, D_{5d} (3)   12, D_5 (4)

32, D_{5d} (5)   4, T_d (6)   5, D_{3h} (7)

8, D_{6h} (8)   8, O_h (9)   20, I_h (10)

5, D_{5h} (11 for $b = 5$)   4, D_{2d} (12)   16, D_{7d} (13)

Figure 9.1 All strictly face-regular ($\{a, b\}, k$)-spheres that are not polyhedra, from Theorem 9.2.3 (first part)

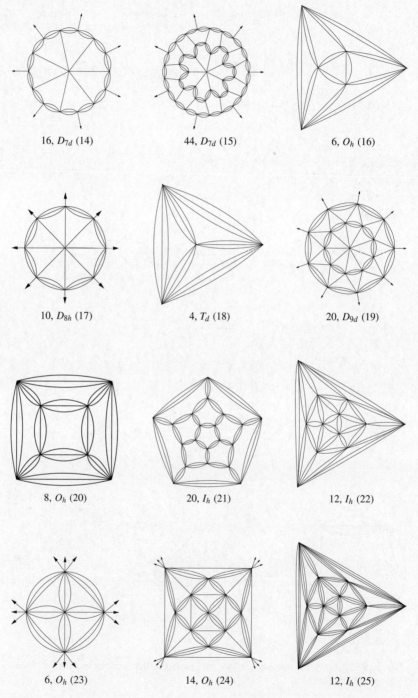

16, D_{7d} (14)   44, D_{7d} (15)   6, O_h (16)

10, D_{8h} (17)   4, T_d (18)   20, D_{9d} (19)

8, O_h (20)   20, I_h (21)   12, I_h (22)

6, O_h (23)   14, O_h (24)   12, I_h (25)

Figure 9.2 All strictly face-regular ($\{a, b\}, k$)-spheres that are not polyhedra, from Theorem 9.2.3 (second part)

9.3 Strictly face-regular ({a, b}, k)-planes

We are now considering the classification of strictly face-regular ($\{a, b\}, k$)-planes.

Following [GLST85, GrSh87a], call a tiling 2-*homeohedral* if the faces form two transitivity orbits under the group of *combinatorial* self-transformations of the Euclidean plane, that maps the tiling onto itself. If the plane is 2-periodic, then it is the universal cover of a torus and we can represent symmetries as isometries of the plane using primal-dual circle representations (see Section 1.3). However, for general planes, we do not have those tools and it may happen that some combinatorial symmetries cannot be represented as isometries.

The list of all 39 2-homeohedral types of k-valent tilings is given in [GLST85, pages 135–136] and in [GrSh87a, Figure 4.6.3]; see also the pioneering thesis [Löc68]. (The tilings with two orbits of *edges* were considered in [GrSh83]; they have at most three orbits of tiles and vertices each.) Every 2-homeohedral type of k-valent plane yields a strictly face-regular ($\{a, b\}, k$)-plane. Therefore, we will find those 39 types in this section.

Amongst strictly face-regular two-faced tilings, only Nrs. 6, 14, 30, 32, 33 (see the last picture of this section), i.e. Archimedean plane tilings with vertex coronas (3.12^2), (4.8^2), $(3.6.3.6)$, $(3^2.4.3.4)$, $(3^3.4^2)$ (with symmetry $p6m$, $p4m$, $p6m$, $p4g$, cmm, respectively; unique remaining two-faced Archimedean tiling $(3^4.6)$ is $6R_0$ but not $3R_i$) are *mosaics*, i.e. tilings of plane $\mathbb{R}^2$ by regular polygons.

We will give in Table 9.3 all 33 parameter set for face-regular ($\{a, b\}, k$)-plane: 21 continua and 12 sporadic cases (when the tiling is unique). The use of continua to describe discrete structures is not new:

1 Kelvin's intermediate structures between cubic closed packings and hexagonal closed packings.
2 3-dimensional sphere packings of the highest density are also described by a layer structure (see, for example, [OKHy96]).
3 In [DeSt00c], four cases of the continuum paradigm in three dimensions are described (partitions 31 to 34 of Table 4). One such example is the tiling of $\mathbb{R}^3$ by regular Tetrahedra and Octahedra.

In chemistry, face-regular spheres and tori are potentially interesting as putative molecular structures, see Figure 9.3.

The consideration of three-faced ($\{a, b, c\}, k$)-polyhedra and tori that are strictly face-regular is also interesting; see on Figure 9.4 an example occurring in chemistry. For such a structure we need a matrix t_{ij} that specifies the number of j-gonal faces adjacent to a i-gonal face. The first instance of such computations was in [BrDe99].

We already mentioned that classification of face-regular maps on a surface of genus greater than 1 is very difficult. However, such surfaces are of interest in

Table 9.3 *All strictly face-regular* $(\{a, b\}.k)$-plane aR_i and bR_j

Nr.	k	a, b	i, j	Nr. of all	Nr. of 2-homeohedral	Description
1	3	3,7	0,6	∞	1	$\frac{1}{6}$-truncated $\{6, 3\}$
2	3	3,8	0,6	∞	1	$\frac{1}{3}$-truncated $\{6, 3\}$
3	3	3,9	0,6	∞	3	$\frac{1}{2}$-truncated $\{6, 3\}$
4	3	3,10	0,6	∞	0	$\frac{2}{3}$-truncated $\{6, 3\}$
5	3	3,11	0,6	∞	0	$\frac{5}{6}$-truncated $\{6, 3\}$
6	3	3,12	0,6	1	1	trunc.$\{6, 3\} = (3.12^2)$
7	3	4,8	2,6	∞	1	4-triakon of Nr.1
8	3	4,10	2,6	∞	1	4-triakon of Nr.2
9	3	4,12	2,6	∞	1	4-triakon of Nr.3
10	3	4,14	2,6	∞	0	4-triakon of Nr.4
11	3	4,16	2,6	∞	0	4-triakon of Nr.5
12	3	4,18	2,6	1	1	4-triakon of Nr.6
13	3	4,7	0,5	∞	2	8-halved (4.8^2)
14	3	4,8	0,4	1	1	trunc.$\{4, 4\} = (4.8^2)$
15	3	4,8	1,5	∞	3	4-halved Nr.13
16	3	4,10	1,4	∞	2	4-halved (4.8^2)
17	3	5,7	1,3	∞	2	a 6-halved $\{6, 3\}$
18	3	5,7	2,4	$\infty + 1$	2	decorated $\{6, 3\}$
19	3	5,8	2,2	$\infty + 2$	2	a 6-halved $\{6, 3\}$
20	3	5,8	3,4	1	0	decorated $\{6, 3\}$
21	3	5,10	3,2	1	0	decorated $\{6, 3\}$
22	3	5,11	3,1	1	0	decorated $\{6, 3\}$
23	3	5,12	3,0	1	1	decorated $\{6, 3\}$
24	4	3,5	2,4	∞	2	decorated $\{4, 4\}$
25	4	3,6	2,4	∞	2	decorated $\{4, 4\}$
26	4	3,7	2,4	∞	0	decorated $\{4, 4\}$
27	4	3,8	2,4	1	1	4-capped (4.8^2)
28	4	3,5	0,2	1	1	decorated $\{4, 4\}$
29	4	3,5	1,3	∞	3	decorated $\{4, 4\}$, (4.8^2)
30	4	3,6	0,0	1	1	Archimedean $(3.6.3.6)$
31	4	3,6	1,2	1	1	decorated $\{4, 4\}$
32	5	3,4	1,0	1	1	Archimedean $(3^2.4.3.4)$
33	5	3,4	2,2	$\infty + 2$	2	decorated $\{4, 4\}$

physics. A *minimal surface* (of mean curvature 0) is a surface that minimizes surface tension (see [Oss69], for definitions and the relation with soap structures). Infinite 3-periodic minimal surfaces in $\mathbb{R}^3$ are of interest in Crystallography (see, for example, [KoFi87, KoFi96]).

Their quotient surface, under the translation group, is a surface of genus greater or equal to three. *Schwarzits*, i.e. $(\{6, 7, 8\}, 3)$-maps on surface of genus three, have been used to model 3-periodic minimal surfaces (see [Ki00]).

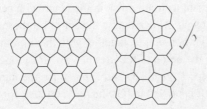

Figure 9.3 Two chemical azulenoid nets ([OKHy96]): the boron net YCrB$_4$ (left) and ThMoB$_4$ (right); they are strictly face-regular $(5, 7)$-planes $5R_1, 7R_3$ (case 17)

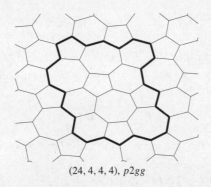

$(24, 4, 4, 4)$, $p2gg$

Figure 9.4 The net Y$_2$ReB$_6$ ([OKHy96]) is three-faced strictly face-reg
$0, 2, 3, t_{6j} = 2, 1, 3, t_{7j} = 3, 3, 1$ for $j = 5, 6, 7$) structure with a delimit
domain

9.3.1 Case determination

We follow here [GLST85] in presenting the notions of normality an

A *normal tiling* $\mathcal{T}$ is a tiling by a set of plane tiles $\mathcal{T} = \{T_1, T$
following properties:

1. Each tile is a closed topological disk.
2. The tiles cover the plane.
3. The intersection of any two tiles $T_i \cap T_j$ is a connected set of measure 0.
4. The tiles are *uniformly bounded*, i.e. there exist two parameters u and U such that every tile T_i contains, as a subset, a closed circular disk of radius u, and is contained in a closed circular disk of radius U.

The crucial hypothesis is the fourth one. It excludes hyperbolic tilings of the plane, like those which we can see in some of Escher art (planar ones but we cannot draw them with given metric constraints).

Let $D(\rho, P)$ represent a circular disk in the plane with center P and radius ρ. Let $A(\rho, P)$ denote the patch of tiles that is the set of all those tiles of T, whose intersection with $D(\rho, P)$ is non-empty, together with the minimum number of additional tiles needed to make the union of the tiles of $A(\rho, P)$ a topological disk (i.e. simply connected). Write $p(\rho, P)$, $e(\rho, P)$ and $v(\rho, P)$ for the number of tiles, edges, and vertices in $A(\rho, P)$. It is proved in [GrSh87a, §3.2] that, for any fixed number σ, we have the limit:

$$\lim_{\rho \to \infty} \frac{p(\rho + \sigma, P) - p(\rho, P)}{p(\rho, P)} = 0. \tag{9.4}$$

This expresses the fact that $A(\rho, P)$ is very near to a ball. If, in addition, the limits:

$$\lim_{\rho \to \infty} \frac{e(\rho, P)}{p(\rho, P)} \quad \text{and} \quad \lim_{\rho \to \infty} \frac{v(\rho, P)}{p(\rho, P)} \tag{9.5}$$

exist and are finite, then the tiling T is called *balanced*. If the values of the limits in (9.5) are denoted by $e(T)$ and $v(T)$, respectively, then *Euler formula for tilings* holds ([GrSh87a, §3.1]), that is:

$$v(T) - e(T) + 1 = 0.$$

Theorem 9.3.1 *A normal balanced* $(\{a, b\}, k)$*-plane* aR_i *and* bR_j *satisfies the equation:*

$$(2k - a(k - 2))(b - j) + (2k - b(k - 2))(a - i) = 0. \tag{9.6}$$

Proof. For $x = a, b$, denote by $p_x(\rho, P)$ the number of x-gonal faces in $A(\rho, P)$. Of course, we have $p_a(\rho, P) + p_b(\rho, P) = p(\rho, P)$. We would like to have $p_a(\rho, P)$ $(a - i) = p_b(\rho, P)(b - j)$ but this relation is not true in general, since the domain $A(\rho, P)$ is not closed, i.e. some of the edges are boundary edges. What we have is only:

$$|p_a(\rho, P)(a - i) - p_b(\rho, P)(b - j)| \le \{e(\rho + U, P) - e(\rho - U, P)\},$$

i.e. all the edges of the difference of the above terms are included in the difference between two disks $A(\rho, P)$. From Limits 9.4 and 9.5, we know that:

$$\lim_{\rho \to \infty} \frac{e(\rho + U, P)}{p(\rho, P)} = \lim_{\rho \to \infty} \frac{e(\rho - U, P)}{p(\rho, P)} = e(T),$$

which yields:

$$\lim_{\rho \to \infty} \frac{p_a(\rho, P)}{p(\rho, P)}(a - i) - \frac{p_b(\rho, P)}{p(\rho, P)}(b - j) = 0.$$

Combining with the exact equation $\dfrac{p_a(\rho, P)}{p(\rho, P)} + \dfrac{p_b(\rho, P)}{p(\rho, P)} = 1$, we obtain that:

$$\lim_{\rho \to \infty} \frac{p_a(\rho, P)}{p(\rho, P)} \quad \text{and} \quad \lim_{\rho \to \infty} \frac{p_b(\rho, P)}{p(\rho, P)}$$

exist and are equal to $p_a(T) = \frac{b-j}{a+b-i-j}$ and $p_b(T) = \frac{a-i}{a+b-i-j}$, respectively. Using the same technique, we prove the equations $kv(T) = 2e(T)$ and $2e(T) = ap_a(T) + bp_b(T)$ (which, like the preceding, are only approximately valid in the patch $A(\rho, P)$) and finally:

$$0 = (2k - a(k-2))p_a(T) + (2k - b(k-2))p_b(T),$$

which yields the required relation. $\qquad\qquad\qquad\qquad\qquad\qquad\qquad\square$

We now proceed to the determination of all possible parameters. Before that, a few remarks are in order. Equation (9.6) is useful only for the relations between k, a, b, i, and j. The difficulty is that an equation, say, $p_a(T) = 0$ does not imply $p_a(\rho, P) = 0$, since a function can be strictly positive and have a zero limit. The global equation (9.6) does not give local information. For local reasoning, our only possibility is to use *corona arguments*, i.e. those based on possible corona of faces.

Theorem 9.3.2 *The list of all possible cases (k, a, b, i, j) of 3-connected normal balanced $(\{a, b\}, k)$-planes aR_i and bR_j is the one given in Table 9.3.*

Proof. We will consider the cases successively.

If $k = 3$, then we have the relation $(6 - a)(b - j) + (6 - b)(a - i) = 0$.

If $a = 3$, then $i = 0$ (otherwise, the plane is not 3-connected) and the relation simplifies to $j = 6$. Since the pattern $b33b$ is forbidden in the corona, every b-gon is adjacent to at most six 3-gons and so, $7 \le b \le 12$.

If $a = 4$, then $0 \le i \le 2$. If $i = 0$, then the relation simplifies to $12 = b + j$, whose solutions are $(b, j) = (7, 5)$ and $(8, 4)$. Those two solutions are realizable. If $i = 1$, then the relation is $18 = b + 2j$, whose solutions are $(b, j) = (18, 0)$, $(16, 1)$, $(14, 2)$, $(12, 3)$, $(10, 4)$, $(8, 5)$. The b-gons contain patterns $b44b$ and $b4b$ in their corona, which implies $j \ge b - 2j$ and leaves only (realizable) solutions $(b, j) = (10, 4)$ and $(8, 5)$. If $i = 2$, then the 4-gons are necessarily organized in triples and come by 4-triakon of $(\{3, b'\}, 3)$-plane $3R0$, $b'R_6$.

If $a = 5$, then $0 \le i \le 3$. If $i = 0$, then $30 = j + 4b$, whose only solution is $(b, j) = (7, 2)$. This solution is not valid since a 5-gon would have corona 7^5 and those 7-gons have corona 7575555, i.e. 5-gons are not isolated. If $i = 1$, then $24 = j + 3b$, whose solution are $(b, j) = (7, 3)$ and $(8, 0)$. Solution $(8, 0)$ is not valid since 8-gons would have corona 5^8 and so the 5-gons would be adjacent to at least two 5-gons. If $i = 2$, then the equation is $18 = j + 2b$ whose solutions are $(7, 4)$, $(8, 2)$, $(9, 0)$. The solution $(9, 0)$ is not valid since a 9-gon would have corona 5^9 and those 5-gons would have corona 59599, i.e. a 9-gon would be adjacent to a 9-gon. If $i = 3$, then $12 = b + j$, whose solutions are $(b, j) = (7, 5)$, $(8, 4)$, $(9, 3)$, $(10, 2)$, $(11, 1)$, $(12, 0)$. The solution $(7, 5)$ is excluded by looking at three possible coronas in Lemma 12.5.5 and seeing that none of them has five 7-gons in it. The solution $(9, 3)$ requires a more detailed analysis. The condition $5R_3$ implies that the 5-gons of the plane are in $(5, 3)$-polycycles E_1 and E_2. We can then see that the boundary of the 9-gon is necessarily of the form:

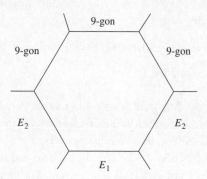

Then we see that the sequence of E_1 and E_2 propagate and we obtain a cycle $(E_1, E_2)^9$. Those are bounded by nine 9-gons and at the center of the structure, there is a 9-gon, which is adjacent to nine 9-gons and not three as requested. We get a contradiction. All other solutions are realizable.

If $k = 4$, the relation takes the form $(4 - a)(b - j) + (4 - b)(a - i) = 0$. The number a is necessarily equal to 3. If $i = 0$, the relation simplifies to $12 = 2b + j$, whose solutions are $(b, j) = (5, 2)$ and $(6, 0)$, which are both realizable. If $i = 1$, then $8 = b + j$, whose solutions are $(b, j) = (5, 3)$, $(6, 2)$, $(7, 1)$, and $(8, 0)$. If cases $(7, 1)$ and $(8, 0)$ occur, then a b-gon having the pattern 3^3 occurs in its corona. The 3-gon T, central to this pattern, is adjacent to another 3-gon T'. Regardless of the choice made, T' will be adjacent to at least two 3-gons, which is impossible. If $i = 2$, then $j = 4$. The 3-gons are organized in quadruples and the corona of 3-gons does not contain the pattern $b33b$. So, b-gons are adjacent to at most four 3-gons and $5 \leq b \leq 8$.

If $k = 5$, then $(10 - 3a)(b - j) + (10 - 3b)(a - i) = 0$ and a is necessarily equal to 3. If $i = 0$, the equation takes the form $30 = j + 8b$, which has no solution for $b \geq 4$. If $i = 1$, the equation is $20 = j + 5b$ and its unique (realizable) solution is $(b, j) = (4, 0)$. If $i = 2$, then the equation is $10 = j + 2b$, whose solutions are $(b, j) = (4, 2)$ and $(5, 0)$. In the case $(5, 0)$, all 5-gons have the corona:

All the edges in the boldfaced boundary must be incident to 5-gons, which is, clearly, impossible.   □

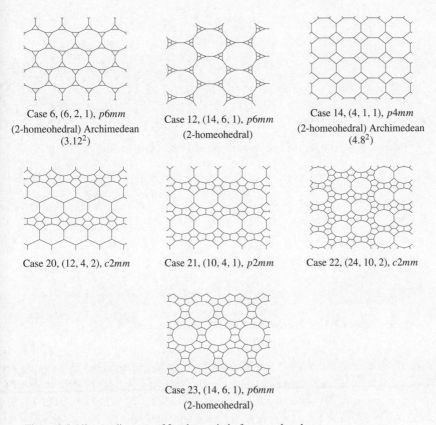

Case 6, (6, 2, 1), *p6mm*
(2-homeohedral) Archimedean
(3.12²)

Case 12, (14, 6, 1), *p6mm*
(2-homeohedral)

Case 14, (4, 1, 1), *p4mm*
(2-homeohedral) Archimedean
(4.8²)

Case 20, (12, 4, 2), *c2mm*

Case 21, (10, 4, 1), *p2mm*

Case 22, (24, 10, 2), *c2mm*

Case 23, (14, 6, 1), *p6mm*
(2-homeohedral)

Figure 9.5 All sporadic cases of 3-valent strictly face-regular planes

Note that in the next section we do the enumeration of planes for every case, without assumption of balancedness and normality. It turns out that, for every case, except possibly Case 29, all obtained planes are balanced and normal, but this was not guaranteed a priori. If we did not restrict ourselves to normal and balanced planes, we would have obtained hyperbolic plane tilings, which, as is well known, cannot be classified easily.

Planes 6, 12, 14, 20–23 (3-valent ones) and 27, 28, 30–32 are unique; those 12 sporadic cases are represented on Figures 9.5 and 9.12, respectively. The notation (v, p_a, p_b) (under picture of plane) denotes the number of vertices, the number of a-gons and the number of b-gons in a minimal torus obtained from the plane.

9.3.2 Proof and description of 33 parameter sets

Cases 1–6, take a ($\{3, b\}$, 3)-plane $3R_0$, bR_6. The 3-gons are isolated; therefore, we can shrink each of them into a single vertex. The obtained plane is $\{6, 3\}$, i.e. 3-valent

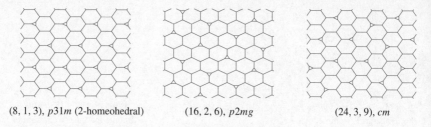

(8, 1, 3), $p31m$ (2-homeohedral) (16, 2, 6), $p2mg$ (24, 3, 9), cm

Figure 9.6 Some ($\{3, 7\}$, 3)-planes that are $3R_0$ and $7R_6$ (Case 1)

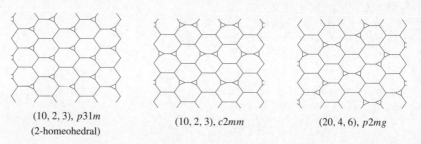

(10, 2, 3), $p31m$
(2-homeohedral) (10, 2, 3), $c2mm$ (20, 4, 6), $p2mg$

Figure 9.7 Some ($\{3, 8\}$, 3)-planes that are $3R_0$ and $8R_6$ (Case 2)

plane tiling by 6-gons. So, ($\{3, b\}$, 3)-planes $3R_0$, bR_6 are obtained by taking a set Y_b (for $7 \le b \le 12$) of vertices in $\{6, 3\}$ such that every face is incident to exactly $b - 6$ vertices in Y_b. By doing the truncation over such set Y_b, we obtain a ($\{3, b\}$, 3)-plane that is $3R_0$ and bR_6. Of course, by complementing the set Y_b, we obtain a set Y_{18-b}, thus establishing a one-to-one mapping between the classes 1 and 5, as well as between the classes 2 and 4. Also, since Y_6 is unique, ($\{3, 12\}$, 3)-plane $3R_0$, $12R_6$ is unique, but there is an infinity of sets Y_b for $7 \le b \le 11$. See Figures 9.6, 9.7, and 9.8 for some ($\{3, b\}$, 3)-planes $3R_0$, bR_6 with $b = 7, 8, 9$, respectively.

Cases 7–12, i.e. ($\{4, b\}$, 3)-planes $4R_2$, bR_6 are obtained from cases 1–6, respectively by 4-*triakon*, i.e. replacing each 3-gon by triple of adjacent 4-gons. The 2-homeohedral ones are shown on Figure 9.9.

Case 13, i.e. ($\{4, 7\}$, 3)-plane $4R_0$, $7R_5$. Given a ($\{4, 8\}$, 3)-plane G that is $4R_1$ and $8R_5$, the removal of the central edge, separating the pair of adjacent 4-gons, yields a ($\{4, 7\}$, 3)-plane that is $4R_0$ and $7R_5$. Clearly, the flipping (defined below in the proof for Case 15) of G does not change the obtained ($\{4, 7\}$, 3)-plane.

Take a ($\{4, 7\}$, 3)-plane G that is $4R_0$ and $7R_5$, and call an edge *isolated* if its four adjacent edges are not included into 4-gonal faces. By a corona argument, i.e. scanning the possible sequences of gonalities of faces, we see easily that every 7-gon contains at most 2 isolated edges. Hence, the possible structures for those isolated edges are triples of isolated edges (with a vertex contained in three such edges), paths, and circuits (with vertices contained in at most 2 edges).

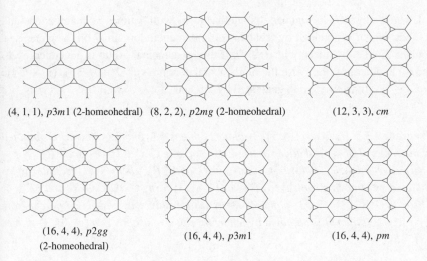

(4, 1, 1), *p*3*m*1 (2-homeohedral) (8, 2, 2), *p*2*mg* (2-homeohedral)   (12, 3, 3), *cm*

(16, 4, 4), *p*2*gg*   (16, 4, 4), *p*3*m*1   (16, 4, 4), *pm*
(2-homeohedral)

Figure 9.8 Some ({3, 9}, 3)-planes that are $3R_0$ and $9R_6$ (Case 3)

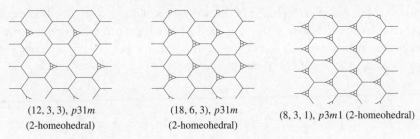

(12, 3, 3), *p*31*m*   (18, 6, 3), *p*31*m*
(2-homeohedral)   (2-homeohedral)   (8, 3, 1), *p*3*m*1 (2-homeohedral)

Figure 9.9 2-homeohedral ({4, *b*}, 3)-planes $4R_2$ and bR_6 with $b = 8, 10, 12$ (cases 7, 8, 9)

If G contains a circuit of isolated edges, then this set of isolated edges form a zigzag, i.e. every two, but not three, consecutive edges are contained in a face. By using local (i.e. corona) arguments, we see easily that the structure can be completed in a unique way. So, the corresponding ({4, 7}, 3)-plane is:

(6, 1, 2), *c*2*mm* (2-homeohedral)

which we denote by ({4, 7}, 3)$_{spec}$.

Given a torus G, its *squaring* is another torus with 4 times as many edges and vertices, which is obtained by replacing a fundamental domain $\mathcal{D}$ on the plane by a domain $2\mathcal{D}$, i.e. by all $2x$ with $x \in \mathcal{D}$. If the fundamental group is generated by v_1 and v_2; another possibility for the fundamental domain is $\mathcal{D} \cup (v_1 + \mathcal{D}) \cup (v_2 + \mathcal{D}) \cup (v_1 + v_2 + \mathcal{D})$, for the group generated by $2v_1$ and $2v_2$.

Theorem 9.3.3 *(i) Take a* $(\{4, 7\}, 3)$-*plane G that is $4R_0$ and $7R_5$; then G is obtained by the removal of central edges of a* $(\{4, 8\}, 3)$-*plane G' that is $4R_1$ and $8R_5$.*

(ii) Take a $(\{4, 7\}, 3)$-*torus G that is $4R_0$ and $7R_5$; then G or its squaring is obtained by removing central edges of a* $(\{4, 8\}, 3)$-*torus G' that is $4R_1$ and $8R_5$.*

(iii) If G is a $(\{4, 7\}, 3)$-*plane or torus whose universal cover is different from* $(\{4, 7\}, 3)_{spec}$, *then G is obtained by removing edges of exactly two* $(\{4, 8\}, 3)$-*planes* $4R_1$, $8R_5$.

Proof. Let us prove (i); we can assume that G is distinct from $(\{4, 7\}, 3)_{spec}$. The first step consists of associating to G another 4-valent plane $skel(G)$, whose vertex-set consists of all 4-gons. Every 7-gonal face is adjacent to two 4-gons; hence, it defines an edge of $skel(G)$ and $skel(G)$ is 4-regular. See below some representations of the local structure of G (in straight lines) and $skel(G)$ (in dashed lines):

Clearly, the faces of $skel(G)$ can be triples of edges, paths of isolated edges, or 2-gonal faces enclosing a single edge. Also, since G is different from $(\{4, 7\}, 3)_{spec}$, $skel(G)$ is connected.

An examination of all possibilities for faces of $skel(G)$ shows that, for every face, we can find two sets of lines cutting the 4-gons realizing the $(\{4, 8\}, 3)$-plane (one of them is shown on the pictures above).

Now we indicate how we can find a *coherent cutting* set for G, i.e. a cutting leaving a $(\{4, 8\}, 3)$-plane. If a path $v_0, \ldots, v_m$ of adjacent vertices in $skel(G)$ and the cutting line of v_0 are chosen, then this defines uniquely the choice of cutting lines of v_i. Assume that a coherent cutting does not exist. Then there exists a closed path $v_0, \ldots, v_m = v_0$, such that the choice of a cutting line on v_0 led us in the end to a different cutting line on $v_m = v_0$. If such a path exists, then we can assume that it does not self-intersect, i.e. it makes a closed circle, say, C.

Consider the set of interior faces of this circle. We can find an ordering $F_0, \ldots, F_q$, such that, for any $0 \leq i \leq q$, the graphs determined by $F_0, \ldots, F_i$, are enclosed by a path P_i, without self-intersections. The face F_i admits a coherent cutting of its 4-gons. Hence, by removing the face F_i, we obtain a path P_{i-1}, which does not admit a coherent cutting. But, in the end, we reach a contradiction since P_0 is a single face of $skel(G)$ and such faces admit coherent cutting sets.

Let us now prove (ii) for G being a $(\{4, 7\}, 3)$-torus $4R_0, 7R_5$. The above argument can be easily generalized to the following result: if P and P' are two closed paths of $skel(G)$, which are homologous, then P admits a coherent cutting set if and only if P' admits a coherent cutting set. The homology group $H_1(G)$ is isomorphic to $\mathbb{Z}^2$, i.e. there are two closed paths P_1 and P_2, such that any other closed path P is homologous to $n_1 P_1 + n_2 P_2$.

Suppose now that P_1 and P_2 admit a coherent cutting set. Then G admits a cutting set. Assume that P_1 or P_2 do not admit a coherent cutting set. If P_1 is of the form $v_0, \ldots, v_m$, then the path P_1', that correspond to P_1 in the squaring G' of G, is of the form $v_0, \ldots, v_m, \ldots, v_{2m}$. Now, $v_{2m} = v_0$ and $v_0 \neq v_m$. Since there are only two possibilities for the cutting of v_0, the cutting of v_{2m} is coherent with the cutting of v_0. The same argument applies to P_2 and so the squaring G' does admit a coherent cutting set. Assertion (iii) is a consequence of the connectedness of $skel(G)$ for G different from $(\{4, 7\}, 3)_{spec}$-torus. $\qquad\square$

See on Figure 9.10 some examples of $(\{4, 7\}, 3)$-planes that are $4R_0$ and $7R_5$.

Case 14, i.e. $(\{4, 8\}, 3)$-plane $4R_0, 8R_4$. Necessarily, the corona of 8-gons is of the form $(48)^4$, so we have only one sporadic plane.

Case 15, i.e. $(\{4, 8\}, 3)$-plane $4R_1, 8R_5$. A *special perfect matching SPM* in $\{3, 6\}$ is a set of edges, such that it holds:

1 every vertex is contained in exactly one edge of SPM and
2 every vertex is contained in exactly one 3-gon whose edge, opposite to the vertex, belongs to SPM.

The *flip* of a $(3, 6)$-plane with a special perfect matching SPM consists of changing the edges of SPM to their opposite according to the diagram below:

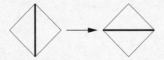

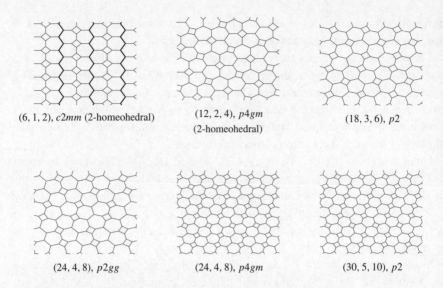

(6, 1, 2), *c2mm* (2-homeohedral)   (12, 2, 4), *p4gm*   (18, 3, 6), *p2*
                                           (2-homeohedral)

(24, 4, 8), *p2gg*   (24, 4, 8), *p4gm*   (30, 5, 10), *p2*

Figure 9.10 Some ({4, 7}, 3)-planes that are $4R_0$ and $7R_5$ (Case 13)

We obtain again {3, 6} but with another special perfect matching.

Given a ({4, 8}, 3)-torus G that is $4R_1$ and $8R_5$, consider the plane $skel(G)$, whose vertex-set is the set of 8-gonal faces and whose edge-set is made of two classes. The first class of edges comes from the edge separating adjacent 8-gons, while the second class comes from the central edge of the (4, 3)-polycycles $P_2 \times P_3$ of G.

Theorem 9.3.4 *(i) Given a ({4, 8}, 3)-plane G that is $4R_1$ and $8R_5$, then skel(G) is {3, 6} with a special perfect matching.*

(ii) Given a special perfect matching of {3, 6}, we can define a ({4, 8}, 3)-plane that is $4R_1$ and $8R_5$.

(iii) The flipping of a (3, 6)-plane with a special perfect matching corresponds to the following transformation:

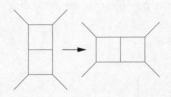

Proof. (i) Given a ($\{4, 8\}$, 3)-plane that is $4R_1$, it is clear from the definition of $skel(G)$, that it is 6-valent and that it has a perfect matching. Clearly, the corona of a 8-gon is of the form 84484888 or 84488488. The pattern 848 means that the vertex corresponding to this 8-gonal face is contained in a 3-gon, whose opposite edge belongs to the perfect matching. Hence, the perfect matching is special.

(ii) If G is $\{3, 6\}$ with a special perfect matching, then take the dual of it and transform every edge, arising from the perfect matching, according to the following scheme:

Clearly, we get a ($\{4, 8\}$, 3)-plane $4R_1$ and $8R_5$.

(iii) This assertion is obvious. $\qquad\qquad\qquad\qquad\qquad\qquad\qquad\qquad\qquad\qquad$ □

See on Figure 9.11 some examples of such planes.

Case 16, i.e. ($\{4, 10\}$, 3)-plane $4R_1$, $10R_4$. The 4-gons are organized by pairs. Therefore, the corona of a 10-gon is either $c_1 = 44b44b4b4b$, $c_2 = 44b4b44b4b$, or $c_3 = 44b44b44bb$ with $b = 10$. Corona c_3 implies a vertex contained in three 10-gons F_1, F_2, F_3. But considering all possibilities of putting the 4-gons around this vertex, we see that we cannot arrange them so that the F_i are adjacent to six 4-gons.

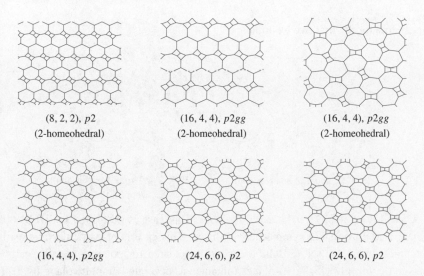

(8, 2, 2), $p2$	(16, 4, 4), $p2gg$	(16, 4, 4), $p2gg$
(2-homeohedral)	(2-homeohedral)	(2-homeohedral)
(16, 4, 4), $p2gg$	(24, 6, 6), $p2$	(24, 6, 6), $p2$

Figure 9.11 Some ($\{4, 8\}$, 3)-planes that are $4R_1$ and $8R_5$ (Case 15)

Therefore, this does not happen and two possibilities for corona are c_1 and c_2. Suppose that every 10-gon is of type c_2. Then the plane is completely determined. Suppose now that one 10-gon is of type c_1. Then we have the following infinite string:

Then we conclude easily that every such plane is described by an infinite word $\ldots \alpha_i \ldots$ with α_i being u or v, with u:

and v

See below for the first two examples:

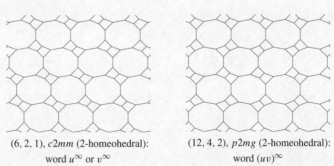

(6, 2, 1), $c2mm$ (2-homeohedral):   (12, 4, 2), $p2mg$ (2-homeohedral):
word u^∞ or v^∞ word $(uv)^\infty$

Case 17, i.e. ($\{5, 7\}, 3$)-planes $5R_1$, $7R_3$. This case is described in [DFSV00], where also all possible symmetries are listed. This case is of particular interest in Organic Chemistry; see [CBCL96], where the search for putative metallic carbon nets in the form of ($\{5, 7\}, 3$)-planes (obtained as decorated graphite plane $\{6, 3\}$) is warranted.

The possible coronas of a given 7-gon are 5575757 and 5577557. The following drawing shows why the second configuration is not possible:

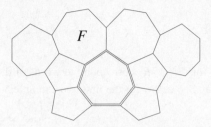

The 7-gon F should have 5-gons on its three remaining edges. This would imply that a 5-gon is adjacent to at least two 5-gons, which is forbidden. The same proof shows that no vertex is contained in three 7-gons.

So, 7-gons have the corona 5575757 and we have only the following three possibilities for those 7-gons and the pair of 5-gons attached to them, up to rotation of the plane:

We can thus represent the plane as a decoration of $\{6, 3\}$, i.e. the addition of edges to this plane.

Now consider how the individual motifs A and B may pack to cover the plane. It is clear, that choosing a given 7-gon as the center of a B motif forces two of its three 7-gonal neighbors to adopt a B' configuration. Therefore, a linear chain of B centers propagates to infinity in both directions. Motifs of type A also have this propagation property:

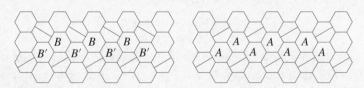

We now introduce the symbols u:

and v:

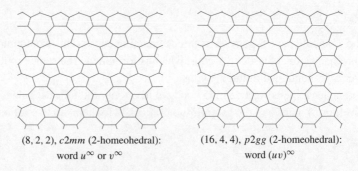

and see that $(\{5, 7\}, 3)$-tori $5R_1, 7R_3$ correspond to words of the form $\ldots \alpha_i \ldots$ with α_i being equal to u or v. See below for the first two examples:

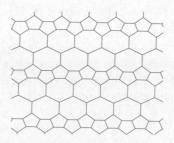

| $(8, 2, 2)$, $c2mm$ (2-homeohedral): | $(16, 4, 4)$, $p2gg$ (2-homeohedral): |
| word u^∞ or v^∞ | word $(uv)^\infty$ |

Case 18, i.e. $(\{5, 7\}, 3)$-plane $5R_2, 7R_4$. Suppose that one 7-gon has the corona 5557777. Then, by looking at the corona, we find that those three 5-gons are part of an infinite string and so; we have the following unique plane:

$(8, 2, 2)$, $p2mg$ (2-homeohedral)

Assume now that 5-gons are organized in triples. Suppose that one 7-gon has the corona 5757577, then one of the adjacent 7-gons is itself adjacent to at least four 5-gons, which is impossible. Therefore, every 7-gon has the pattern 7557 and 757 in its corona. So, if we replace a triple of 5-gons by a 6-gon, we obtain $\{6, 3\}$. This means that those planes are obtained by putting triples of 5-gons in $\{6, 3\}$. Simple arguments yield that they are described by infinite words $\ldots \alpha_i \ldots$ with α_i being u or v; here u denotes:

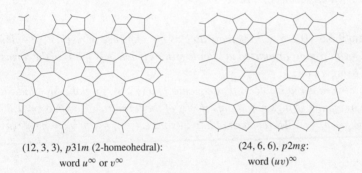

and v denotes:

See below for the first two examples:

$(12, 3, 3)$, $p31m$ (2-homeohedral):
word u^∞ or v^∞

$(24, 6, 6)$, $p2mg$:
word $(uv)^\infty$

Case 19, i.e. ($\{5, 8\}$, 3)-plane $5R_2$, $8R_2$. Such a plane will have its 5- and 8-gons organized in infinite strings, or we will have one triple of 5-gons or 8-gons. If we have one such triple, then the plane is completely determined: around this triple one have concentric circles of 5- and 8-gons and they are uniquely defined. Assume now that 5- and 8-gons are organized in lines and so, necessarily, in infinite strings.

Given a 5-gon, its corona is of the form 58588. Up to orientation of the infinite string of 5-gons, this makes two possible choices, which we write as u or v:

u v

Hence, we can write the infinite word representing the plane in the form of a 5-word (p-words are infinite words describing the orientation of the p-gon in one of the infinite string of p-gons) $\ldots \alpha_i \ldots$ with α_i being u or v. Another viewpoint is possible by considering the infinite string of 8-gons. Up to orientation of the infinite strings of 8-gons, this makes three possible choices, which we write as L, S or R:

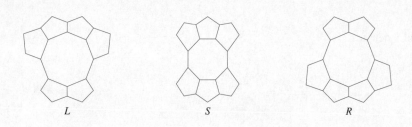

Hence, we can write the infinite word representing the plane in the form of a 8-word $\ldots \beta_i \ldots$ with β_i being L, S or R.

Theorem 9.3.5 *(i) A 5-word is realizable as the sequence of a ($\{5, 8\}$, 3)-plane $5R_2$ and $8R_2$ with p-gons organized in infinite strings if and only if it is of the form $(\gamma_i)_{i \in \mathbb{Z}}$ with γ_i being uv or vu.*

(ii) A 8-word is realizable as the sequence of a ($\{5, 8\}$, 3)-plane $5R_2$ and $8R_2$ with p-gons in infinite strings if and only if it is of the form $(\alpha_i)_{i \in \mathbb{Z}}$ with $\alpha_i = LS^{m_i} RS^{n_i}$ for some $m_i, n_i \geq 0$. The corresponding 5-word is $(\gamma_i)_{i \in \mathbb{Z}}$ with $\gamma_i = (vu)^{m_i} (uv)^{n_i}$.

Proof. Let us first prove (ii). It suffices to show that no two L or R can appear in a sequence, even with S between them. Suppose that the pattern $LS^m L$ appear in a sequence. The corresponding infinite string of 8-gons has two adjacent infinite strings of 8-gons. It is easy to see that one contains the pattern $LS^{m-1}L$ and the other contains the pattern $LS^{m+1}L$. By iterating this construction, we find an infinite string that contains LL. But this is excluded since it corresponds to having a triple of 5-gons.

It is easy to see that the 5-word corresponding to the above 8-word is $\ldots (vu)^{m_i} (uv)^{n_i} \ldots$. Clearly, any 5-word of the form $(\gamma_i)_{i \in \mathbb{Z}}$ with γ_i being uv or vu, can be realized in this form. □

Note that there is still another description of those planes. Consider a *step* to be two 5-gons being put together; then put the steps together to form an infinite stairway (the infinite string of 5-gons), which can go up or down. The infinite word $\ldots \gamma_i \ldots$ with γ_i being uv or vu corresponds to the stairway, for example by assigning uv to "up" and vu to "down" directions.

See below for the first two examples:

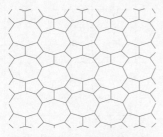 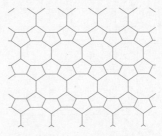

(12, 4, 2), *p2mg* (2-homeohedral):   (6, 2, 1), *c2mm* (2-homeohedral):
5-word $(uuvv)^\infty$ or 8-word $(LR)^\infty$ 5-word $(uv)^\infty$ or 8-word S^∞

Cases 20–23, i.e. ($\{5, b\}$, 3)-planes $5R_3$ and bR_{12-b} for $b = 8$, 10, 11, and 12. Those cases are solved in Theorems 12.5.7, 12.5.13 and Lemmas 12.5.4(ii), 12.5.3(ii).

Cases 24–27, i.e. ($\{3, b\}$, 4)-plane $3R_2\, bR_4$ with $5 \leq b \leq 8$. Suppose that some 3-gons are not organized in quadruples but instead in sequences like the one that appears in $APrism_m$ for some $m \geq 4$. Then, clearly, there are no pending edges left and the plane is reduced to it, which is impossible.

So, all 3-gons are organized in quadruples. Every b-gon is adjacent to $b-4$ quadruples. If we collapse the quadruples to single points, then the obtained plane is still 4-valent and the b-gons have been reduced to 4-gons. Therefore, we have the quadrilateral $\{4, 4\}$-plane. Combinatorially, such planes correspond to selecting a set S of vertices in $\{4, 4\}$ such that every 4-gons is incident to $b-4$ elements in S. Clearly, if we take the complement of S in the vertex-set of $\{4, 4\}$, we obtain a bijection between case 24 and case 26. Note, however, that case 24 admits two 2-homeohedral planes, while case 26 admits none.

Let us consider Case 24: every 4-gon should be incident to one vertex in S. All 4-gons, incident to a fixed vertex, form a block of four 4-gons. Clearly, such blocks are organized in infinite strings. Two consecutive infinite strings are in front of each other or not and we denote this by u or v:

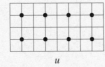 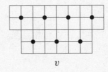

u v

Therefore, the planes, in case 24, correspond to infinite words composed of u and v.

See below for the first two examples:

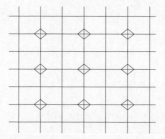

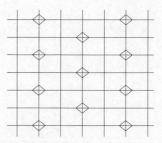

$(8, 4, 4)$, *p4mm* (2-homeohedral): u^∞ $(8, 4, 4)$, *c2mm* (2-homeohedral): v^∞

Let us consider case 25: every 4-gon should be incident to two vertices in S. For a given 4-gon, either those vertices are on opposite sides of a diagonal or they are on one of the edges. We will prove that one has a structure of infinite strings, in which the choices are uniform. Denote by u, v the edge, diagonal cases, respectively:

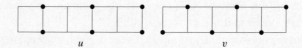

If every 4-gon has its vertices in S on a diagonal, then, clearly, it corresponds to the infinite word v^∞. Suppose that one 4-gon has two vertices in S along an edge, then this edge defines an infinite string in which only the edge choice is taken. Then this plane belongs to the ones described by an infinite word.

See below for the first two examples:

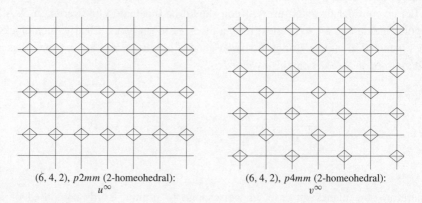

$(6, 4, 2)$, *p2mm* (2-homeohedral): $(6, 4, 2)$, *p4mm* (2-homeohedral):
u^∞ v^∞

Case 26 is obtained by taking the complement of the set S in Case 24.

See below for the first two examples:

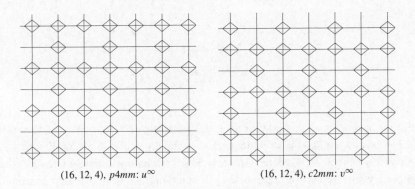

$(16, 12, 4)$, $p4mm$: u^∞ $(16, 12, 4)$, $c2mm$: v^∞

Case 27 is easy: every 4-gon should be incident to four vertices in S; so, every vertex of $\{4, 4\}$ is in S. See Figure 9.12 for a representation.

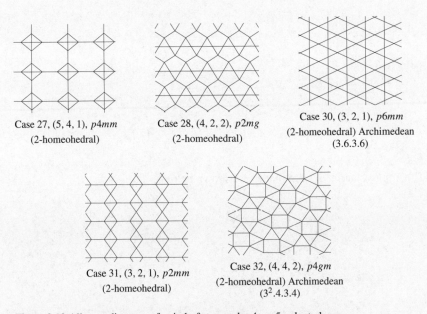

Case 27, $(5, 4, 1)$, $p4mm$
(2-homeohedral)

Case 28, $(4, 2, 2)$, $p2mg$
(2-homeohedral)

Case 30, $(3, 2, 1)$, $p6mm$
(2-homeohedral) Archimedean
$(3.6.3.6)$

Case 31, $(3, 2, 1)$, $p2mm$
(2-homeohedral)

Case 32, $(4, 4, 2)$, $p4gm$
(2-homeohedral) Archimedean
$(3^2.4.3.4)$

Figure 9.12 All sporadic cases of strictly face-regular 4- or 5-valent planes

Case 28, i.e. $(\{3, 5\}, 4)$-planes $3R_0$ and $5R_2$. Since 3-gons are adjacent only to 5-gons, we have the following possibilities for the vertex corona: 5555, 5553, and 5353. Let us prove that 5555 does not occur. If it were the case, then we would have a quadruple of 5-gons around a vertex v. Any of four vertices adjacent to v should have also corona 5533, which is impossible by property $3R_0$. Take a 5-gon P_1; since it is adjacent to a 5-gon, we have a vertex with corona 5553. This extends to a sequence of 5-gons as below with all vertices 5553 being black circles:

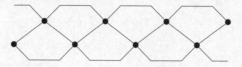

Since we have a plane, this extends to an infinite string of 5-gons. We can, therefore, add 3-gons and see that the structure extends uniquely and is shown on Figure 9.12.

Case 29, i.e. $(\{3, 5\}, 4)$-planes $3R_1$ and $5R_3$. This case is the only one that is not completely solved, but we present two different continua in it. The first continuum comes as infinite words $\dots \alpha_i \dots$ with α_i being r or s; where r is as follows:

and s is:

The first two cases of such planes are shown below:

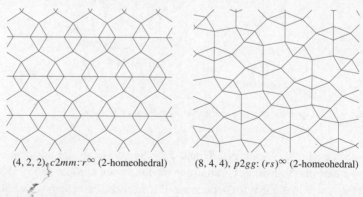

$(4, 2, 2), c2mm\!:\!r^\infty$ (2-homeohedral)　　　　$(8, 4, 4), p2gg\!:\!(rs)^\infty$ (2-homeohedral)

Another class comes by continuum $\dots \alpha_i \dots$ with α_i being u or v, where u is as follows:

and v is:

Two such planes are shown below (the plane corresponding to $(uv)^\infty$ is the same as the one corresponding to $(rs)^\infty$):

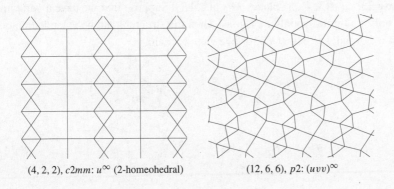

$(4, 2, 2)$, $c2mm$: u^∞ (2-homeohedral) $(12, 6, 6)$, $p2$: $(uvv)^\infty$

Case 30, i.e. ($\{3, 6\}$, 4)-plane $3R_0$ and $6R_0$. The condition $3R_0$, $6R_0$ implies that every vertex has corona 3636. It is easy to see that there is a unique plane tiling, called the *Kagome tiling*, shown on Figure 9.12.

Case 31, i.e. ($\{3, 6\}$, 4)-plane $3R_1$ and $6R_2$. Clearly, vertices have coronas $\mathbf{a} = 3636$, $\mathbf{b} = 3366$ or $\mathbf{c} = 6666$. Every 3-gon is adjacent to another 3-gon. The vertices of the common edge have corona $\mathbf{b}$, while the vertices that are not on a common edge have corona $\mathbf{a}$. Every 6-gon is adjacent to four 3-gons. Easy considerations yield the following local structure around every 6-gon:

It is then easy to see that there is a unique way to extend this structure (shown on Figure 9.12).

Case 32, i.e. ($\{3, 4\}$, 5)-planes $3R_1$ and $4R_0$. Take a 4-gon Q. Since it is adjacent to only 3-gons, there are four 3-gons around them. The vertex corona cannot contains 333, since it would imply that a 3-gon is adjacent to at least two 3-gons. Therefore, the corona of a vertex incident to Q is, up to rotation and change of rotation order, 33434. Since a 3-gon is adjacent to a unique 3-gon, we can find only two possible cases:

When we make a choice, it is easy to see that it extends uniquely. Therefore, we have a unique $(\{3, 4\}, 5)$-plane $3R_1, 4R_0$ shown on Figure 9.12.

Case 33, i.e. $(\{3, 4\}, 5)$-planes $3R_2$ and $4R_2$. Suppose that we have a vertex incident only to 3-gons or only to 4-gons. Then the structure can be extended uniquely and it is normal; see below a birds eye view of them.

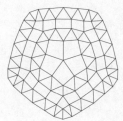 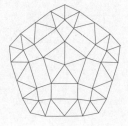

Assume in the following that no vertex is contained in only 3-gons or 4-gons. It is impossible that a vertex has the corona 34444, since this 3-gon would be adjacent only to one 3-gon. The 4-gons are organized in sequence (F_i) of adjacent 4-gons. Orient the sequence and then we have a choice: either the sequence turns left L, right R, or straight S. From the forbidding of 34444, we know that patterns LL and RR cannot occur. Assume that the pattern $LS^m L$ appears in the sequence. Then by taking two bounding sequences of 3-gons and two bounding sequences of 4-gons, we obtain the pattern $LS^{m+2}L$ and $LS^{m-2}L$ in them. So, we can decrease the number of patterns S between two consecutive L. In the end, we are left with a single vertex contained in five 3- or 4-gons. We excluded that possibility; so, the patterns $LS^m L$ and $RS^m R$ do not occur in the sequence. Therefore, the sequence is of the form $(\gamma_i)_{i \in \mathbb{Z}}$ with γ_i being of the form $LS^{m_1} RS^{n_1}$. We must now prove that the sequence is not a circuit. If this were the case, then the neighboring circuits, obtained as above by adding a layer of 3-gons and a layer of 4-gons, would have the same sequence. So, both sides of the circuit would contain an infinity of faces. This is impossible since a closed circuit in the plane contains a finite number of faces in its interior.

So, we have a sequence (γ_i) and see that such planes are obtained by adding edges to $\{4, 4\}$.

See below for the first two examples:

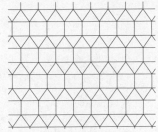 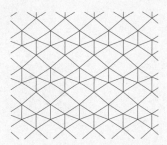

$(2, 2, 1)$, *c2mm* (2-homeohedral): S^∞ $(4, 4, 2)$, *p2mg* (2-homeohedral): $(LR)^\infty$

Archimedean $(3^3.4^2)$

10

Parabolic weakly face-regular spheres

In this chapter and the next, Nr. x means the xth polyhedron described in Table 9.1. We consider here $(\{a, b\}, k)$-spheres that are bR_j or aR_i and satisfy to $2k = b(k-2)$, see Chapter 2 for introduction.

10.1 Face-regular ($\{2, 6\}$, 3)-spheres

See on Figure 10.1 the first examples of $GC_{k,l}(Bundle_3)$ (the Goldberg–Coxeter operation $GC_{k,l}(G)$ is defined in Section 2.1). By Theorem 2.2.1, any $(\{2, 6\}, 3)$-sphere comes as $GC_{k,l}(Bundle_3)$ for some integers $0 \leq l \leq k$.

Any $(\{2, 6\}, 3)$-sphere, which is not a $Bundle_3$, is $2R_0$. $Bundle_3$ is $6R_j$ for any j. If a $(\{2, 6\}, 3)$-sphere distinct from $Bundle_3$ is $6R_j$, then $(6 - j)p_6 = 2p_2 = 6$. So, the cases $j = 0, 1, 2$ are impossible. For $j = 3, 4, 5$, the unique solution is $GC_{1,1}(Bundle_3)$, $GC_{2,0}(Bundle_3)$, $GC_{2,1}(Bundle_3)$, respectively, with 3, 4, 5 6-gons neighboring any 6-gon.

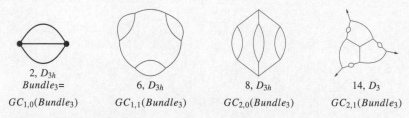

2, D_{3h}			
Bundle_3=	6, D_{3h}	8, D_{3h}	14, D_3
$GC_{1,0}(Bundle_3)$	$GC_{1,1}(Bundle_3)$	$GC_{2,0}(Bundle_3)$	$GC_{2,1}(Bundle_3)$

Figure 10.1 First examples of ($\{2, 6\}$, 3)-spheres

10.2 Face-regular ({3, 6}, 3)-spheres

By Theorem 2.0.2, a ({3, 6}, 3)-sphere that is not 3-connected, is one of T_n for some $n \geq 1$. Hence, a ({3, 6}, 3)-sphere that is not Tetrahedron, is either $3R_1$ and belongs to this infinite series, or is $3R_0$.

A ({3, 6}, 3)-sphere that is $6R_j$ and distinct from Tetrahedron, satisfies to $(6-j)p_6 \leq 3p_3 = 12$. Therefore, we have an upper bound on the number of vertices, which allow us to find the complete list of such spheres:

1 the first member T_1 of the infinite series (of 2-connected spheres, see Theorem 2.0.2) that is $3R_1$ and $6R_2$,
2 the second member T_2 of this infinite series that is $3R_1$ and $6R_4$,
3 Nr. 3, i.e. $GC_{1,1}(Tetrahedron)$ that is $3R_0$ and $6R_3$,
4 Nr. 4, i.e. $GC_{2,0}(Tetrahedron)$ and Nr. 5 (twist of Nr. 4) that are $3R_0$ and $6R_4$,
5 Nr. 6, i.e. $GC_{2,1}(Tetrahedron)$ that is $3R_0$ and $6R_5$.

10.3 Face-regular ({4, 6}, 3)-spheres

There is an infinity of ({4, 6}, 3)-spheres that are $4R_0$; in fact, as the number of vertices goes to infinity, the proportion of such spheres amongst all ({4, 6}, 3)-spheres tends to 1.

Take a ({4, 6}, 3)-sphere that is $4R_1$. Insert on every edge, separating two 4-gons, a 2-gon. The resulting map is a ({2, 6}, 3)-sphere with at most one 2-gon being adjacent to each 6-gon. Such spheres are described by $GC_{k,l}(Bundle_3)$. Hence, there exists a ({4, 6}, 3)-sphere $4R_1$ with v vertices if and only if $v = 2(k^2 + kl + l^2) - 6 \geq 18$, and it has symmetry D_3 or D_{3h}. In fact, there is a bijection between ({4, 6}, 3)-spheres $4R_1$ and all but the five smallest ({2, 6}, 3)-spheres (i.e. four on Figure 10.1 and one with $(k, l) = (3, 0)$). So, the smallest ({4, 6}, 3)-sphere $4R_1$ has 18 vertices and corresponds to $(k, l) = (2, 2)$.

Theorem 10.3.1 *The only* ({4, 6}, 3)-*spheres that are* $4R_2$, *are either* $Prism_6$, *or the series of* ({4, 6}, 3)-*spheres with* $8 + 6t$ *vertices*, $t \geq 1$, *called t-hex-elongated cubes (see Figure 10.2 for the first three spheres of this infinite series).*

Proof. Recall that $p_4 = 6$. Let F_0 be a 4-gon of a ({4, 6}, 3)-sphere that is $4R_2$. Then F_0 is adjacent to two 4-gons F_1 and F_2. These 4-gons are adjacent to other 4-gons. There are two cases: either F_1 and F_2 are adjacent or not. In the first case, we obtain a configuration of three 4-gons surrounded by three 6-gons. This configuration generates the infinite family. In the second case, we obtain a ring of six 4-gons that uniquely gives $Prism_6$. □

There is no ({4, 6}, 3)-sphere that is $4R_3$.

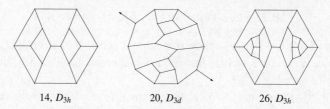

14, D_{3h} 20, D_{3d} 26, D_{3h}

Figure 10.2 (({4, 6}, 3)-spheres of the infinite series of t-hex elongated cubes, for $t = 1$, 2 and 3

A (({4, 6}, 3)-sphere that is $6R_j$ and distinct from Cube, satisfies to $(6 - j)p_6 \leq 4p_4 = 24$. Hence, it has at most 56 vertices. For $j = 0$, there is only $Prism_6$. There is no such sphere for $j = 1$.

1 For $j = 2$, there is Nr. 19 and the following sphere:

2 For $j = 3$, there are Nrs. 21, 22.
3 For $j = 4$, there are Nrs. 20, 23, 24, 25.
4 For $j = 5$, there is Nr. 26.

10.4 Face-regular (({5, 6}, 3)-spheres (fullerenes)

If P is a (({5, 6}, 3)-sphere that is $6R_j$, then we have, clearly, $(6 - j)p_6 \leq 5p_5 = 60$. For $j = 0, 1, 2, 3, 4, 5$, this yields an upper bound (on the number of vertices $v = 20 + 2p_6$) of 30, 32, 50, 60, 80, 140. The complete enumeration was done by computer for $j \leq 4$ and the results are presented on Figures 10.3, 10.4, 10.5, 10.6, and 10.7. For $j = 5$, such spheres are also $5R_0$ by Lemma 11.1.3 and so unique such sphere is Nr. 55.

There is an infinity of (({5, 6}, 3)-spheres that are $5R_0$; in fact, as the number of vertices goes to infinity, the proportion of such spheres amongst all (({5, 6}, 3)-spheres tends to 1. Such fullerenes are said to satisfy the *isolated pentagon rule* in Chemistry ([FoMa95]).

The number of (({5, 6}, 3)-spheres $5R_1$ is also infinite and also there is no simple description of them (all 130 of such $5R_1$ fullerenes with $v \leq 72$ are listed in [Fow93]; the two smallest ones have $(v, Aut) = (50, D_3)$ and $(52, T)$.)

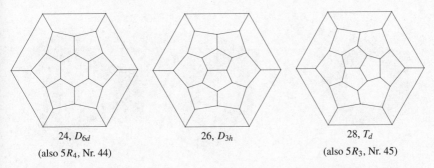

24, D_{6d} 26, D_{3h} 28, T_d

(also $5R_4$, Nr. 44)   (also $5R_3$, Nr. 45)

Figure 10.3 All ($\{5, 6\}, 3$)-spheres that are $6R_0$ (all classical dual *Frank – Kasper* spheres), besides Dodecahedron

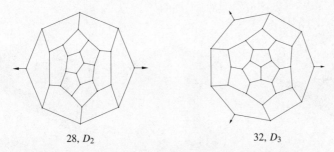

28, D_2 32, D_3

Figure 10.4 All ($\{5, 6\}, 3$)-spheres that are $6R_1$, besides Dodecahedron

The ($\{5, 6\}, 3$)-spheres that are $5R_2$, are enumerated in Theorem 10.4.3 below. The only ($\{5, 6\}, 3$)-spheres that are $5R_3$, are Nrs. 45 and 46; see Theorem 12.5.6. Snub *Prism*$_6$, i.e. the smallest Nr. 44, is a unique ($\{5, 6\}, 3$)-sphere that is $5R_4$.

Lemma 10.4.1 *All ($\{5, 6\}, 3$)-spheres with 5-gons organized in a ring are four spheres (all but the second one) drawn on Figure 10.8.*

Proof. First, such a ring ought to have 12 5-gons. The ring splits the 6-gons into two regions, which are ($6, 3$)-polycycles P and P'. Let us consider the polycycle P, its boundary sequence is of the form $b_1 \ldots b_m$, with $b_i = 2$ or 3. Denote by v_i the number of boundary vertices of degree i for $i = 2$ or 3. There are no two consecutive vertices of degree 3 in the sequence, since it would imply a vertex contained in three 5-gons. Denote by p_j the number of 5-gons of the ring adjacent to P on j edges for $j = 1, 2$. We have, clearly, $p_2 = v_3$ and $p_1 + p_2 = v_2$. If we consider the polycycle P', then we have, as previously, $p'_2 = v'_3$ and $p'_1 + p'_2 = v'_2$ with the relations $p'_1 = p_2$ and $p'_2 = p_1$. Combining with the relation $v_2 = 6 + v_3$, $v'_2 = 6 + v'_3$ of Theorem 5.2.1, we obtain $v_2 = v'_2 = 12$ and $v_3 = v'_3 = 6$. Therefore, the number of edges of P and P' is 18.

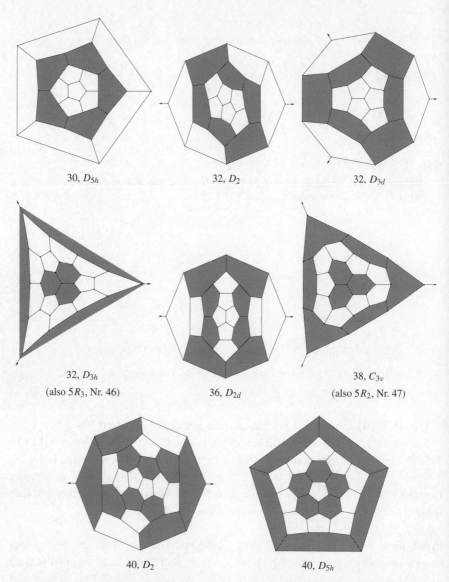

30, D_{5h} 32, D_2 32, D_{3d}

32, D_{3h}

(also $5R_3$, Nr. 46)   36, D_{2d} 38, C_{3v}

(also $5R_2$, Nr. 47)

40, D_2 40, D_{5h}

Figure 10.5 All $(\{5, 6\}, 3)$-spheres that are $6R_2$, besides Dodecahedron

It is proved in [HaHa76] (see also [BBG03, Gre01]) that for a given number h of 6-gons, the minimal perimeter of a $(6, 3)$-polycycle is $2\lceil\sqrt{12h - 3}\rceil$. Therefore, the maximal number h of 6-gons of a $(6, 3)$-polycycle with perimeter 18 is 7. So, the maximal number of 6-gons of a $(\{5, 6\}, 3)$-sphere with a ring of 5-gons is 14 and the maximal number of vertices of this $(\{5, 6\}, 3)$-sphere is 48. The result follows by computer enumeration.   □

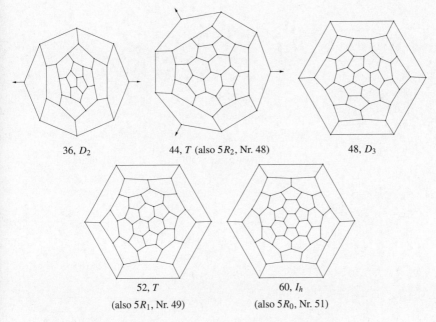

| 36, D_2 | 44, T (also $5R_2$, Nr. 48) | 48, D_3 |

52, T

(also $5R_1$, Nr. 49)

60, I_h

(also $5R_0$, Nr. 51)

Figure 10.6 All $(\{5, 6\}, 3)$-spheres that are $6R_3$, besides Dodecahedron

Lemma 10.4.2 *The possible graphs G_5 of 12 5-gons of a $(\{5, 6\}, 3)$-sphere that is $5R_2$, belong to following five cases:*

$$G_5 = 4C_3, G_5 = 2C_3 + C_6, G_5 = C_3 + C_9, G_5 = 2C_6, G_5 = C_{12}.$$

Proof. If G is a $(\{5, 6\}, 3)$-sphere $5R_2$, then its graph G_5 is an union of cycles and we must determine the possible lengths. Suppose that a ring of 5-gons is not reduced to a triple of 5-gons incident to a vertex, then the ring encloses a $(6, 3)$-polycycle P. Denote by v_2, v_3 the number of boundary vertices of P of degree 2, 3. By Theorem 5.2.1, we have $v_2 = 6 + v_3$ and the number of 5-gons adjacent to P is equal to v_2 and is, therefore, greater or equal to 6. If the length of the cycle is 6, then $v_3 = 0$ and P is reduced to a 6-gon.

It is easy to see that cycles of length 2 and 3, not enclosing a vertex, do not exist, since, otherwise, we would get only a 1-connected graph, while $(\{5, 6\}, 3)$-spheres are 3-connected (see Theorem 2.0.2).

If G has a single cycle, then G_5 is equal to C_{12}. If G has at least two cycles, then it is easy to see that there exist at least two cycles enclosing a vertex or a $(6, 3)$-polycycle. If both those cycles are C_3, then either we have a C_6, or we have two C_3 for the remaining 5-gons. If only one of those cycles is C_3, then either one has a C_6 and another C_3, or one has, simply, a C_9. If none of those cycles is C_3, then both have length at least 6, so both have length 6. $\quad\square$

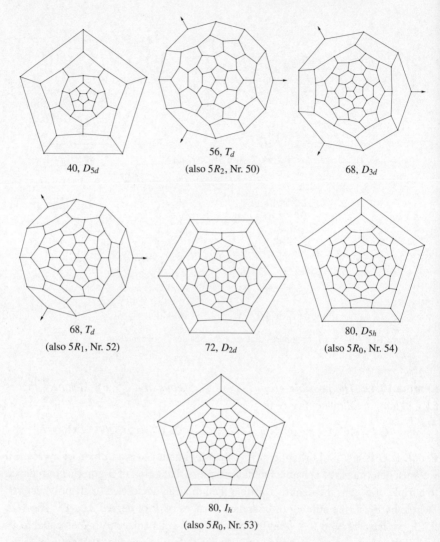

Figure 10.7 All ({5, 6}, 3)-spheres that are $6R_4$, besides Dodecahedron

Theorem 10.4.3 *([DGr02]) All ({5, 6}, 3)-spheres that are $5R_2$, are:*

(i) The sporadic ones on Figure 10.8.

(ii) An infinite series of $(12t + 24)$-vertex (for any $t > 0$, and of symmetry D_{6d}, if t is even, D_{6h} if t is odd) fullerenes with 5-gons organized into two 6-cycles. They are obtained from snub $Prism_6$ by inserting t more 6-cycles of 6-gons.

(iii) An infinite series of (symmetry D_2, D_{2d}, D_{2h}, T or T_d) v-vertex (for any $v \equiv 0$ (mod 4) with $v \geq 40$) fullerenes with 5-gons organized into four 3-cycles. They

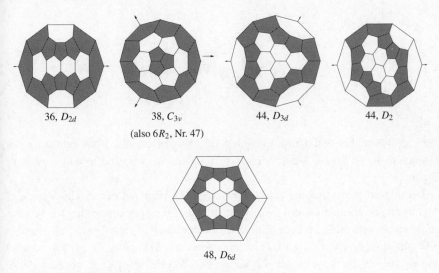

36, D_{2d} 38, C_{3v} 44, D_{3d} 44, D_2

(also $6R_2$, Nr. 47)

48, D_{6d}

Figure 10.8 All sporadic ($\{5, 6\}$, 3)-spheres that are $5R_2$

are obtained, by collapsing to four points all four 3-gons, from any ($\{3, 6\}$, 3)-sphere, such that no 6-gon is adjacent to more than one 3-gon (see Section 2.2 for a description of ($\{3, 6\}$, 3)-spheres).

Proof. Let G be a ($\{5, 6\}$, 3)-sphere $5R_2$. We will consider all possible graphs G_5 of 5-gons listed in the above lemma, one by one. The case $G_5 = C_{12}$ is settled in Lemma 10.4.1.

Let us consider the case $G_5 = C_9 + C_3$. The (6, 3)-polycycle P, bounded by C_9, has v_2, v_3 vertices on its boundary. Denote again, by p_1, p_2 the number of 5-gons in the 9-cycle bounding P, sharing one, respectively, 2 edges with P. We have $v_2 = 6 + v_3$ (see Theorem 5.2.1) and the equalities:

$$p_1 + p_2 = 9, \ p_2 = v_3, \ p_1 + p_2 = v_2,$$

which yields $v_2 = 9$, $v_3 = 3$ and P having 12 edges. (6, 3)-polycycles with h 6-gons have at least $2\lceil\sqrt{12h - 3}\rceil$ boundary edges; so, P has at most three 6-gons (see [BBG03, Gre01]). All (6, 3)-polycycles with at most three 6-gons are drawn below:

The third possibility is the only one that fits. We add the 9-cycle of 5-gons around it and then a cycle of 6-gons:

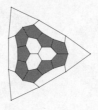

We can then either add a ring of 5-gons or a ring of 6-gons. If we add a ring of 6-gons, then we have a 3-gon, which is excluded. So, we need to add a cycle of 5-gons.

Let us now prove that the case $G_5 = 2C_3 + C_6$ does not occur. The 6-cycle of 5-gons can be considered as a cycle with six *tails*, i.e. edges connecting the vertices of the cycle with other vertices. Similarly, the boundary of each 3-cycle of 5-gons is a circuit of nine vertices, six of which are endpoints of six tails. We have to connect the six tails of the 6-cycle with 12 tails of two 3-cycles in order to obtain a net of 6-gons.

The 6-cycle C_6 has two domains: outer and inner. There are two cases: either two 3-cycles lie in distinct domains, or both lie in the same, say, outer domain. In the first case, by symmetry, we can consider only the outer domain. The Euler formula shows that the boundary circuit of the cycle of 5-gons should have three tails. It is easy to verify that it is not possible to form a net of 6-gons using three tails of the 6-cycle and six tails of the 3-cycle.

In the second case, by the Euler formula, we have the 6-cycle with six tails and two 3-cycles C_3^A and C_3^B each with six tails. Suppose there is a fullerene containing this configuration. Then, in this fullerene, there are chains of 6-gons connecting a 5-gon of the 6-cycle and a 5-gon of a 3-cycle. Consider such a chain of minimal, say, q, length. In this case, the 6-cycle is surrounded by q rings each containing six 6-gons. If we dissect the qth ring of 6-gons into two 6-cycles each with six tails, we obtain a 6-cycle surrounded by $q - 1$ rings. The boundary of the $(q - 1)$th ring contains six tails.

Let the chain of q 6-gons connect the 6-cycle with C_3^A. At least two tails of the $(q - 1)$th ring correspond (are connected) to tails of C_3^A. Since the boundary of the $(q - 1)$th ring with six tails is similar to the boundary of the 6-cycle with six tails, our problem is reduced to the case when two tails of the 6-cycle are connected to two tails of C_3^A. There are two cases: either endpoints of two tails of C_3^A are separated on the boundary of C_3^A by a vertex, or not. We obtain two configurations, each consisting of a circuit with vertices that have or do not have tails. Both these configurations have the unique extension by 6-gons, which cannot be glued with the cycle C_3^B having six tails.

If $G_5 = 4C_3$, then G comes by collapsing into a point of all four 3-gons of a $(\{3, 6\}, 3)$-sphere vertices, such that no 6-gon has more than one 3-gonal neighbor.

If $G_5 = 2C_6$, then G has two 6-cycles of 5-gons. Both of those cycles are encircling $(6, 3)$-polycycles, which are reduced to a single 6-gon. It is easy to see that one has t 6-rings of 6-gons separating two 6-cycles of 5-gons. So, we are in case (ii) of the theorem. $\qquad\square$

10.5 Face-regular ($\{3, 4\}$, 4)-spheres

All ($\{3, 4\}$, 4)-spheres that are $4R_j$, satisfy $e_{3-4} = (4 - j)p_4$. But, clearly, $e_{3-4} \leq 3p_3 = 24$; so, we get $v = 6 + p_4 \leq 6 + \frac{24}{4-j}$, which yields, for $j = 0, 1, 2, 3$, the upper bounds 12, 14, 18, 30 on the number v of vertices. Direct computation, using ENU, of the ($\{3, 4\}$, 4)-spheres yields the list presented on Figures 10.9, 10.10, 10.11, and 10.12.

It looks too hard to describe all ($\{3, 4\}$, 4)-spheres that are $3R_0$; for example, the medial graph of any ($\{3, 4\}$, 4)-sphere with v vertices is a ($\{3, 4\}$, 4)-sphere with $2v$ vertices that are $3R_0$. Actually, as the number of vertices goes to infinity, the proportion of ($\{3, 4\}$, 4)-spheres that are $3R_0$ goes to 1. There are "much less" (still, an infinity) ($\{3, 4\}$, 4)-spheres $3R_1$, but a classification seems difficult also.

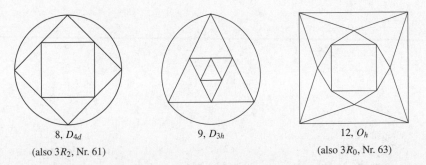

| 8, D_{4d} | 9, D_{3h} | 12, O_h |
| (also $3R_2$, Nr. 61) | | (also $3R_0$, Nr. 63) |

Figure 10.9 All ($\{3, 4\}$, 4)-spheres that are $4R_0$, besides Octahedron

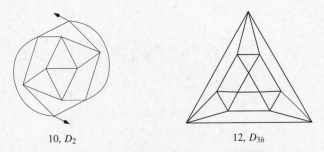

10, D_2 12, D_{3h}

Figure 10.10 All ($\{3, 4\}$, 4)-spheres that are $4R_1$, besides Octahedron

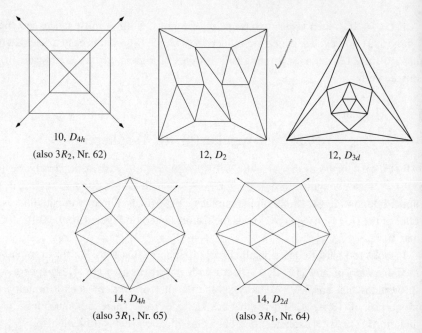

10, D_{4h}

(also $3R_2$, Nr. 62)

12, D_2

12, D_{3d}

14, D_{4h}

(also $3R_1$, Nr. 65)

14, D_{2d}

(also $3R_1$, Nr. 64)

Figure 10.11 All ($\{3, 4\}$, 4)-spheres that are $4R_2$, besides Octahedron

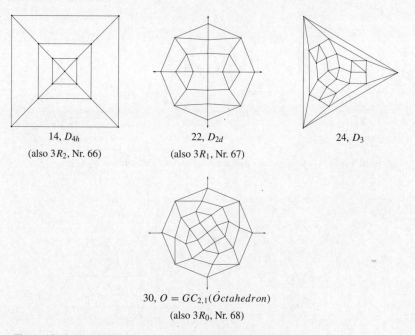

14, D_{4h}

(also $3R_2$, Nr. 66)

22, D_{2d}

(also $3R_1$, Nr. 67)

24, D_3

30, $O = GC_{2,1}(Octahedron)$

(also $3R_0$, Nr. 68)

Figure 10.12 All ($\{3, 4\}$, 4)-spheres that are $4R_3$, besides Octahedron

But the case of $3R_2$ is much simpler; we have the following:

Theorem 10.5.1 *The only* $(\{3, 4\}, 4)$-*spheres that are* $3R_2$ *are either* $APrism_4$, *or the infinite family of* t-*elongated octahedra (first spheres on Figures 10.11, 10.12 correspond to cases* $t = 1, 2$).

Proof. Let F_0 be a 3-gon of a $(\{3, 4\}, 4)$-sphere with $3R_2$. Then F_0 is adjacent to two 3-gons F_1 and F_2. These 3-gons are adjacent to other 3-gons. There are two cases: either F_1 and F_2 have, or have not, a common second adjacent 3-gon. In the first case, we obtain a configuration of four 3-gons surrounded by four 4-gons. This configuration generates the family of t-elongated octahedra, $t \geq 1$. In the second case, one obtains uniquely $APrism_4$. □

10.6 Face-regular ($\{2, 3\}$, 6)-spheres

A $(\{2, 3\}, 6)$-sphere that is $3R_j$ has $e_{2-3} = (3 - j)p_3$, so we get $v = 2 + \frac{1}{2}p_3 \leq 2 + \frac{6}{3-j}$, which yields, for $j = 0, 1, 2$, the respective upper bounds 4, 5, 8 on v. We obtain the list presented in Figures 10.13, 10.14, and 10.15.

If a $(\{2, 3\}, 6)$-sphere G is $2R_1$, then its dual G^* is a sphere with vertices of degree two or three and faces of gonality 6. The vertices of degree two of G^* are organized in pairs. We replace the edge between them by a 2-gon and obtain a $(\{2, 6\}, 3)$-sphere. This establishes a bijection between v-vertex $(\{2, 3\}, 6)$-spheres $2R_1$ and

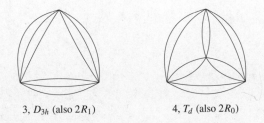

3, D_{3h} (also $2R_1$)   4, T_d (also $2R_0$)

Figure 10.13 All $(\{2, 3\}, 6)$-spheres that are $3R_0$, besides $Bundle_6$

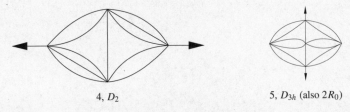

4, D_2 5, D_{3h} (also $2R_0$)

Figure 10.14 All $(\{2, 3\}, 6)$-spheres that are $3R_1$, besides $Bundle_6$

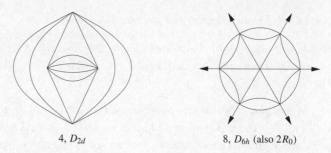

$4, D_{2d}$ $8, D_{6h}$ (also $2R_0$)

Figure 10.15 All $(\{2, 3\}, 6)$-spheres that are $3R_2$, besides $Bundle_6$

$(2v + 2)$-vertex $(\{2, 6\}, 3)$-spheres different from $Bundle_3$ and $GC_{1,1}(Bundle_3)$, here $v = k^2 + kl + l^2$ for some $0 \leq l \leq k$. For example, a unique 3-vertex $(\{2, 3\}, 6)$-sphere is $2R_1$ and it corresponds to the 8-vertex $(\{2, 6\}, 3)$-sphere.

If we take a $(\{5, 6\}, 3)$-sphere G that is $5R_1$, then its dual G^* is a sphere with vertices of degree 5 or 6 and faces of gonality three. The vertices of degree 5 of G^* are in pairs and so, after we replace the edge between them by a 2-gon, we obtain a $(\{2, 3\}, 6)$-sphere $2R_0$. It shows that there is no reasonable hope of obtaining a description of $(\{2, 3\}, 6)$-spheres $2R_0$.

11

General properties of 3-valent face-regular maps

In this chapter we address the following problem: characterize all *weakly* face-regular $(\{a, b\}, 3)$-maps on sphere or torus (see Chapter 9 for strictly face-regular ones).

We used computer methods (by the consideration of all possibilities), when this approach worked, and theoretical, otherwise. The computer approach cannot work in the torus case, since, given a $(\{a, b\}, 3)$-torus that is bR_j or aR_i, we can obtain a $(\{a, b\}, 3)$-torus with the same property and arbitrary large number v of vertices.

However, for many subcases, the torus case is simpler, since the Euler formula takes the form $v - e + f = 0$, instead of more complicated $v - e + f = 2$. To illustrate this point, a $(\{5, 7\}, 3)$-sphere that is $7R_2$, satisfies to $x_0 + x_3 + p_7 = 20$ (see Theorem 15.1.1), which allows us to have an upper bound on v and enumerate such spheres, while a $(\{5, 7\}, 3)$-torus that is $7R_2$ satisfies to $x_0 + x_3 + p_7 = 0$ and, hence, does not exist at all.

The shape of the results is also interesting. If the bR_j $(\{a, b\}, 3)$-tori admit classification, then, usually, there is more freedom for the bR_j $(\{a, b\}, 3)$-spheres (compare, for example, Theorems 13.2.2 and 13.2.3). However, if the aR_i $(\{a, b\}, 3)$-tori admit classification, then, usually, the possibilities for aR_i $(\{a, b\}, 3)$-sphere are more restricted (see, for example, Theorem 12.5.2).

Here is a summary of the results and conjectures on finiteness of the number of $(\{4, b\}, 3)$-spheres:

- The number of $(\{4, b\}, 3)$-spheres bR_j is 0 for $j \geq b - 2 \geq 6$ (Theorem 18.2.2); it is finite for $j \leq 3$ (all such spheres are given in Theorem 14.1.2 for $j \leq 2$ and in Theorem 16.1.1(ii) for $j = 3$); it is infinite for $j = 4$ (all such spheres with $b = 8$ are conjecturally listed in 17.1.3 and infiniteness is conjectured for $b \geq 9$).
- The number of $(\{4, b\}, 3)$-spheres $4R_0$ is infinite for $b = 6$ and 7 only (see Theorems 12.0.1(i), 12.1.2, and 12.0.1).
- The number of $(\{4, b\}, 3)$-spheres $4R_1$ is infinite for $b = 6, 7, 8,$ and 9 (see Theorems 12.2.2, 12.2.3, 12.2.4, and 12.0.1).

- The number of $(\{4, b\}, 3)$-spheres $4R_2$ (different from $Prism_b$) is infinite for $6 \leq b \leq 13$ and $b = 15$ and such spheres do not exist for other values of b (see Theorems 12.3.1(ii), 12.3.2, 12.3.3, 12.3.4, 12.3.5, 12.3.7, 12.3.9, 12.3.6, and 12.3.8).

Here is a summary of results and conjectures on the existence of $(\{4, b\}, 3)$-tori:

- A $(\{4, b\}, 3)$-torus bR_j does not exist if $j \leq 3$ (Theorems 14.1.2 and 16.1.1(i)). For $j = 4$, it exists if and only if $b \geq 8$; moreover, it is $4R_0$ for $b = 8$ (Theorem 17.1.2 (i), (ii)) and characterized (Theorem 17.1.2 (iii), (iv)) for $b = 9$. For $j = 5$, it exists if and only if $b \geq 7$; moreover, it is $4R_0$ for $b = 7$ (Theorem 18.1.1). For $j = 6$ and if the $(\{4, b\}, 3)$-torus is 3-connected, it is also $4R_2$ (Theorem 18.2.1).
- A $(\{4, b\}, 3)$-torus $4R_i$ can exist only for $(i, b) = (0, 7)$, $(1, 7)$, $(1, 8)$, $(1, 9)$, $(2, 7 \leq b \leq 16)$, $(2, 18)$ (Theorems 12.0.1(ii) and 12.3.1(i)). See Figure 12.4 for an example with $(i, b) = (1, 9)$.

Here is the summary of results and conjectures on the finiteness of the number of $(\{5, b\}, 3)$-spheres:

- The number of $(\{5, b\}, 3)$-spheres bR_0 is finite (and equal 4, 2, 3, 3, 5, and 4) if and only if $6 \leq b \leq 11$ (Theorem 13.2.3 lists them for $b \leq 12$ and proves it for $b = 12$; we conjecture it for $b > 12$).
- The number of $(\{5, b\}, 3)$-spheres bR_1 is finite (and equal 3, 4, 7, and 22) for $6 \leq b \leq 9$ (Figures 10.4, 14.1, 14.2, 14.3, and 14.4 lists them for $6 \leq b \leq 9$; we conjecture infiniteness for $b > 9$).
- The number of $(\{5, b\}, 3)$-spheres bR_2 is finite (and equal 9 and 27) if and only if $6 \leq b \leq 7$ (Theorem 15.1.3 proves it for $b \geq 8$; for $b = 6$ and $b = 7$ all are listed in Figures 10.5, 15.6, and 15.7).
- The number of $(\{5, b\}, 3)$-spheres bR_3 is finite (and equal 6) if and only if $b = 6$ (it is Conjecture 16.2.3; Theorem 16.2.4 proves it for $b = 9, 10, 12$).
- The number of $(\{5, b\}, 3)$-spheres bR_4 is finite (and equal 8) if and only if $b = 6$ (see Figure 10.7, Theorems 17.2.1, 17.2.2, 17.2.4 prove it for $b = 7, 8, 10, 13, 16$).
- The number of $(\{5, b\}, 3)$-spheres bR_5 is finite (and equal 2) if and only if $b = 6$ (Theorems 18.1.6, 18.1.7 prove it for $b \leq 21$, except for undecided cases $b = 7, 10, 13, 16, 19$; we conjecture it for all b and Figures 19.2, 18.1 give examples for $b = 7, 8, 9$, respectively).
- By Theorem 12.0.1, if $b \geq 7$, then a $(\{5, b\}, 3)$-sphere $5R_i$ can exist only for $i = 2$ (see [DGr02]) or $i = 3$. A $(\{5, b\}, 3)$-sphere $5R_3$ exists if and only if $6 \leq b \leq 10$ (Theorem 12.5.2 lists them for $b = 7$ and proves it for $b \leq 10$, Theorem 12.5.9 gives an infinite series and seven examples for $b = 8$, Theorem 12.5.11 gives a unique case for $b = 9$, Theorem 12.5.12 gives three examples for $b = 10$).

Here is a summary of the results and conjectures on the existence of $(\{5, b\}, 3)$-tori:

- $(\{5, b\}, 3)$-torus bR_0 exists if and only if $b \geq 12$; for $b = 12$ it is bR_0 if and only if it is $5R_3$ (Theorems 13.2.2 and 12.5.2(ii)).
- $(\{5, b\}, 3)$-torus bR_1 exists if and only if $b \geq 10$; for $b = 10$ it corresponds to a special perfect matching of a 6-valent tiling of the torus by 3-gons (Theorems 14.2.2(ii), 14.2.5, and 14.2.3).
- $(\{5, b\}, 3)$-torus bR_2 exists if and only if $b \geq 8$; for $b = 8$ it is bR_2 if and only if it is $5R_2$ (Theorems 15.2.1 and 15.2.2).
- $(\{5, b\}, 3)$-torus bR_3 exists if and only if $b \geq 7$; for $b = 7$ it is bR_3 if and only if it is $5R_1$ (Theorems 16.2.2 and 16.2.1).
- $(\{5, b\}, 3)$-torus bR_4 exists if and only if $b \geq 7$ (Theorem 17.2.3).
- By Theorem 12.0.1, a $(\{5, b\}, 3)$-torus $5R_i$ can exist only for $i = 2$ and $i = 3$. A $(\{5, b\}, 3)$-torus $5R_2$ exists if and only if $b = 7$ or 8 (see Theorem 12.4.1). A $(\{5, b\}, 3)$-torus $5R_3$ exists if and only if $b = 8, 10, 11, 12$; moreover, it is bR_{12-b} for $b = 10, 11, 12$ (Theorems 12.5.2, 12.5.11(ii), Lemma 12.5.5(ii), and Conjecture 12.5.14 for $b = 10$).

In view of above summaries for spheres and tori, we conjecture:

Conjecture 11.0.1 *The number of* $(\{a, b\}, 3)$-*spheres* bR_j *is infinite if and only if a* $(\{a, b\}, 3)$-*torus* bR_j *exists.*

But the similar conjecture for aR_i does not hold, for example for $(a, b; i) = (5, 11; 3)$ (Lemma 12.5.4) or $(5, 12; 3)$ (Lemma 12.5.3).

We present some pictures of $(\{a, b\}, 3)$-maps, especially when full classification was undertaken. However, space constraints prevent us from showing all that we would have liked and we refer to [DuDe06] for more information.

Face-regular maps are of interest for chemistry and physics, because many of them already appear there. For example, many of known polyhedral (energy) minimizers in the Thomson problem (for given number of particles on sphere) or Tammes problem of minimal distance between n points on the sphere or Skyrme problem (for given integer barionic number) are face-regular $(\{5, 6\}, 3)$-polyhedra. Face-regular $(\{5, 7\}, 3)$-planes are related to a putative "metallic carbon" deformation of the graphite lattice (see [DFSV00]). Also, for example, all known polyhedra P, such that their skeleton is an isometric subgraph of a hypercube or a half-hypercube, have either P, or its dual P^* face-regular.

Theorem 11.0.2 *If the set of 5-gonal faces of a* $(\{5, b\}, 3)$-*sphere* bR_j *contains at least two* $(5, 3)$-*polycycles* A_2 *or at least two* $(5, 3)$-*polycycles* A_3, *then there exists an infinity of* $(\{5, b\}, 3)$-*spheres that are* bR_j.

Proof. The polycycle A_2 (see Figure 7.2) has a central edge; by removing it, we obtain a $(5, 3)_{gen}$-polycycle (see Section 4.5) that is the union of two $(5, 3)$-polycycles E_2 (see Figure 7.2). This $(5, 3)_{gen}$-polycycle has two boundaries, which have the same boundary sequence $(23^3)^2$. Hence, both sides can be filled by the same structure, which again has at least two $(5, 3)$-polycycles A_2 and is again bR_j. This construction can, obviously, be repeated and we obtain an infinite series.

Take an elementary $(5, 3)$-polycycle A_3 (see Figure 7.2) and remove its central vertex. The result is a $(5, 3)_{gen}$-polycycle with two boundary sequences. It turns out that those boundary sequences are identical, namely, $(3^2 2)^3$. Hence, we can fill both those boundaries by the same structure, which is again bR_j. So, we obtain a larger $(\{5, 8\}, 3)$-sphere, which is bR_j. This operation can, obviously, be repeated and we obtain larger $(\{5, b\}, 3)$-spheres, which are bR_j. By creating a chain of such spheres, we get an infinity of them. $\square$

The following theorem, which is a slight generalization of Theorem 5.2.1, is very helpful in deriving classification results.

Theorem 11.0.3 *Let P be a finite $(a, 3)_{gen}$-polycycle with t boundaries. Denote by v_2 and v_3 the number of vertices of degree 2, 3 on the boundary. Let x and p_a be the number of interior vertices and a-gonal interior faces. Then, we have:*

$$\begin{cases} ap_a - 3x = v_2 + 2v_3 \\ p_a - \frac{x}{2} = (2 - t) + \frac{v_3}{2}. \end{cases}$$

If $a \neq 6$, then the system of equations has the solution:

$$\begin{cases} p_a = \frac{1}{a-6}\{v_2 - v_3 - 6(2 - t)\} \\ x = \frac{1}{a-6}\{2v_2 - (a - 4)v_3 - 2a(2 - t)\}. \end{cases}$$

Proof. Consider this set of 5-gonal faces as a plane graph with 5-gons and t other faces.

By counting in two different ways the number e of edges, we obtain $2e = ap_a + v_2 + v_3 = 2v_2 + 3v_3 + 3x$, which implies $ap_a - 3x = v_2 + 2v_3$.

On the other hand, Euler formula $v - e + f = 2$ implies, by writing $v = v_2 + v_3 + x$ and $f = t + p_a$, the relation $p_a - \frac{x}{2} = (2 - t) + \frac{v_3}{2}$. The solution comes by solving the linear system. $\square$

11.1 General $(\{a, b\}, 3)$-maps

Definition 11.1.1 *Given an $(\{a, b\}, 3)$-map G (on sphere or torus) that is bR_j, associate to it a map $b(G)$ (on sphere or torus, respectively) formed by the b-gonal faces of G and their adjacencies. It is an induced map of the dual map of G.*

Note that the map $b(G)$ can be non-connected. All $(\{5, b\}, 3)$-maps that are bR_0 or bR_1 are such. By the classification of $(4, 3)$-polycycles in Section 4.2 the only $(\{4, b\}, 3)$-map that is bR_j and with $b(G)$ non-connected is $Prism_b$.

Corollary 11.1.2

(i) If a $(\{a, b\}, 3)$-polyhedron G is bR_j, then $j \leq 5$.
(ii) If a 3-connected $(\{a, b\}, 3)$-torus G is bR_j, then $j \leq 6$.

Proof. (i) The sphere $b(G)$ does not contain any 2-gon, since, otherwise, it would imply that G is not 3-connected. Hence, we can apply Theorem 1.2.3(ii) and obtain $j \leq 5$.

(ii) In the toroidal case, the 3-connectedness implies that the gonalities of faces of $b(G)$ is at most 3. From Theorem 1.2.3(ii), we get $j \leq 6$. $\qquad\square$

If we remove the hypothesis of 3-connectedness, then there is no upper bound on j. For example, by Theorem 13.2.4, there exists a $(\{5, 5j\}, 3)$-sphere that is $5jR_j$, for any $j \geq 2$.

Lemma 11.1.3 *If an $(\{a, b\}, k)$-map is bR_j for $j = b - 1$, then it is aR_0.*

Proof. It suffice to see that any $a - a$ adjacency implies $j < b - 1$. $\qquad\square$

Theorem 11.1.4 *Let M be a $(\{a, b\}, 3)$-sphere or torus that is bR_j.*

(i) If $a = 3$, then:

- *If M is a torus, then $j \geq 6$. If $j = 6$, then M is $3R_0$.*
- *If M is a sphere, then $j < 6$ implies $p_b \leq \frac{12}{6-j}$. If $j = 6$, then $e_{3-3} = 6$.*

(ii) If $a = 4$, then:

- *If M is a torus, then $j \geq 12 - b$. If $j = 12 - b$, then M is $4R_0$.*
- *If M is a sphere, then $j < 12 - b$ implies $p_b \leq \frac{24}{12-b-j}$. If $j = 12 - b$, then $e_{4-4} = 12$.*

(iii) If $a = 5$, then:

- *If M is a torus, then $j \geq 30 - 4b$. If $j = 30 - 4b$, then M is $5R_0$.*
- *If M is a sphere, then $j < 30 - 4b$ implies $p_b \leq \frac{60}{30-4b-j}$. If $j = 30 - 4b$, then $e_{5-5} = 30$.*

Proof. Euler formula (1.1) is written as:

$$p_b(b - 6) = p_a(6 - a) - 6\chi$$

with $\chi = 2$ for the sphere and 0 for the torus. We have furthermore the following equality:

$$e_{a-b} = p_b(b - j) = ap_a - 2e_{a-a}$$

If we eliminate the term p_a from those equations then we obtain the following equations for $a = 3, 4, 5$:

$$p_b(j - 6) = 2e_{3-3} - 6\chi; \quad p_b(j + b - 12) = 2e_{4-4} - 12\chi;$$

$$p_b(4b + j - 30) = 2e_{5-5} - 30\chi$$

Let us consider the case $a = 4$, the others being similar. For tori the equation is $p_b(b + j - 12) \geq 0$, which implies $j \geq 12 - b$. If $j = 12 - b$, then we have $e_{4-4} = 0$. For spheres, if $j < 12 - b$, then $p_b = \frac{24 - 2e_{4-4}}{12 - b - j} \leq \frac{24}{12 - b - j}$, while if $j = 12 - b$, the equation simplifies to $e_{4-4} = 12$.   □

The $(\{4, b\}, 3)$-spheres bR_3 are listed in Theorem 16.1.1, the $(\{4, b\}, 3)$-spheres bR_2 are listed in Theorem 15.1.4, there is no $(\{4, b\}, 3)$-spheres bR_1 by Theorem 14.1.2 and the only $(\{4, b\}, 3)$-sphere bR_0 is $Prism_b$. The $(\{4, 7\}, 3)$-spheres $7R_4$ are listed in Figure 17.1, A conjectural list of $(\{4, 8\}, 3)$-spheres $8R_4$ is given in Conjecture 17.1.3; there is an infinity of them. Three examples of $(\{4, 7\}, 3)$-spheres that are $7R_5$ are known (see Figure 18.3), but finiteness or not is undecided for them.

By Theorem 12.0.1 we know that there is no $(\{5, 7\}, 3)$-spheres $5R_i$ for $i \leq 1$. All $(\{5, 7\}, 3)$-spheres that are $7R_2$ are enumerated in Chapter 15 (see Figures 15.6 and 15.7).

11.2 Remaining questions

We list here some remaining interesting problems for 3-valent face-regular two-faced maps:

1 Decide finiteness or not for $(\{4, 7\}, 3)$-spheres $7R_5$ and $(\{5, 7\}, 3)$-spheres $7R_5$.
2 Decide finiteness or not for $(\{5, 7\}, 3)$-spheres $7R_3$ and $(\{5, 10\}, 3)$-spheres $5R_3$.
3 Decide existence of $(\{5, b\}, 3)$-tori bR_5

One of the most interesting questions that arise in this research is to check our conjecture that an infinity of $(\{a, b\}, 3)$-spheres bR_j exists if and only if an $(\{a, b\}, 3)$-torus bR_j exists.

For every pair aR_i, bR_j and fixed genus $g \geq 2$, the number of strictly face-regular possibilities is, clearly, finite. But it is, certainly, extremely large. An interesting question would be to decide, what type of strict face-regularity can appear on surfaces of genus $g \geq 2$. Another direction is to study all weakly face-regular $(\{a, b\}, k)$-maps with $k \geq 4$.

We can also permit $a = b$, i.e. we can distinguish two classes of a-gons in the main problem. This is similar to (R, q)-polycycles considered before, where faces were partitioned into holes and proper ones.

12

Spheres and tori that are aR_i

We start with the following general result.

Theorem 12.0.1 *Let G be a 3-connected* $(\{a, b\}, 3)$*-sphere or torus that is* aR_i *with* $b \geq 7$. *Then the following hold:*

(i) aR_i *is in one of following cases:* $3R_0$, $4R_0$, $4R_1$, $4R_2$, $5R_1$, $5R_2$, $5R_3$, $5R_4$.

(ii) *If it is* $5R_4$, *then it is snub* $Prism_b$.

(iii) *If it is* $5R_1$, *then* $b = 7$ *it is Case 17 of strictly face-regular planes in Table 9.3.*

Proof. We have $a = 3$, 4, or 5. If a $(\{3, b\}, 3)$-map is 3-connected, then it is $3R_0$, which excludes $3R_1$ and $3R_2$.

Property aR_{a-1} implies bR_0 (see Theorem 11.1.3). There is no $(\{4, b\}, 3)$-maps $4R_3$ and bR_0 and the only $(\{5, b\}, 3)$-map that is $5R_4$ and bR_0 is snub $Prism_b$ (see Table 9.1).

Take a $(\{5, b\}, 3)$-sphere or torus that is $5R_1$. Euler formula reads $12\chi = p_5 - (b - 6)p_b$. There are $n_P = \frac{p_5}{2}$ pairs of 5-gons. Every such pair defines $4n_P$ patterns $b5b$ and $2n_P$ patterns $b55b$ in the boundary sequences of b-gons. A packing argument yields:

$$2(4n_P) + 3(2n_P) = 7p_5 \leq bp_b.$$

Since G is on sphere or torus, we have $\chi \geq 0$, i.e. $p_5 \geq (b - 6)p_b$. Therefore, $b = 7$, $p_5 = p_7$ and $\chi = 0$; so, G is a torus. Since $p_5 = p_7$, the boundary sequence of 7-gons is of the form $\alpha_1 \ldots \alpha_m$ with $\alpha_i = 57$ or 557. It is easy to see that the only possibility is two terms of the form 57 and one term of the form 557. So, G is $7R_3$. $\qquad\square$

12.1 Maps aR_0

Theorem 12.1.1 *Let G be a* $(\{a, b\}, 3)$*-sphere or torus* aR_0 *with* $b > 6$; *then it holds:*

(i) *If* $a = 3$, *then* $b \leq 12$.

    1 If $b = 12$, then G is Case 6 of strictly face-regular torus.

    2 If $b = 11$, then G is Case 5 of strictly face-regular torus.

    3 If $b = 7, 8, 9, 10$ and G is a sphere, then G has at least $20, 24, 52, 60$ vertices, respectively.

(ii) If $a = 4$, then $b \leq 8$.

    1 If $b = 8$, then G is Case 14 of strictly face-regular torus.

    2 If $b = 7$, and G is a sphere, then G has at least 80 vertices.

The case $a = 5$ does not occur.

Proof. $a = 3, 4$ or 5 by Euler formula for $(\{a, b\}, 3)$-maps:

$$p_a(6 - a) - p_b(b - 6) = 6\chi$$

with $\chi = 2$ or 0 for sphere or torus. By the property aR_0, the number e_{a-b} of a-b edges satisfies to:

$$e_{a-b} = ap_a \leq p_b \lfloor b/2 \rfloor,$$

because the corona of any b-gon cannot have a-gons on 2 neighboring edges. So, we obtain:

$$6\chi = (6 - a)p_a - p_b(b - 6) \leq p_b\Psi_a(b) \text{ with } \Psi_a(b) = \frac{6 - a}{a}\lfloor b/2 \rfloor - (b - 6).$$

If $\Psi_a(b) < 0$, then it excludes the existence of both the sphere and torus. If $\Psi_a(b) = 0$, then $\chi = 0$, i.e. G is a torus and it is bR_j for some j. If $\Psi_a(b) > 0$, then both the torus and sphere cases are possible. Furthermore, if G is a torus, we have $p_b \geq \frac{6\chi}{\Psi_a(b)}$, thereby yielding the lower bound on v.

    The result follows by considering $a = 3, 4, 5$ in the above relation. $\qquad\square$

Theorem 12.1.2 *There exist two infinite series of $(\{4, 7\}, 3)$-spheres that are $4R_0$. They have $140 + 84i$ vertices (see two examples on Figure 12.1). For i even, they are distinct, one is of symmetry D_{7h}, the other of symmetry D_{7d}. For i odd, they are isomorphic and of symmetry D_7.*

Proof. From the drawing on Figure 12.1, it is clear that such spheres exist. Now we will show the existence of an infinity of them.

    We have the following band structure of 4- and 7-gons:

The left- and right-hand sides of this band can be closed, in order to obtain a structure with 14 4-gons.

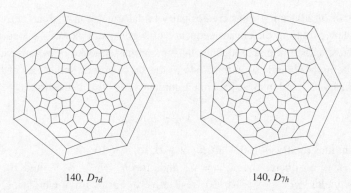

140, D_{7d} 140, D_{7h}

Figure 12.1 First examples of two infinite series of ($\{4, 7\}$, 3)-spheres that are $4R_0$ (see Theorem 12.1.2)

This structure can be inserted along one of the cutting lines, indicated below, and we gets ($\{4, 7\}$, 3)-sphere that is again $4R_0$ and has 84 more vertices.

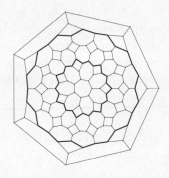

Obviously, the above operation can be repeated, again and again.   □

12.2 Maps $4R_1$

Theorem 12.2.1 *Let G be a ($\{4, b\}$, 3)-sphere or torus that is $4R_1$. Then it holds that:*

(i) $b \leq 10$.

(ii) If $b = 7, 8, 9$ and G is a sphere, then it has at least 32, 48, 108 vertices.

(iii) If $b = 10$, then G is a torus and it is also $10R_4$ (i.e. Case 16 of Table 9.3)

Proof. (i) ($\{4, b\}$, 3)-maps $4R_1$ satisfy to Euler formula $6\chi = 2p_4 - (b - 6)p_b$ with χ being equal to 2 for spheres and 0 for tori.

We have $p_4 = 3\chi + \frac{b-6}{2}p_b$.

Every pair of adjacent 4-gons creates pair of adjacent 4-gons, which corresponds to subsequence $b44b$ in the corona sequence of b-gons, and pair of isolated 4-gons, which corresponds to the subsequence $b4b$ in corona sequence of b-gons. So, there are p_4 patterns $b44b$ and p_4 patterns $b4b$ in the set of corona sequences of b-gonal faces of the considered map. A packing argument yields the inequality:

$$2p_4 + 3p_4 \leq bp_b,$$

which simplifies to $30\chi \leq (30 - 3b)p_b$. $\chi \geq 0$; so, $b \leq 10$.

(ii) If G is a sphere, then $\chi = 2$ and, for $b = 7, 8, 9$, and the above inequality yields $p_b \geq \frac{20}{3}$, 10, 20, respectively. Parity arguments yield $p_b \geq$ 8, 10, 20 and the corresponding lower bounds 32, 48, 108 on the number of vertices.

(iii) For $(\{4, 10\}, 3)$-torus $4R_1$, the inequality reads $30\chi \leq 0$, which yields $\chi = 0$, i.e. G is a torus and $2p_4 + 3p_4 = 10p_{10}$. But this means that in the corona of 10-gons, the pattern 10^2 does not occur. So, their corona is of the form $\alpha_1 \ldots \alpha_r$ with α_u being 10.4 or 10.4^2. Denote by y_2 and y_3 the number of α_i being equal to 10.4, 10.4^2, respectively, for a given 10-gonal face F. We have, clearly, $2y_2 + 3y_3 = 10$, whose solutions are $(y_2, y_3) = (5, 0)$ or $(2, 2)$. So, for all solutions we get $y_3 \leq y_2$. But, on average over all 10-gonal faces, we have $y_3 = y_2$; this is possible only if $y_2 = y_3$ for every 10-gonal face. So, the torus is $10R_4$.   □

The lower bound on the number of vertices of $(\{4, b\}, 3)$-spheres is met only for $b = 8$ (see Figures 12.2, 12.3, 12.5 and Theorem 12.2.4).

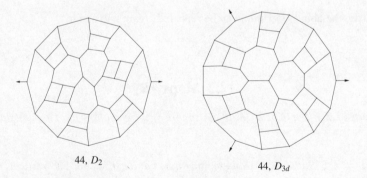

44, D_2

44, D_{3d}

Figure 12.2 Two smallest $(\{4, 7\}, 3)$-spheres that are $4R_1$

Theorem 12.2.2 *There exist an infinity of $(\{4, 7\}, 3)$-spheres that are $4R_1$.*

Proof. The proof consists of building the following initial example with 140 vertices and symmetry D_{7d}:

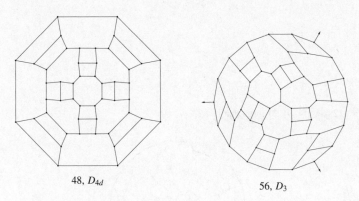

48, D_{4d} 56, D_3

Figure 12.3 Two $(\{4, 8\}, 3)$-spheres that are $4R_1$

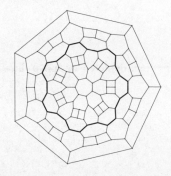

We will cut along the over-lined path and insert the following structure, which consists of 7 units:

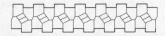

Obviously, the obtained map is again a $(\{4, 7\}, 3)$-sphere that is $4R_1$, and the construction can be repeated. □

Theorem 12.2.3 *There is an infinity of $(\{4, 8\}, 3)$-spheres that are $4R_1$.*

Proof. We construct the following example of a $(\{4, 8\}, 3)$-sphere that is $4R_1$:

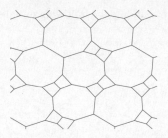

Figure 12.4 A ({4, 9}, 3)-torus that is $4R_1$

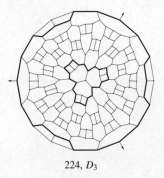

224, D_3

The idea is to cut along the line, depicted in above drawing, and insert inside the following band structure:

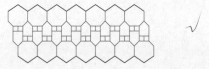

Obviously, the above operation can be repeated.   □

There exist ({4, 9}, 3)-tori that are $4R_1$, see an example on Figure 12.4, with $(v, p_4, p_9) = (20, 6, 4)$.

Take a ({4, 9}, 3)-sphere G that is $4R_1$, and map every pair of adjacent 4-gons to a single edge. The obtained *reduced sphere*, denoted by $Red(G)$, is still 3-valent. The set of pairs of adjacent 4-gons of G yields an edge-set $\mathcal{E}S(G)$ in $Red(G)$, which satisfies the following properties:

1. It is a *matching*, i.e. no vertex belong to two edges of $\mathcal{E}S(G)$.
2. For every face F of G, denote by $h(F)$ the number of edges in $\mathcal{E}S(G)$ that are incident to a vertex of F (those edges contain either an edge, or just a vertex

of G). We have the equation $h(F) + l(F) = 9$ with $l(F)$ being the gonality of F.

If a set of faces of a plane graph satisfies the above conditions, then we call it a *special* $(\{4, 9\}, 3)$-*matching*.

Theorem 12.2.4 *(i) If G is a v-vertex $(\{4, 9\}, 3)$-sphere that is $4R_1$, and $G' = Red(G)$ is its associated graph, then $v = \frac{5}{2}v' + 18$.*

(ii) The smallest $(\{4, 9\}, 3)$-sphere that is $4R_1$ is the one with 128 vertices depicted on Figure 12.5.

(iii) There is an infinity of $(\{4, 9\}, 3)$-spheres that are $4R_1$.

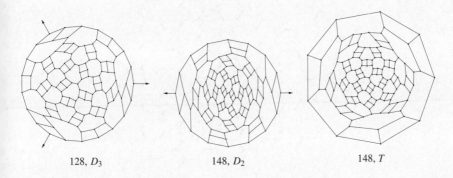

128, D_3 148, D_2 148, T

Figure 12.5 Some $(\{4, 9\}, 3)$-spheres that are $4R_1$

Proof. (i) Take a $(\{4, 9\}, 3)$-sphere that is $4R_1$ and has v vertices. We have the equations $3v = 4p_4 + 9p_9$ and $2p_4 - 3p_9 = 12$. The number of pairs of 4-gons is $\frac{p_4}{2}$. The sphere $Red(G)$ has $v' = v - 2p_4$ vertices. We obtain easily $v' = 2p_9 - 4$ and $v = 8 + 5p_9$ from which the result follows.

(ii) Take a $(\{4, 9\}, 3)$-sphere that is $4R_1$, and consider its reduced sphere $Red(G)$ that is a $(\{5, 6, 7, 8, 9\}, 3)$-sphere. We enumerate $(\{5, 6, 7, 8, 9\}, 3)$-spheres up to 44 vertices and, for every one of them, we search for special $(\{4, 9\}, 3)$-matchings. We found one graph with 44 vertices that is a $(\{5, 6\}, 3)$-sphere and has a unique special $(\{4, 9\}, 3)$-matching. It defines a $(\{4, 9\}, 3)$-sphere that is $4R_1$, and part (i) above proves that it is the smallest one.

(iii) Consider the following $(\{5, 6\}, 3)$-sphere with its special $(\{4, 9\}, 3)$-matching:

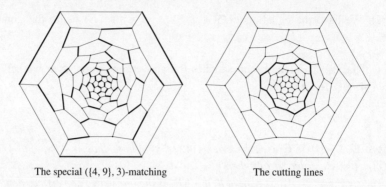

The special ({4, 9}, 3)-matching   The cutting lines

This sphere is cut along the cutting lines and in place is inserted the following structure:

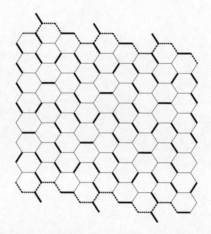

It is easy to see that the obtained structure is still a ({5, 6}, 3)-sphere with a special ({4, 9}, 3)-matching. Furthermore, the operation can be repeated indefinitely.   □

Although the above proof is very easy to check, the way to obtain the example is interesting. First, restrict to fullerenes. Second, restrict the search to cylindrical structures, since almost all infinite series, so far, were of that form. Then search amongst those of symmetry D_3, since it is the maximal possible symmetry. The requirement of special symmetry allow us to restrict ourselves to *equivariant special* ({4, 9}, 3)-*matchings*, i.e. special ({4, 9}, 3)-matchings, which have the same symmetry group as the fullerene, so as to prune the search tree. We obtained the fullerene with the special ({4, 9}, 3)-matching drawn above, set a cutting line, and considered the problem of finding a possible structure to insert as a torus problem. We found 14 different possibilities and selected the one of maximal symmetry.

If we search for special ({4, 9}, 3)-matchings in fullerenes, then this leads to ({4, 9}, 3)-spheres with special ({4, 9}, 3)-matchings:

1 one with 128 vertices of symmetry D_3,
2 two with 148 vertices $((1, D_2)$ and $(1, T))$,
3 10 with 168 vertices $((5, C_2)$ and $(5, D_3))$,
4 23 with 188 vertices $((9, C_1), (10, C_2), (3, D_2)$, and $(1, D_3))$,
5 66 with 208 vertices $((44, C_1), (19, C_2), (2, C_3)$, and $(1, C_s))$.

12.3 Maps $4R_2$

Theorem 12.3.1 below gives the necessary conditions for existence of $(\{4, b\}, 3)$-spheres or tori that are $4R_2$. The existence part of an infinity of $(\{4, b\}, 3)$-spheres $4R_2$ for $b = 6 - 13, 15$ is proved in Theorems 10.3.1, 12.3.2, 12.3.3, 12.3.4, 12.3.5, 12.3.6, 12.3.7, 12.3.8, and 12.3.9. The existence of $(\{4, b\}, 3)$-tori $4R_2$ bR_6 for $b = 8$, 10, 12, 14, 16, and 18 proves, a fortiori, the existence of $(\{4, b\}, 3)$-tori $4R_2$ for those values (see Table 9.3). For $b = 7, 9, 11, 13, 15$ the existence of $(\{4, b\}, 3)$-tori $4R_2$ is proved in Theorems 12.3.2, 12.3.4, 12.3.6, 12.3.8, and 12.3.9.

Theorem 12.3.1 *(i) A $(\{4, b\}, 3)$-torus $4R_2$ exist only for $b = 7 - 16, 18$. For $b = 14$, 16, 18 such tori are strictly face-regular.*

(ii) A $(\{4, b\}, 3)$-sphere $4R_2$, which is not $Prism_b$, exist only for $b = 6 - 13, 15$. The number of vertices of such spheres should be at least 20 (for $b = 7$), 32 (for $b = 8$), 28 (for $b = 9$), 44 (for $b = 10$), 92 (for $b = 11$), 56 (for $b = 12$), 116 (for $b = 13$), 140 (for $b = 15$). If it has this minimal number of vertices, then it is strictly face-regular.

Proof. A cycle of 4-gons cannot exist either in case (i) or in case (ii), due to the exclusion of $Prism_b$. So, all 4-gons are part of triples of 4-gons $\{4, 3\} - v$. Denote by n_t the number of such triples. We have the relations $p_4 = 3n_t$. Furthermore, by a packing argument, we obtain the inequality:

$$6n_t = e_{4-b} \leq p_b 2 \lfloor \frac{b}{3} \rfloor.$$

The Euler formula (1.1) is $6\chi = 2p_4 - (b - 6)p_b$ with χ being 2 for the sphere and 0 for the torus. So, we obtain:

$$6\chi = 2p_4 - (b - 6)p_b \leq p_b \Psi(b) \text{ with } \Psi(b) = 2 \lfloor \frac{b}{3} \rfloor - (b - 6).$$

The function Ψ satisfies to:

- $\Psi(b) > 0$ for $b \in \{6, \ldots, 13, 15\}$;
- $\Psi(b) = 0$ for $b = 14, 16, 18$;
- $\Psi(b) < 0$ for $b = 17$ or $b \geq 19$.

If $\Psi(b) < 0$, then it excludes the existence of a sphere or a torus. If $\Psi(b) = 0$, then $\chi = 0$. Also, all b-gonal faces should be adjacent to exactly $2\lfloor\frac{b}{3}\rfloor$ 4-gons, i.e. the torus is strictly face-regular.

If $\Psi(b) > 0$, then we have the condition $p_b \geq \frac{6\chi}{\Psi(b)}$, which gives announced lower bounds.   □

The $(\{4, 3\} - v)$-*replacement* of a map G by a set S of vertices consists of replacing every vertex in S by a $(4, 3)$-polycycle $\{4, 3\} - v$.

Theorem 12.3.2 *(i) Every $(\{4, 7\}, 3)$-map is obtained from a $(\{5, 7\}, 3)$-map by selecting a set S, such that every 5-gon is incident to exactly one vertex of S, and doing $(\{4, 3\} - v)$-replacement of all vertices in S.*

(ii) There exists a $(\{4, 7\}, 3)$-torus $4R_2$.

(iii) Given a v-vertex $(\{4, 7\}, 3)$-sphere $4R_2$, we obtain a $(v + 36)$-vertices $(\{4, 7\}, 3)$-sphere $4R_2$, by replacing the central vertex of this triple by the following structure:

(iv) There is an infinity of $(\{4, 7\}, 3)$-spheres that are $4R_2$.

Proof. (i) Take a $(\{4, 7\}, 3)$-map and replace every $(4, 3)$-polycycle $\{4, 3\} - v$, appearing in it, by a vertex. We get a map with 3-, 5-, and 7-gons. We need to prove that 3-gons cannot occur. If there is a 3-gon, then, in the original map, a 7-gonal face was incident to two $(4, 3)$-polycycles $\{4, 3\} - v$. So, one of adjacent 7-gons is also incident to those two $\{4, 3\} - v$. This implies that those two 7-gons are adjacent to two common 7-gonal faces, say, F_1 and F_2. Those two faces are incident to one $(4, 3)$-polycycle $\{4, 3\} - v$. We get the contradiction by seeing that those faces, F_1 and F_2, are adjacent to a 2-gonal face.

(ii) There exists a $(\{5, 7\}, 3)$-torus $7R_4$ with 5-gons organized in triples (see Section 9.3, Case 18). So, we can apply the operation, given in (i), and get the torus.   □

(iii) This result is obvious and (iv) follows by repeated applications of (iii).   □

Theorem 12.3.3 *There is an infinity of $(\{4, 8\}, 3)$-spheres $4R_2$.*

Proof. Take a $(4, 3)$-polycycle $\{4, 3\} - v$ and add $2t$ rings of three 6-gons around it. Then make three vertices of degree 2 adjacent to one other vertex (this forms another

(4, 3)-polycycle $\{4, 3\} - v$). We obtain a 3-valent plane graph G, which contains two triples of 4-gons. In order to obtain a ($\{4, 8\}$, 3)-sphere $4R_2$, we should find a subset S of the vertex-set such that:

- every 4-gon is incident to two vertices of S and
- every 6-gon is incident to one vertex of S.

For the elements of S, we take first the vertices of G that are incident to just one 4-gon of G. Then we need to find vertices that are incident to three 6-gons and cover the remaining 6-gons. It is easy to see that this is, indeed, possible.

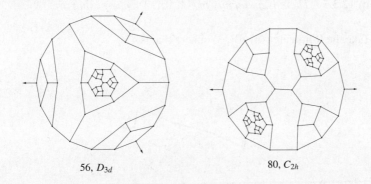

56, D_{3d} 80, C_{2h}

See above first examples of the infinite series.   □

Theorem 12.3.4 *(i) There is an infinity of ($\{4, 9\}$, 3)-spheres $4R_2$.*
    (ii) There exists a ($\{4, 9\}$, 3)-torus $4R_2$.

Proof. (i) Take the following graph:

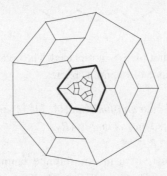

cut it along the boldfaced edges and insert the following structure:

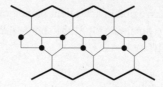

with the circled vertices being ($\{4, 3\} - v$)-replaced. The operation can, clearly, be repeated and we get an infinite series.

(ii) The above drawing in (i) is, clearly, a part of a plane tiling by such structures. Its quotient is the required torus.   □

Theorem 12.3.5   *There is an infinity of* ($\{4, 10\}, 3$)*-spheres that are* $4R_2$.

Proof. Take the following ($\{4, 10\}, 3$)-sphere that is $4R_2$:

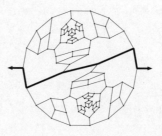

and insert, along the boldfaced edges, the following structure:

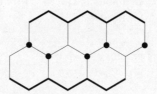

with the circled vertices being $\{4, 3\} - v$-replaced. The operation can be repeated and we obtain an infinite sequence of required spheres.   □

Theorem 12.3.6 *(i) There exist an infinity of* ($\{4, 11\}, 3$)*-spheres that are* $4R_2$.
   (ii) There exist a ($\{4, 11\}, 3$)*-torus that is* $4R_2$.

Proof. (i) The proof consists of using the infinite families of ($\{5, 7\}, 3$)-spheres that are $7R_4$, constructed in Theorem 17.2.2. The 5-gons of those polycycles are organized into two polycycles A_3 and bands (of length 6) of 5-gons. Define a

($\{4, 3\} - v$)-replacement set S by assigning the ($\{4, 3\} - v$)-vertices in the following way:

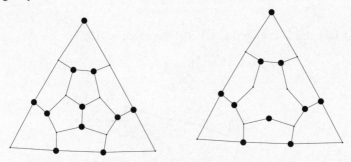

So, every 5-, 7-gon is incident to three, two, respectively, vertices in S. This means that, by doing the ($\{4, 3\} - v$)-replacement, we obtain a ($\{4, 11\}, 3$)-sphere $4R_2$. Since the series of Theorem 17.2.2 is infinite, we have an infinite series.

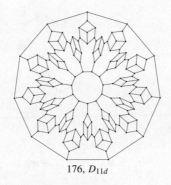

176, D_{11d}

Figure 12.6 A ($\{4, 11\}, 3$)-sphere that is $4R_2$

(ii) The figure below gives a ($\{5, 7\}, 3$)-torus.

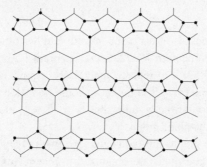

By doing ($\{4, 3\} - v$)-replacement of the circled vertices, we get a ($\{4, 11\}, 3$)-torus $4R_2$.   □

Theorem 12.3.7 *(i) There is an infinity of* $(\{4, 12\}, 3)$*-spheres that are* $4R_2$.
(ii) There is a $(\{4, 12\}, 3)$*-sphere* $4R_2$ *of symmetry* O.

Proof. (i) Take the following $(\{4, 12\}, 3)$-sphere that is $4R_2$,

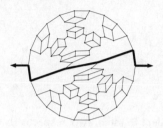

and insert along the boldfaced edges the following structure:

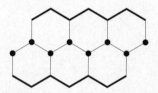

with the circled vertices being $\{4, 3\} - v$-replaced. The operation can be repeated and we obtain an infinite series of required spheres.

(ii) Take $GC_{2,1}(Cube)$, i.e. Nr. 26 and triple it along the set of vertices, which are incident to a 4-gon or to a 3-fold axis of symmetry.   □

Theorem 12.3.8 *(i) There exist an infinity of* $(\{4, 13\}, 3)$*-spheres that are* $4R_2$.
(ii) There exist a $(\{4, 11\}, 3)$*-torus that is* $4R_2$.

Proof. (i) The proof consists of using the infinite families of $(\{5, 7\}, 3)$-spheres that are $7R_4$ constructed in Theorem 17.2.2. As in Theorem 12.3.6, we will define a set S, which defines the $(\{4, 3\} - v)$-replacement. Every 5-gonal face should be incident to four elements of S and every 7-gonal face should be incident to three elements of S. Hence, it is easier to use the complement $\overline{S} = \{1, \dots, v\} - S$ with v being the number of vertices of the plane graph.

The 5-gons of those graphs are organized into two polycycles A_3 and bands, of length 6, of 5-gons. The 7-gons are organized in a series of two parallel rings of length 3. There exists a simple zigzag that separates those two rings. Assign all those vertices to $\overline{S}$. For the 5-gons, assign vertices in the following way:

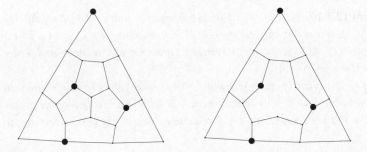

So, every 5-, 7-gon is incident to one, four, respectively, vertices in $\overline{S}$. This means that by doing the $(\{4, 3\} - v)$-replacement on S, we obtain a $(\{4, 13\}, 3)$-sphere $4R_2$. Since the series of Theorem 17.2.2 is infinite, we have an infinite series.

(ii) The figure below gives a $(\{5, 7\}, 3)$-torus.

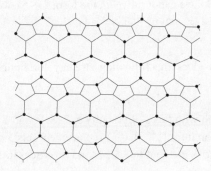

By doing $(\{4, 3) - v)$-replacement of the non-circled vertices, we get a $(\{4, 13\}, 3)$-torus $4R_2$.   □

Theorem 12.3.9 *(i) Given a $(\{5, 7\}, 3)$-map that is $7R_4$ and such that 7-gons have the corona $7^4 5^3$, we can obtain a $(\{4, 15\}, 3)$-map that is $4R_2$, by $(\{4, 3\} - v)$-replacement of the set of vertices incident to 5-gonal faces.*

(ii) There exists a $(\{4, 15\}, 3)$-torus that is $4R_2$.

(iii) There exists an infinity of $(\{4, 15\}, 3)$-spheres that are $4R_2$.

Proof. (i) This result is obtained by considering the local structure.

(ii) Take the unique $(\{5, 7\}, 3)$-torus that is $5R_2$ and $7R_4$, and use (i).

(iii) The $(\{5, 7\}, 3)$-spheres, constructed in Theorem 17.2.2, give, using (i), an infinite series.   □

Theorem 12.3.10 *(i) If G is a $(\{4, 11\}, 3)$-sphere that is $4R_2$ and different from $Prism_{11}$, then it is obtained by the $(\{4, 3\} - v)$-replacement of a set S in a 3-valent plane graph G'. G' has $-4 + 12x$ vertices, G has $8 + 42x$ vertices, and S has $2 + 5x$ vertices, for some $x \geq 1$.*

(ii) If G is a $(\{4, 13\}, 3)$-sphere that is $4R_2$ and different from $Prism_{13}$, then it is obtained by the $(\{4, 3\} - v)$-replacement in a set S of a 3-valent plane graph G'. G' has $-4 + 12x$ vertices, G has $8 + 54x$ vertices, and S has $2 + 7x$ vertices, for some $x \geq 1$.

Proof. We prove only (i), since the proof of (ii) is very similar. Denote by n_1 the number of vertices of G', which are not $(\{4, 3\} - v)$-replaced, and by n_2 the number of vertices of G', which are $(\{4, 3\} - v)$-replaced. Then we have $p_4 = 3n_2$ and $-5p_{11} + 2p_4 = 12$. Denote by v the number of vertices of G. We have $v = n_1 + 7n_2$ and $3v = 11p_{11} + 4p_4$. Eliminating the unknown p_{11}, v and p_4, we obtain the relation $n_1 = \frac{7n_2 - 44}{5}$. So, we can write n_2 in the form $2 + 5x$ with $x \in \mathbb{Z}$. This yields $n_1 = 7x - 6$, $v = 8 + 42x$, and $n_1 + n_2 = -4 + 12x$. $\square$

The above theorem gives the following strategy for finding some $(\{4, 11\}, 3)$-spheres that are $4R_2$:

1 Take a 3-valent graph with faces of gonality 5, 7, 8, 9, and 11.
2 Do an exhaustive search for $(\{4, 3\} - v)$-replacement sets S such that every 5-, 7-, 9-, and 11-gonal face is incident to 3, 2, 1, and 0 vertices in S respectively.

If we consider all 3-valent graphs G' up to $12m - 4$ vertices, and manage to do an exhaustive search, then we can get the complete list of $(\{4, 11\}, 3)$-spheres up to $42m + 8$ vertices.

The same applies for $(\{5, 13\}, 3)$-spheres. But, in this case, more than half of the vertices are in the $(\{4, 3\} - v)$-replacement set S. Hence, it is better, from the computational viewpoint, to do an exhaustive enumeration of their complements.

The $(\{5, 7, 9, 11\}, 3)$-spheres can be enumerated up to 56 vertices. By applying an exhaustive enumeration procedure, we obtained 87 spheres. This means that the enumeration of $(\{4, 11\}, 3)$-spheres $4R_2$ has been completed up to 218 vertices. Besides two strictly face-regular ones, the remaining spheres have 176 vertices (Figure 12.6). The repartition by symmetry is the following: $(38, C_1)$, $(31, C_2)$, $(4, C_{2h})$, $(1, C_3)$, $(10, C_i)$, $(2, C_s)$, and $(1, D_{11d})$. If we limit ourselves to $(\{5, 7\}, 3)$-spheres, then we can extend the enumeration up to 68 vertices. Using the exhaustive enumeration technique, we found 27276 graphs, whose repartition by symmetry is the following: $(26299, C_1)$, $(895, C_2)$, $(3, C_{2h})$, $(1, C_{2v})$, $(9, C_3)$, $(16, C_i)$, $(28, C_s)$, $(16, D_2)$, $(8, D_3)$, and $(1, S_4)$.

The $(\{5, 7, 9, 11, 13\}, 3)$-spheres can, again, be enumerated up to 56 vertices. By applying the exhaustive enumeration procedure, we obtain 12 spheres. This means that the enumeration of $(\{4, 13\}, 3)$-spheres $4R_2$ has been completed up to 278 vertices. Besides two strictly face-regular ones, the remaining spheres have 224 vertices. The repartition by symmetry is the following: $(3, C_1)$, $(4, C_2)$, $(2, C_i)$, $(2, D_3)$, and $(1, S_6)$. If we limit ourselves to $(\{5, 7\}, 3)$-spheres, then we can extend the enumeration up to 68 vertices. Using the exhaustive enumeration technique, we found 805 graphs, whose repartition by symmetry is the following: $(707, C_1)$, $(86, C_2)$, $(4, C_3)$, $(1, C_s)$, $(3, D_2)$, and $(4, D_3)$.

12.4 Maps $5R_2$

Theorem 12.4.1 *(i) A $(\{5, b\}, 3)$-sphere that is $5R_2$ has $b = 7$.*
(ii) A $(\{5, b\}, 3)$-torus that is $5R_2$ has $b = 7$ or 8.

Proof. Euler formula in Theorem 13.1.1(iii) implies the relation:

$$(6 - b)(x_0 + x_3) + (8 - b)p_5 = 4b\chi$$

with χ being 2 and 0 for the sphere and torus, respectively.

If $b > 8$, we get an impossibility. If $b = 8$, then $-2(x_0 + x_3) = 2\chi$, which implies $\chi = 0$ (i.e. a torus) and $x_0 = x_3 = 0$. $\qquad\square$

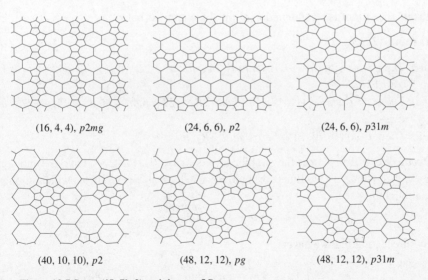

(16, 4, 4), *p2mg*   (24, 6, 6), *p2*   (24, 6, 6), *p31m*

(40, 10, 10), *p2*   (48, 12, 12), *pg*   (48, 12, 12), *p31m*

Figure 12.7 Some $(\{5, 7\}, 3)$-tori that are $5R_2$

Note that an infinity of $(\{5, 7\}, 3)$-spheres $5R_2$ is constructed in [HaSo07]. It is proved in Theorem 15.2.1 that a $(\{5, 8\}, 3)$-torus is $5R_2$ if and only it is $8R_2$.

See on Figures 12.7, 17.15 some $(\{5, 7\}, 3)$-tori that are $5R_2$, respectively, $7R_4$.

12.5 Maps $5R_3$

The following simple theorem is key to this section.

Theorem 12.5.1 *The set of 5-gonal faces of a* $(\{5, b\}, 3)$-*map that is* $5R_3$, *is partitioned into polycycles* E_1 *and* E_2.

Proof. Take a $(\{5, b\}, 3)$-map $5R_3$ and a 5-gon F of this map. F is adjacent to 5-gons on either three consecutive edges, or two consecutive edges and one isolated edge. The first case gives E_2, while the second case gives E_1 or E_2.   □

In this section n_1 and n_2 are the number of $(5, 3)$-polycycles E_1 and E_2.

Theorem 12.5.2 *Let G be a* $(\{5, b\}, 3)$-*torus that is* $5R_3$; *then it holds:*

  (i) $b \leq 12$.
 (ii) *If $b = 12$, then it is also* $12R_0$.
(iii) *If $b = 11$, then it is also* $11R_1$.

Let G be a $(\{5, b\}, 3)$-*sphere that is* $5R_3$; *then it holds:*

  (i) $b \leq 10$.
 (ii) *If $b = 7$, then there are exactly two such spheres, shown on Figure 12.8.*

Proof. The proof is a combination of the three lemmas below: □

Lemma 12.5.3

  (i) *There is no* $(\{5, b\}, 3)$-*torus that is* $5R_3$, *for $b > 12$; for $b = 12$ such a torus is also* $12R_0$.
 (ii) *There is a unique* $(\{5, 12\}, 3)$-*plane* $5R_3$, $12R_0$ *represented on Figure 9.5 in Case 23.*
(iii) *There is no* $(\{5, b\}, 3)$-*sphere that is* $5R_3$, *for $b \geq 12$.*

Proof. We will first treat the simpler toroidal case (i). First, we have the relation $p_5 = 3n_1 + 4n_2$. By the Euler formula, we also have $p_5 = (b - 6)p_b$, which implies the relation:

$$e = 3(b - 5)p_b = 3\frac{b - 5}{b - 6}p_5.$$

By direct counting, we have:

$$e_{5-5} = \left(3 + \frac{3}{2}\right)n_1 + 6n_2 \quad \text{and} \quad e_{5-b} = 6n_1 + 8n_2.$$

We then obtain the relation:

$$e_{b-b} = 3\frac{b-5}{b-6}(3n_1 + 4n_2) - \left(\left(3+\frac{3}{2}\right)n_1 + 6n_2\right) - (6n_1 + 8n_2)$$

$$= \left(\frac{9(b-5)}{b-6} - \left(9+\frac{3}{2}\right)\right)n_1 + \left(\frac{12(b-5)}{b-6} - 14\right)n_2$$

$$= \frac{18-\frac{3}{2}b}{b-6}n_1 + \frac{24-2b}{b-6}n_2 = \frac{12-b}{b-6}\left\{\frac{3}{2}n_1 + 2n_2\right\}.$$

If $b > 12$, then $e_{b-b} = n_2 = n_1 = 0$, which is impossible. If $b = 12$, then $e_{12-12} = 0$, i.e. the torus is $12R_0$.

In order to prove assertion (ii) take a $(\{5, 12\}, 3)$-plane $5R_3$, $12R_0$. The $(5, 3)$-polycycle E_2 cannot occur in the decomposition of the 5-gons, since it will imply that two 12-gons are adjacent. Therefore, every 12-gon is bounded by six $(5, 3)$-polycycles E_1 and the major skeleton, formed by $(5, 3)$-polycycles E_1, is 3-valent and has faces of gonality six; it is $\{6, 3\}$. So, G is the sporadic plane represented on Figure 9.5.

In order to prove assertion (iii), enumeration of $(\{5, b\}, 3)$-spheres is essentially a remake, with some additional constants, of (i). We obtain first $p_5 = 12 + (b-6)p_b$ and then it holds that:

$$e = 3\left(10 + (b-5)\frac{p_5 - 12}{b-6}\right) \quad \text{and} \quad v = 2\left(10 + (b-5)\frac{p_5 - 12}{b-6}\right)$$

and, finally:

$$e_{b-b} = \left(30 - 36\frac{b-5}{b-6}\right) + \frac{12-b}{b-6}\left(\frac{3}{2}n_1 + 2n_2\right).$$

If $b \geq 12$, then e_{b-b} becomes negative, which is impossible. $\qquad\square$

Lemma 12.5.4

(i) A $(\{5, 11\}, 3)$-torus that is $5R_3$ is also $11R_1$.

(ii) There is a unique $(\{5, 11\}, 3)$-plane $5R_3$, $11R_1$ represented on Figure 9.5 in Case 22.

(iii) There is no $(\{5, 11\}, 3)$-sphere that is $5R_3$.

Proof. Any corona of an 11-gon, in a $(\{5, 11\}, 3)$-map $5R_3$, is one of the following six *types*:

Type 1	$E_1 E_2(11\text{-gon})E_2 E_1 E_1$	Type 4	$E_2(11\text{-gon})E_2 E_2(11\text{-gon})^2 E_2$
Type 2	$E_1 E_2(11\text{-gon})^3 E_2 E_1$	Type 5	$E_2(11\text{-gon})^7 E_2$
Type 3	$E_1 E_2(11\text{-gon})^5 E_2$	Type 6	$(11\text{-gon})^{11}$

Denote by $p_{11,i}$, with $1 \leq i \leq 6$, the number of 11-gonal faces of type i.

In torus case (i), consider the number x_i of vertices, which are contained in exactly i b-gonal faces. We have, clearly:

$$x_0 = n_1 + 2n_2, \quad x_1 = 6n_1 + 6n_2 \quad \text{and} \quad x_2 = 2n_2.$$

Let us consider first the toroidal case. The number of vertices of our torus is equal to $2\frac{b-5}{b-6}(3n_1 + 4n_2)$. From this we get:

$$x_3 = 2\frac{6}{5}(3n_1 + 4n_2) - (n_1 + 2n_2) - (6n_1 + 6n_2) - 2n_2$$

$$= \left(\frac{36}{5} - 7\right) n_1 + \left(8\frac{6}{5} - 10\right) n_2 = \frac{1}{5}n_1 - \frac{2}{5}n_2.$$

By direct counting, we get also:

$$\begin{cases} 3x_3 = 2p_{11,2} + 4p_{11,3} + p_{11,4} + 6p_{11,5} + 11p_{11,6}, \\ 3n_1 = 3p_{11,1} + 2p_{11,2} + p_{11,3}, \\ 4n_2 = 2p_{11,1} + 2p_{11,2} + 2p_{11,3} + 4p_{11,4} + 2p_{11,5}. \end{cases}$$

Using the previous equations, we obtain:

$$x_3 = \frac{2}{3}p_{11,2} + \frac{4}{3}p_{11,3} + \frac{1}{3}p_{11,4} + 2p_{11,5} + \frac{11}{3}p_{11,6},$$

$$\frac{1}{5}n_1 - \frac{2}{5}n_2 = -\frac{1}{15}p_{11,2} - \frac{2}{15}p_{11,3} - \frac{2}{5}p_{11,4} - \frac{1}{5}p_{11,5},$$

which, clearly, implies $p_{11,2} = p_{11,3} = p_{11,4} = p_{11,5} = p_{11,6} = 0$. Hence, assertion (i) is true.

In order to prove (ii), take a $(\{5, 11\}, 3)$-plane G that is $5R_3$ and $11R_1$. Clearly, only type 1 can occur for the 5-gons. Consider now the 3-valent plane $\widetilde{G}$ defined by the $(5, 3)$-polycycles E_1 of G with two polycycles E_1 being adjacent if they share an edge or are linked by an $(5, 3)$-polycycle E_2. $\widetilde{G}$ is a 3-valent plane with faces of gonality 6, i.e. it is $\{6, 3\}$. Every pair of adjacent 11-gons is adjacent to six $(5, 3)$-polycycles E_1 and two $(5, 3)$-polycycles E_2. Those two E_2 define opposite edges in the 6-gons of $\{6, 3\}$. The set of those edges form a set of parallel edges of $\{6, 3\}$; the tiling $\{6, 3\}$ has three such sets of parallel edges but they are all isomorphic. Therefore, the $(\{5, 11\}, 3)$-plane G is unique and it is the one depicted in Figure 9.5.

For spheres in assertion (iii), we have:

$$x_3 = 2\left(10 + 6\frac{p_5 - 12}{5}\right) - x_0 - x_1 - x_2 = \frac{-22}{5} + \frac{1}{5}n_1 - \frac{2}{5}n_2.$$

Using the same expression of x_3, n_1, and n_2 in terms of $p_{11,i}$, we obtain the equality:

$$x_3 = -\frac{22}{5} - \frac{1}{15}p_{11,2} - \frac{2}{15}p_{11,3} - \frac{2}{5}p_{11,4} - \frac{1}{5}p_{11,5},$$

i.e. x_3 is negative, an impossibility.   □

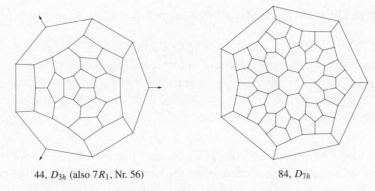

44, D_{3h} (also $7R_1$, Nr. 56)   84, D_{7h}

Figure 12.8 All ($\{5, 7\}$, 3)-spheres that are $5R_3$

Lemma 12.5.5

(i) A ($\{5, 7\}$, 3)-sphere that is $5R_3$ is one of two spheres shown on Figure 12.8.
(ii) There is no ($\{5, 7\}$, 3)-torus that is $5R_3$.

Proof. By analogy with the previous lemma, we can split the set of 7-gonal faces into the following types:

Type 1	$E_1 E_2(\text{7-gon}) E_2$
Type 2	$E_2(\text{7-gon})^3 E_2$
Type 3	$(\text{7-gon})^7$

Suppose that a sphere contains a face of type 1, then this face is necessarily adjacent to another face of type 1. Those two faces are bordered by four $(5, 3)$-polycycles: two E_1 and two E_2. Each of two $(5, 3)$-polycycles E_1 has two vertices of degree 2. Since the polycycle E_1 is adjacent only to 7-gons of type 1, we have only one way of filling the structure: by adding faces of type 1. Hence, the obtained sphere is strictly face-regular.

Assume now that there is no faces of type 1; then $n_1 = 0$. Hence, the only appearing $(5, 3)$-polycycles are E_2. Those polycycles are adjacent by pairs, so they make cycles in the sphere. Since this is a ($\{5, 7\}$, 3)-sphere, there exist at least one $(7, 3)$-polycycle, say P_7', bordered by a ring of $(5, 3)$-polycycles E_2.

From the structure of the $(5, 3)$-polycycle E_2, we know that P_7' has boundary sequence $(32^3)^h$ for some h. If we remove the 7-gons on the boundary, then we obtain the boundary sequence 2^h, i.e. a simple h-gon. Hence, $h = 7$. On the other side of the structure, we do the same analysis and obtain the second ($\{5, 7\}$, 3)-sphere that is $5R_3$.

In addition, the above proof shows that there is no ($\{5, 7\}$, 3)-torus that is $5R_3$. $\square$

Theorem 12.5.6 *All* $(\{5, 6\}, 3)$*-spheres that are* $5R_3$ *are Nrs. 45 and 46.*

Proof. Take a $(\{5, 6\}, 3)$-sphere, i.e. a fullerene G that is $5R_3$. A given 6-gon can be adjacent to two $(5, 3)$-polycycles E_2 or three $(5, 3)$-polycycles E_1, or no $(5, 3)$-polycycles at all. The sphere G has at least one 6-gon adjacent to 5-gons. If this 6-gon is adjacent to two $(5, 3)$-polycycles E_2, then it is Nr. 46. If it is adjacent to three $(5, 3)$-polycycles E_1, then it is Nr. 45.   □

Consider now, for $b = 8, 9, 10$, $(\{5, b\}, 3)$-maps that are $5R_3$. The set of 5-gonal faces of such maps admits a partition into $(5, 3)$-polycycles E_1 and E_2. The polycycle E_1 has two open edges and E_2 has three open edges. If we consider the graph formed by all those polycycles, then it has 2- and 3-valent vertices. This graph is not necessarily connected. The connected components of this graph (i.e. the $(b, 3)_{gen}$-polycycles formed by the b-gonal faces) are bounded by 5-gons on one or several boundaries. We will obtain some classification results of those $(b, 3)_{gen}$-polycycles only when they are $(b, 3)$-polycycles. This will allow us to obtain some examples of $(\{5, 8\}, 3)$-maps $5R_3$ and $(\{5, 10\}, 3)$-maps $5R_3$. Also, we will completely classify the $(\{5, 9\}, 3)$-spheres or tori that are $5R_3$.

It is known (see [DDS05a] and Chapter 5) that the $(b, 3)$-boundary sequence of a finite $(b, 3)$-polycycle does not characterize it, in general; however, this phenomenon will not occur for the polycycles considered in this section and we will use the $(b, 3)$-boundary sequence to denote the $(b, 3)$-polycycles used.

Take such a $(b, 3)$-polycycle, it is bounded by elementary $(5, 3)$-polycycles E_1 and E_2, which we represent, in a symbolic way, by $E_1 E_2^{n_1} \ldots E_1 E_2^{n_u}$. This *symbolic sequence* of E_1 and E_2 corresponds to the boundary sequence:

$$b(n_1, \ldots, n_u) = 22(2232)^{n_1} \ldots 22(2232)^{n_u}.$$

Theorem 12.5.7 *There is a unique* $(\{5, 8\}, 3)$*-plane* $5R_3$, $8R_4$, *represented on Figure 9.5 in Case 20.*

Proof. A 8-gon of such a plane should be adjacent to four 5-gons and so to two $(5, 3)$-polycycles E_1 and E_2. If E_1 occurs, then this 8-gon is adjacent to at least three $(5, 3)$-polycycles E_1 and E_2. So, E_2 does not occur and the neighborhood of the 8-gons are of the form:

The plane is, therefore, uniquely defined and is the one on Figure 9.5.   □

Lemma 12.5.8 *(the (8, 3)-case)*

(i) *Consider the symbolic sequence $(n_1, \ldots, n_u)$.*

- *If $n_i = 0$, 1 or 2 for some i, then the boundary sequence $b(n_1, \ldots, n_u)$ is (8, 3)-fillable if and only if $(n_1, \ldots, n_u)$ is equal to $(0, 0, 0, 0)$, $(1, 1, 1)$, or $(2, 2, 2)$, respectively.*

- *If $n_i \geq 3$ for all i, then the boundary sequence $b(n_1, \ldots, n_u)$ is (8, 3)-fillable if and only if the boundary sequence:*

$$22(23)^{x_1} \ldots 22(23)^{x_u} \text{ with } x_i = n_i - 3$$

is (8, 3)-fillable.

(ii) *For any given u, there is a finite number of symbolic sequences $(n_1, \ldots, n_u)$, such that the boundary sequence $b(n_1, \ldots, n_u)$ is (8, 3)-fillable. Up to isomorphism, the list consists, for $t \leq 11$, of:*

u	symbolic sequences	p_8
3	$(1, 1, 1)$ *and* $(2, 2, 2)$	3 *and* 6
4	$(0, 0, 0, 0)$ *and* $(3, 3, 3, 3)$	1 *and* 13
6	$(3, 4, 3, 4, 3, 4)$	24
8	$(3, 4, 3, 5, 3, 4, 3, 5)$	35
9	$(3, 4, 4, 3, 4, 4, 3, 4, 4)$	39
10	$(3, 4, 3, 5, 3, 5, 3, 4, 3, 6)$	46
11	$(3, 4, 3, 5, 4, 3, 4, 4, 3, 4, 5)$	50

(iii) *The boundary sequence $b(n_1, \ldots, n_u)$ is (8, 3)-fillable if and only if the boundary sequence $b(3, 4^{n_1}, \ldots, 3, 4^{n_u})$ is (8, 3)-fillable.*

(iv) *There is an infinity of boundary sequences of the form $b(n_1, \ldots, n_u)$ that are (8, 3)-fillable.*

Proof. (i) Clearly, if some $n_i = 0$, then the only way to close the structure is by obtaining an isolated 8-gon and its symbolic sequence is $(0, 0, 0, 0)$. So, assume $n_i \geq 1$. If $n_i = 1$ for some i, then there is a unique way of closing the structure and we obtain a triple of 8-gons associated to the symbolic sequence $(1, 1, 1)$. Hence, assume $n_i \geq 2$. If $n_i = 2$ for some i, then there is a unique way of closing the structure and we obtain six 8-gons associated to the symbolic sequence $(2, 2, 2)$. Hence, assume $n_i \geq 3$. Clearly, the set of edges of the boundary, which are incident to an 8-gonal face of a possible (8, 3)-filling, is a path or the empty set, i.e. the face cannot be incident to two different segments of the boundary. So, the boundary sequence $b(n_1, \ldots, n_u)$ admits an (8, 3)-filling if and only if the boundary sequence, which is obtained by filling all faces incident to the boundary, i.e. $22(23)^{x_1} 22(23)^{x_2} \ldots 22(23)^{x_u}$ with $x_i = n_i - 3$ also admits a (8, 3)-filling.

(ii) Using Theorem 11.0.3, we can see that the numbers p_8 and x (of 8-gonal faces and, respectively, of interior vertices of an possible $(8, 3)$-filling) are:

$$\begin{cases} p_8 = u - 3, \\ x = 2u - 8 - \sum_{i=1}^{u} x_i. \end{cases}$$

Therefore, for a fixed u, we have $x_i \le 2u - 8$ and there is a finite number of possible boundary sequences and so, a finite number of possible $(8, 3)$-fillings. The enumeration is then done by computer.

(iii) Taking the initial boundary sequence $b(3, 4^{n_1}, 3, 4^{n_2}, \dots, 3, 4^{n_u})$ and applying the transformation of the second item of (i), we obtain the boundary sequence $b(n_1, \dots, n_u)$.

(iv) Using the transformation in (iii), we can, from a given boundary sequence $b(n_1, \dots, n_u)$, obtain another one. So, we get an infinity of such boundary sequences. $\square$

From the above analysis, it seems likely that, for any $u \ge 4$, there exists at least one boundary sequence $b(n_1, \dots, n_u)$ that is $(8, 3)$-fillable.

Theorem 12.5.9 *(i) There exists a $(\{5, 8\}, 3)$-torus that is $5R_3$ and not $8R_4$.*

(ii) There exists a sequence $(F_i)_{i \ge 0}$ of $(\{5, 8\}, 3)$-spheres $5R_3$ with $1640 + 1152i$ vertices. Their symmetry is O_h if $i = 0$ and D_{4h}, otherwise.

(iii) There exist $(\{5, 8\}, 3)$-spheres $5R_3$ with the following $(v, Aut(G))$: $(56, O_h)$, $(92, T_d)$, $(164, T_d)$, $(488, O_h)$, $(3944, C_{2v})$, $(4196, T_d)$, $(6248, D_{4h})$.

Proof. (i) The $(\{5, 8\}, 3)$-torus $5R_3$ is obtained by taking the $(8, 3)$-polycycles, whose symbolic sequence is $(3, 3, 3, 3)$ and $(3, 4, 3, 5, 3, 4, 3, 5)$, and gluing them together according to the drawing below:

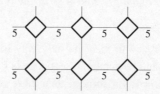

Value 3 assigned to boldfaced edges,
other non-notated edges are assigned the value 4

Clearly, we get the needed torus from this structure.

(ii) An infinity of $(\{5, 8\}, 3)$-spheres $5R_3$ is obtained by a variation of (i). This time we take the symbolic sequences $(3, 3, 3, 3)$, $(3, 4, 3, 4, 3, 4)$, and $(3, 4, 3, 5, 3, 4, 3, 5)$. For $i = 0$, we form a truncated Octahedron of symmetry O_h. For $i = 1, 2$, we form the structure according to the following drawings:

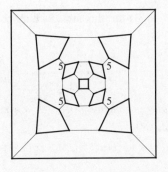 

Values 3 or 4 are assigned (for boldfaced or not) non-notated edges

For $i \geq 3$, we have an obvious generalization.

(iii) The relative wealth of above examples of $(8, 3)$-polycycles allow us to build a variety of examples of $(\{5, 8\}, 3)$-spheres that are $5R_3$. They are described by the assignment of values to the edges of a 3-valent plane graph such that every circuit $(n_1, \ldots, n_t)$ appearing on a face is $(8, 3)$-fillable. Namely, we can do the following:

1. Take Cube and put $(0, 0, 0, 0)$ on each face. It will be Nr. 58.
2. Take Tetrahedron and put $(1, 1, 1)$ on each face. It will be Nr. 59.
3. Take Tetrahedron and put $(2, 2, 2)$ on each face. It will be a $(\{5, 8\}, 3)$-sphere $5R_3$ of symmetry T_d with 164 vertices.
4. Take Cube and put $(3, 3, 3, 3)$ on each face. It will be a $(\{5, 8\}, 3)$-sphere $5R_3$ of symmetry O_h with 488 vertices.
5. Take three graphs on Figure 12.9 and assign values to their edges accordingly. We get three $(\{5, 8\}, 3)$-spheres with 3944, 4196, and 6248 vertices.

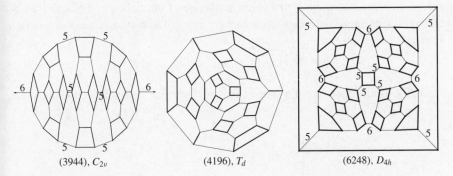

(3944), C_{2v} (4196), T_d (6248), D_{4h}

Figure 12.9 Some 3-valent spheres, which we used as skeletons of $(\{5, 8\}, 3)$-spheres $5R_3$; the boldfaced edges are assigned the value 3, while non-notated edges are assigned the value 4

See on Figure 17.16 some examples of $(\{5, 8\}, 3)$-tori that are $8R_4$ but not $5R_3$.

Lemma 12.5.10 *(the $(9, 3)$-case)*

(i) *Consider the symbolic sequence $(n_1, \ldots, n_u)$.*

- *If $n_i = 0$ for some i, then the boundary sequence $b(n_1, \ldots, n_u)$ is $(9, 3)$-fillable if and only if $(n_1, \ldots, n_u) = (0, 1, 0, 1)$.*
- *If $n_i \geq 1$ for all i, then the boundary sequence $b(n_1, \ldots, n_u)$ is $(9, 3)$-fillable if and only if the boundary sequence:*

$$2(233)^{x_1} \ldots 2(233)^{x_u} \text{ with } x_i = n_i - 3$$

is $(9, 3)$-fillable.

(ii) *For any given u, there is a finite number of symbolic sequences $(n_1, \ldots, n_u)$, such that the boundary sequence $b(n_1, \ldots, n_u)$ is $(9, 3)$-fillable. Up to isomorphism, the list consists, for $u \leq 30$, of two following sequences:*

u	symbolic sequences	p_9
4	$(0, 1, 0, 1)$	2
9	$(1, 1, 1, 1, 1, 1, 1, 1, 1)$	10

Proof. (i) If $n_i = 0$, then the boundary sequence contains at least seven consecutive 2. In order to be fillable, it should contain exactly seven consecutive 2. The only possibility is, clearly, $(n_1, \ldots, n_u) = (0, 1, 0, 1)$.

If $n_i \geq 1$, then, as in the case of $(8, 3)$-polycycles, a face, which is incident to the boundary, is incident on only one segment of edges. Hence, there is a unique way of adding 9-gons on the boundary, so as to form a ring. The boundary sequence of this filling is:

$$2(233)^{x_1} 2(233)^{x_2} \ldots 2(233)^{x_u} \text{ with } x_i = n_i - 1.$$

(ii) By using Euler formula (see 11.0.3), we get that the numbers p_9 and x (of 9-gonal faces and interior vertices of an possible $(9, 3)$-filling) are:

$$\begin{cases} p_9 = \dfrac{1}{3}\left(u - 6 - \sum_i x_i\right), \\[2mm] x = \dfrac{2}{3}\left(u - 9 - 4\sum_i x_i\right). \end{cases}$$

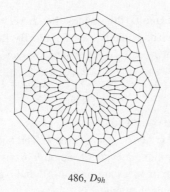

486, D_{9h}

Figure 12.10 The only $(\{5, 9\}, 3)$-sphere that is $5R_3$

So, for a fixed u, we have $x_i \leq \frac{u-9}{4}$ and there is a finite number of possible boundary sequences, i.e. a finite number of possible $(9, 3)$-fillings. The enumeration is then done by computer.   □

It seems likely that the only symbolic sequences $(n_1, \ldots, n_u)$, such that $b(n_1, \ldots, n_u)$ is $(9, 3)$-fillable, are: $(0, 1, 0, 1)$ and $(1, 1, 1, 1, 1, 1, 1, 1, 1)$.

Theorem 12.5.11 *(i) The only $(\{5, 9\}, 3)$-sphere that is $5R_3$, is the one with* 486 *vertices and symmetry D_{9h}; it is obtained by taking $Prism_9$ and assigning the values* 1 *to the edges, which are incident to the 9-gons, and* 0, *otherwise (see Figure 12.10).*
(ii) There is no $(\{5, 9\}, 3)$-torus that is $5R_3$.

Proof. If the $(9, 3)$-polycycle with boundary sequence $b(0, 1, 0, 1)$ appears in the decomposition of the set of 9-gonal faces, then we are done. This is so, since an edge of value 0 can belong only to a $(9, 3)$-polycycle with boundary sequence $b(0, 1, 0, 1)$. Hence, a path of such faces appear. By considering the adjacent 9-gons, we see that the structure should close and obtain the announced graph.

(i) Take such a sphere G and consider the graph $E_1(G)$, whose vertex-set consists of the $(5, 3)$-polycycles E_1 of G with two vertices being adjacent if they are linked by a sequence of $(5, 3)$-polycycles E_2. The graph $E_1(G)$ can be considered as a sphere, except that it is not necessarily connected, i.e. some "faces" of $E_1(G)$ are, possibly, bounded by several cycles. Denote by $C_1, \ldots, C_t$ the connected components of $E_1(G)$ that are plane graphs in the original sense, and by $F_1, \ldots, F_l$ the faces of $E_1(G)$ that have several cycles. Denote by $Conn(G)$ the graph, whose vertex-set consists of both, C_i and F_j, and whose edge, connecting C_i and F_j, corresponds to a cycle of C_i belonging to F_j. Since G is a sphere, $Conn(G)$ is a tree. Hence, it has at least one vertex of degree at most 1.

If the vertex has degree 0, this means that $E_1(G)$ is connected. Denote by p_i the number of faces of gonality i of $E_1(G)$. By Lemma 12.5.10 and Euler formula, only p_4, p_9, or p_i with $i \geq 30$ can be non-zero. So, Euler formula (1.1):

$$2p_4 - 3p_9 - \sum_{i \geq 30}(i - 6)p_i = 12$$

implies $p_4 > 0$ and the conclusion.

Assume that a vertex, say C_i, has degree 1. The connected component C_i is incident to the face F_j along a cycle of length h. Consider now the plane graph formed by C_i only. So, its faces are only 4-, 9- and i-gons with $i \geq 30$ and the h-gon. The Euler formula then reads:

$$(6 - h) + 2p_4 - 3p_9 - \sum_{i \geq 30}(i - 6)p_i = 12.$$

This implies $p_4 > 0$ and the conclusion.

(ii) The proof for the torus uses the same principle, as for spheres, but with additional complications. First, we cannot exclude from the beginning that several boundary sequences can be filled by some $(9, 3)_{gen}$-polycycles, as shown below.

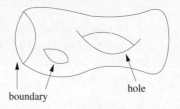

boundary   hole

But, if this happens, then we have two cycles that are not homologous to 0. Then we consider the graph $Conn(G)$, whose vertex-set consists of connected components $C_1, \ldots, C_t$ of $E_1(G)$ and of "faces" $F_1, \ldots, F_s$ having several cycles. The graph $Conn(G)$ has to be a tree, since a cycle would imply that the dimension of $H_1(G)$ is greater than 2. So, the proof for spheres works just as well, and it cannot be a torus.

Suppose now, that one of the maps $E_1(G)$ is a torus. Then the graph $Conn(G)$ is a tree and the proof for the spheres applies. So, this is impossible.

From now on, all connected components of the map $E_1(G)$ are plane graphs and the set of 9-gons is partitioned into $(9, 3)_{gen}$-polycycles with one or more boundaries. The graph $Conn(G)$ is no longer a tree and the fact that it is a torus is encapsulated in the cycles of $Conn(G)$.

Denote by $DE(G)$ the set of directed edges of $Conn(G)$. Denote by $V(G)$ the vector space with canonical basis $(e_d)_{d \in DE(G)}$. For every cycle c of $Conn(G)$, choose an orientation of it and denote by $f(c)$ its representation in $V(G)$. Denote by $H(G)$ the vector space of $V(G)$, generated by all cycles c of $Conn(G)$. It is easy to see that the dimension of the homology group $H_1(G)$ of our map G is equal to $2dim(H(G))$. Since our map is a torus, it holds $dim(H(G)) = 1$. So, there is a single cycle in $Conn(G)$.

Suppose that $Conn(G)$ contains a vertex of degree 3 (either a component C_i, or a face F_j), then there is a vertex of degree 1, which should be a component $C_{i'}$ and so we reach a contradiction by the same method as in the proof for spheres. So, vertices of $Conn(G)$ are of degree 2. This means that $Conn(G)$ is of the form:

$$\ldots - C_1 - F_1 - C_2 - \ldots - C_t - F_t - C_1 - \ldots$$

for some t.

Every connected component C_i is incident to two faces F_i and F_{i+1} (mod t) along cycles of length l_i and k_i, which are the numbers of $(5, 3)$-polycycles E_1 in those cycles. The Euler formula for the plane graph C_i reads:

$$(6 - l_i) + (6 - k_i) + 2p_4 - 3p_9 - \sum_{i \geq 30}(i - 6)p_i = 12,$$

or, in other terms:

$$2p_4 = l_i + k_i + 3p_9 + \sum_{i \geq 30}(i - 6)p_i.$$

If $p_4 > 0$, then we have the pattern $(0, 1, 0, 1)$ and a contradiction is reached. So, $p_4 = 0$. This implies $l_i = k_i = 0$ and also $p_i = 0$. It means that C_i are just rings of $(5, 3)$-polycycles E_2.

Now consider the $(9, 3)_{gen}$-polycycles F_i with two boundary sequences of the form $(2223)^{h_i}$ and $(2223)^{g_i}$. As for the case of $(9, 3)$-polycycles, there is a unique way of filling the faces on the boundaries. But the presence of two boundaries creates several complications, which were not present in the case of $(9, 3)$-polycycles.

We use the same strategy, as for $(9, 3)$-polycycles, whose boundary is of the form $b(\ldots)$: we fill all faces, which are adjacent to 5-gons. The first case is when the added 9-gonal faces do not share an edge in common. In that case, the $(9, 3)_{gen}$-polycycle, obtained by filling previous boundaries, has two boundaries of the form $(323)^{h_i}$ and $(323)^{g_i}$. Denote by p_9 the number of interior 9-gons and by x the number of interior vertices. Application of Theorem 11.0.3 yields:

$$\begin{cases} p_9 = \dfrac{v_2 - v_3}{3} = -\dfrac{h_i + k_i}{3}, \\ x = \dfrac{2v_2 - 5v_3}{3} = -\dfrac{8}{3}(h_i + k_i), \end{cases}$$

which are both strictly negative, an impossibility.

The other case is when the added 9-gons share some edges in common. The possibilities are, locally, the following:

Let p be the number of appearances of such common sequence of edges. This means that the problem of filling the $(9, 3)_{gen}$-polycycles with two boundaries is reduced to the problem of filling p $(9, 3)$-polycycles given by their boundary sequences.

Those boundary sequences are of the form $(323)^{h_i} \alpha (323)^{k_i} \beta$ with α, β being equal to 223, 322, or 32223. By applying again Theorem 11.0.3, we obtain:

$$\begin{cases} p_9 = \dfrac{v_2 - v_3}{3} - 2, \\ x = \dfrac{2v_2 - 5v_3}{3} - 6. \end{cases}$$

Those formulas give:

$$x = -\frac{8}{3}(h_i + k_i) + u_\alpha + u_\beta - 6$$

with u_α, u_β being equal to $-\frac{1}{3}$ or $-\frac{4}{3}$, according to the value of α or β. In either case, $x < 0$ and this is impossible. So, there is no $(\{5, 9\}, 3)$-torus that is $5R_3$. □

In the case $b = 10$, the only known symbolic sequences $(n_1, \dots, n_t)$, such that the boundary sequence $b(n_1, \dots, n_t)$ is $(10, 3)$-fillable, are, up to isomorphism:

t	symbolic sequences	p_{10}
5	$(0, 0, 0, 0, 0)$	1
6	$(0, 1, 0, 1, 0, 1)$	3
9	$(0, 1, 1, 0, 1, 1, 0, 1, 1)$	6

Note that the unique sphere $(\{5, 10\}, 3)$-sphere that is $10R_j$ and $5R_3$ is Nr. 60.

Theorem 12.5.12 *At least two $(\{5, 10\}, 3)$-spheres $5R_3$ exist:*

1 A sphere with 740 vertices and symmetry I_h.
2 A sphere with 7940 vertices and symmetry I_h.

Proof. The first example is constructed using the truncated Icosahedron (the smallest $(\{5, 6\}, 3)$-sphere with isolated 5-gons): its set of edges is partitioned into $(5 - 6)$-edges and $(6 - 6)$-edges. After assigning value 0 to the first kind of edges and value 1 to the second one, all 5-gons have symbolic sequence (0^5) and all 6-gons have symbolic sequence $(0, 1, 0, 1, 0, 1)$. So, we obtain a 740-vertex sphere of symmetry I_h.

The second example is obtained by taking the following $(\{5, 6\}, 3)$-polycycle

Boldfaced edges are assigned value 0, while other edges are assigned value 1 (this graph is an isometric subgraph of truncated Icosahedron). If we take the Icosahedron and substitute every 5-valent vertex with the above $(\{5, 6\}, 3)$-polycycle and glue along the open edges of value 0, then we obtain a sphere with 5-, 6- and 9-gonal faces. The symbolic sequence of the 9-gonal faces is $(1, 0, 1)^3$. So, the structure can be filled. $\qquad\square$

Theorem 12.5.13 *There is a unique $(\{5, 10\}, 3)$-plane $5R_3$, $10R_2$ represented on Figure 9.5 in Case 21.*

Proof. Any 10-gon of such a plane is adjacent to eight 5-gons and so to four $(5, 3)$-polycycles E_1 and E_2. Clearly, we have the following 2-possibilities for coronas of 10-gons:

Type 1	$E_2(\text{10-gon})E_2 E_2(\text{10-gon})E_2$
Type 2	$E_1 E_2(\text{10-gon})^2 E_2 E_1$

If type 2 occurs, then E_1 occurs and we have an $E_1 - E_2$ adjacency. The "edge" $E_1 - E_2$ is incident to two 10-gons, necessarily of type 2. So, we have a sequence of three $(5, 3)$-polycycles E_1, which cannot occur either in type 1 or in type 2. So, type 2 does not occur. Hence, the 10-gons are in type 1 and occur in infinite sequence or in cycles. Every such chain is bounded by two identical chains. So, a cycle of 10-gons cannot occur, since, otherwise, its presence would imply that a bounded region of the plane contains an infinity of faces. So, 10-gons are organized in infinite sequences and they are represented on Figure 9.5. $\qquad\square$

Conjecture 12.5.14 *A $(\{5, 10\}, 3)$-torus that is $5R_3$ is also $10R_2$.*

13

Frank-Kasper spheres and tori

We call the *Frank-Kasper* $(\{5, b\}, 3)$-*map* any $(\{5, b\}, 3)$-map that is bR_0 (in chemistry and crystallography Frank-Kasper polyhedra are just four polyhedra, dual to all four $(\{5, 6\}, 3)$-polyhedra that are $6R_0$). Note that the only oriented $(\{a, b\}, 3)$-maps with $a \leq 4$ that is bR_0 are $Prism_b$, Cube, Tetrahedron, and $Bundle_3$.

13.1 Euler formula for $(\{a, b\}, 3)$-maps bR_0

Recall that p_b is the number of b-gonal faces of a map and denote by x_i with $i = 0, 1, 2, 3$ the number of vertices contained exactly in i a-gonal faces.

Theorem 13.1.1 *Let P be a $(\{a, b\}, 3)$-sphere or torus that is bR_0. It holds that:*

$$\begin{cases} (6 - a)x_3 + (2(a - b) + (6 - a)b)p_b = 4a \text{ on sphere,} \\ (6 - a)x_3 + (2(a - b) + (6 - a)b)p_b = 0 \quad \text{on torus.} \end{cases}$$

Proof. Clearly, $v = x_0 + x_1 + x_2 + x_3$.

By counting the number e of edges in two different ways, we get:

$$2e = 3v = bp_b + ap_a.$$

The Euler formula for an oriented map of genus g is $2 - 2g = v - e + p_a + p_b$; it can be rewritten as:

$$2 - 2g = -\frac{v}{2} + p_a + p_b.$$

Eliminating p_a in the above two equations, we obtain:

$$(6 - a)v - 2(2 - 2g)a = 2(b - a)p_b.$$

We have $x_0 = x_1 = 0$ and $x_2 = bp_b$. Hence, the above relation takes the form $(6 - a)x_3 - 2(2 - 2g)a = (2(b - a) - (6 - a)b)p_b$. $\qquad\square$

The above formula yield finiteness of the number of $(\{5, b\}, 3)$-polycycles for $b \leq 9$ but we will see in the next section that more general finiteness results hold.

13.2 The major skeleton, elementary polycycles, and classification results

A $(5, 3)$-polycycle is called 0-*elementary* if it is elementary and if its boundary sequence is of the form $223^{q_1} \ldots 223^{q_m}$. By inspection of the list of finite elementary $(5, 3)$-polycycles (see Figure 7.2), we obtain that there are exactly three 0-elementary polycycles: E_1, C_1, and C_3.

The classification of all Frank-Kasper $(\{5, b\}, 3)$-spheres can be done using the elementary polycycles decomposition exposed in Chapter 7. If G is a Frank-Kasper $(\{5, b\}, 3)$-sphere, then we remove all its b-gonal faces and obtain a $(5, 3)_{gen}$-polycycle. This $(5, 3)_{gen}$-polycycle is decomposed into elementary $(5, 3)_{gen}$-polycycles along bridges (see Chapter 7 for definition of those notions). In this chapter a *bridge* is an edge, where the two vertices are contained in different b-gonal faces.

Lemma 13.2.1 *Take a Frank-Kasper $(\{5, b\}, 3)$-map that is not snub $Prism_b$. Then the set of all bridges, together with edges incident to b-gonal faces, establish a partition of the set of 5-gonal faces into 0-elementary $(5, 3)$-polycycles.*

Proof. Suppose that a b-gonal face is not incident to any bridge. It means that this b-gonal face is bounded by two concentric rings of 5-gonal faces, i.e. that the map is snub $Prism_b$. So the $(5, 3)_{gen}$-polycycles, appearing from the partition by bridges, are $(5, 3)$-polycycles. Those $(5, 3)$-polycycles are necessarily 0-elementary. $\square$

Now, given a Frank-Kasper map, consider the map, defined by taking, as vertices, all 0-elementary $(5, 3)$-polycycles forming it. The b-gonal faces and bridges of the Frank-Kasper map correspond to faces, and, respectively, edges, of this map. Then we remove vertices of degree 2 and obtain a 3-valent map, which will be called *major skeleton*.

Theorem 13.2.2 *(i) There exist $(\{5, b\}, 3)$-tori that are bR_0, if and only if $b \geq 12$.*
(ii) $(\{5, 12\}, 3)$-tori that are $12R_0$ are also $5R_3$.

Proof. (i) In order to prove the existence of $(\{5, b\}, 3)$-tori that are bR_0, for $b \geq 12$, it suffices to give an example of a periodic $(\{5, b\}, 3)$-planes that is bR_0.

Our basic example is the graphite lattice sheet, i.e. the 3-valent tiling $\{6, 3\}$ of the plane by 6-gons. At every vertex of this tiling, we can substitute a 0-elementary $(5, 3)$-polycycles, either E_1 or C_3. If we substitute only E_1, we obtain a $(\{5, 12\}, 3)$-plane that is $12R_0$. In order to obtain a $(\{5, 13\}, 3)$-plane, we need to substitute a part of the E_1, by some C_3, such that every 6-gon is incident to exactly one C_3. It is easy to see that this is, indeed, possible; see below an example of such a choice.

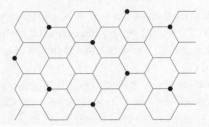

Furthermore, we can partition the set of vertices of the graphite lattice $\{6, 3\}$ into 6 sets O_i, such that every 6-gon contains exactly one vertex in the set O_i. So, by putting C_3 into vertices of sets $O_1, \ldots, O_i$, we obtain a $(\{5, 12 + i\}, 3)$-plane that is $(12 + i)R_0$.

If we insert the $(5, 3)$-polycycle C_1 into edges of the graphite lattice according to the drawing below:

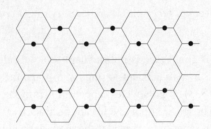

then from a $(\{5, b\}, 3)$-plane obtained by the above procedure we obtain a $(\{5, b + 5\}, 3)$-plane, which is still $(b + 5)R_0$. This procedure can, obviously, be repeated, so we get the existence result for $b \geq 12$.

Assume $b \leq 11$. Given a $(\{5, b\}, 3)$-plane, the gonality of a face in the major skeleton is equal to the number of $(5, 3)$-polycycles E_1 and C_3 to which it is incident. Clearly, there are at most five such incidences for each face. Since the major skeleton is 3-valent, we reach a contradiction by Euler formula (1.1) and (i) holds.

(ii) If M is a $(\{5, 12\}, 3)$-plane that is $12R_0$, and if F is a 12-gonal face, then F is incident to 0-elementary $(5, 3)$-polycycles. Clearly, the gonality of F in the major skeleton is equal to the number of 0-elementary $(5, 3)$-polycycles E_1 and C_3, in which it is contained. So this gonality is at most 6. But a 3-valent torus, whose faces have gonality at most 6, is possible only if all faces have gonality 6, i.e. if all 12-gonal faces are adjacent only to polycycles E_1. Such a structure is unique and it is $5R_3$.   □

Theorem 13.2.3 *(i) For $b \leq 11$, the number of Frank-Kasper ($\{5, b\}$, 3)-polyhedra is finite; they are (besides Dodecahedron):*

- *for $b = 6$, three classical ones,*
- *for $b = 7$, snub $Prism_7$,*
- *for $b = 8$, snub $Prism_8$ and strictly face-regular Nr. 58,*
- *for $b = 9$, 10, and 11, snub $Prism_b$ and the ones indicated on Figures 13.1, 13.2, and 13.3, respectively.*

(ii) For $b = 12$ (besides Dodecahedron and snub $Prism_{12}$), three sporadic spheres and one infinite series $(FK_i)_{i \geq 0}$ with $104 + 56i$ vertices (the symmetry is O_h if $i = 0$, D_{4d} if i is odd, and D_{4h} otherwise) indicated on Figure 13.4.

Proof. Take a ($\{5, b\}$, 3)-sphere G for $b \leq 11$. Then every face of the major skeleton is incident to at most five vertices corresponding to E_1 or C_3. So, the major skeleton is a 3-valent sphere with faces of gonality at most 5. There is a finite number of possibilities, which can be dealt with by computer and, hence, we have (i).

If $b = 12$, then, by the same reasoning, the gonality of faces of the major skeleton is at most 6. Since it is a plane graph, there exists a face of gonality at most 5. Such a face is incident to a vertex v, corresponding to a (5, 3)-polycycle C_3 or C_1. So, the original ($\{5, 12\}$, 3)-sphere contains the pattern below:

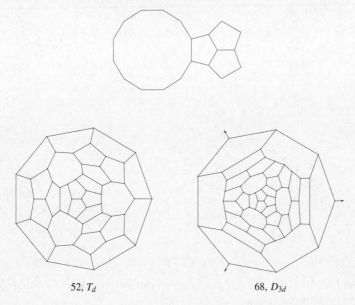

52, T_d 68, D_{3d}

Figure 13.1 All face-regular ($\{5, 9\}$, 3)-spheres that are $9R_0$, besides Dodecahedron and snub $Prism_9$

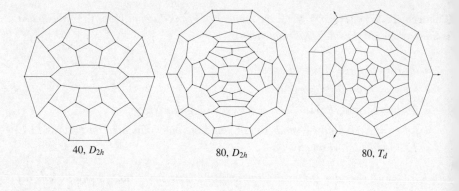

40, D_{2h}　　　　　　80, D_{2h}　　　　　　80, T_d

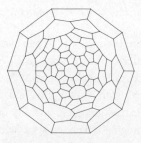

140, I_h (also $5R_3$, Nr. 60)

Figure 13.2 All face-regular ({5, 10}, 3)-spheres that are $10R_0$, besides Dodecahedron and snub $Prism_{10}$

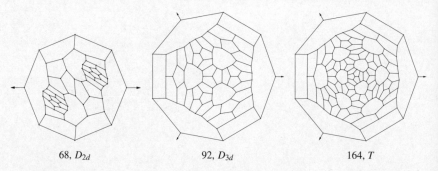

68, D_{2d}　　　　　　92, D_{3d}　　　　　　164, T

Figure 13.3 All face-regular ({5, 11}, 3)-spheres that are $11R_0$, besides Dodecahedron and snub $Prism_{11}$

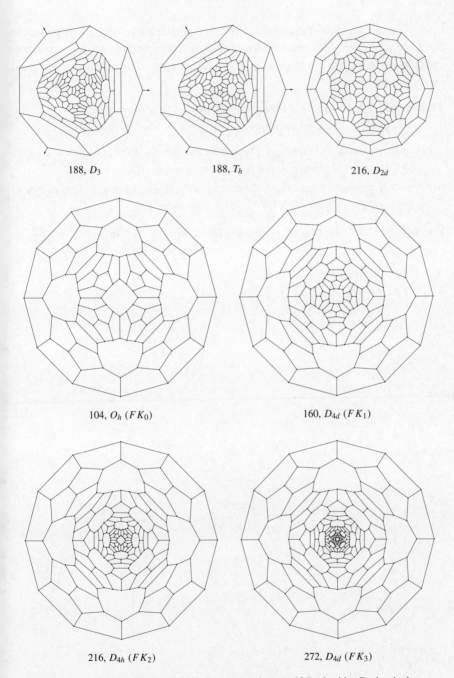

188, D_3

188, T_h

216, D_{2d}

104, O_h (FK_0)

160, D_{4d} (FK_1)

216, D_{4h} (FK_2)

272, D_{4d} (FK_3)

Figure 13.4 All face-regular $(\{5, 12\}, 3)$-spheres that are $12R_0$, besides Dodecahedron and snub $Prism_{12}$: 3 sporadic cases and the series FK_i, illustrated here for $0 \leq i \leq 3$

We then run the enumeration procedure by taking the above $(\{5, 12\}, 3)$-polycycle as the starting point. This enumeration procedure, after creating three spheres, indicated above, goes into a infinite loop, which creates the infinite sequence of maps; hence, (ii) follows. $\qquad\square$

Theorem 13.2.4 *All* $(\{5, b\}, 3)$-*spheres with* $p_b = 2$ *have* $4b$ *vertices.*
   Besides snub $Prism_b$, *all such spheres have* $b \equiv 0$ *(mod 5); they are:*

(i) *A sphere formed by splitting* $\frac{b}{5}$ *Dodecahedra with one edge split into two edges and gluing them together; it is* bR_0 *and has symmetry* $D_{\frac{b}{5}h}$.

(ii) *A sphere formed by splitting* $\frac{b}{5}$ *Dodecahedra with one edge split into two half-edges and gluing them; it is* $bR_{\frac{b}{5}}$ *and has symmetry* $D_{\frac{b}{5}h}$.

Conjecture 13.2.5 *There is an infinity of* $(\{5, b\}, 3)$-*spheres* bR_0 *for any* $b > 12$.

14

Spheres and tori that are bR_1

We consider here $(\{5, b\}, 3)$-spheres that are bR_1. By Theorem 14.1.2, there are no $(\{4, b\}, 3)$-maps bR_1.

Finiteness of the number of $(\{5, b\}, 3)$-spheres bR_1 is proved for $b \leq 9$ in Theorem 14.2.2 with the complete list of such maps shown on Figures 10.4, 14.1, 14.2, 14.3, and 14.4. Infiniteness is expected for $b \geq 10$.

14.1 Euler formula for $(\{a, b\}, 3)$-maps bR_1

Recall that p_b is the number of b-gonal faces of a $(\{a, b\}, 3)$-map and denote by x_i with $i = 0, 1, 2, 3$ the number of vertices contained exactly in i a-gonal faces.

Theorem 14.1.1 *Let P be a $(\{a, b\}, 3)$-sphere or torus that is bR_1, i.e. b-gons are organized into isolated pairs, then p_b is even and it holds that:*

$$\begin{cases} (6 - a)x_3 + (2(a - b) + (6 - a)(b - 1))p_b = 4a & \text{on sphere,} \\ (6 - a)x_3 + (2(a - b) + (6 - a)(b - 1))p_b = 0 & \text{on torus.} \end{cases}$$

Proof. Clearly, $v = x_0 + x_1 + x_2 + x_3$.

By counting the number e of edges in two different ways, we get:

$$2e = 3v = bp_b + ap_a.$$

The Euler formula for an oriented map of genus g is $2 - 2g = v - e + p_b + p_a$; it can be rewritten as:

$$2 - 2g = -\frac{v}{2} + p_b + p_a.$$

Eliminating p_a in above two equations, we obtain:

$$(6 - a)v - 2(2 - 2g)a = 2(b - a)p_b.$$

We have $x_0 = 0$, $x_1 = p_b$ and $x_2 = (b - 2)p_b$. So we get:

$$(6 - a)x_3 - 2(2 - 2g)a = (2(b - a) - (6 - a)(b - 1))p_b,$$

i.e. the desired relation. $\qquad\qquad\square$

225

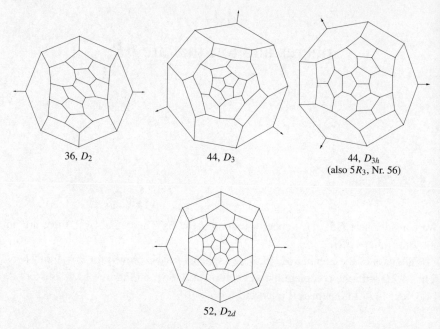

Figure 14.1 All face-regular ({5, 7}, 3)-spheres $7R_1$, besides Dodecahedron

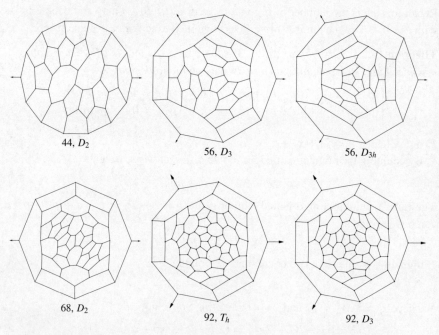

Figure 14.2 All face-regular ({5, 8}, 3)-spheres that are $8R_1$, besides Dodecahedron

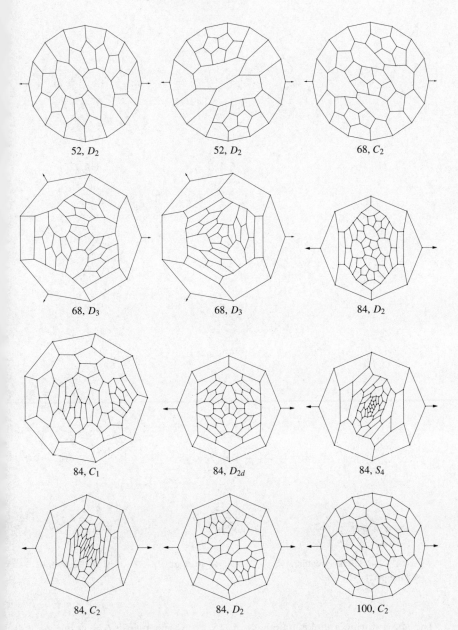

52, D_2 52, D_2 68, C_2

68, D_3 68, D_3 84, D_2

84, C_1 84, D_{2d} 84, S_4

84, C_2 84, D_2 100, C_2

Figure 14.3 All face-regular ($\{5, 9\}$, 3)-spheres that are $9R_1$, besides Dodecahedron (first part)

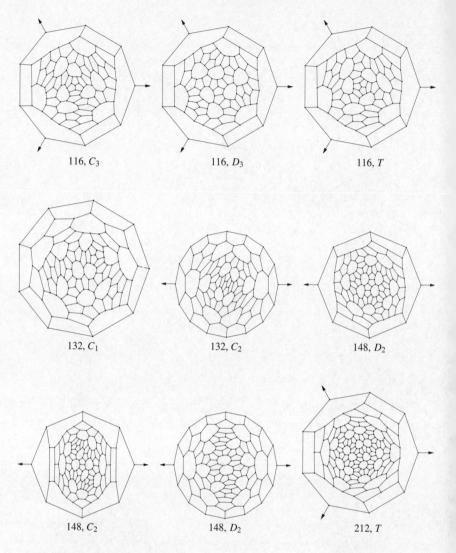

Figure 14.4 All face-regular ($\{5, 9\}$, 3)-spheres that are $9R_1$, besides Dodecahedron (second part)

The above formula yields finiteness of ($\{5, b\}$, 3)-spheres for $b \leq 8$ but we will see in the next section that we can obtain finiteness for $b = 9$ also.

Theorem 14.1.2 *There is no* ($\{4, b\}$, 3)-*sphere or torus that is* bR_1.

Proof. Take a ($\{4, b\}$, 3)-sphere that is bR_1, and two b-gonal faces that are adjacent along an edge. Those two b-gons are encircled by a circuit $F_1, \ldots, F_m$ of 4-gons.

Assume F_i is adjacent to both b-gons, then F_{i-1} and F_{i+1} are adjacent and this implies that F_{i-1} and F_{i+1} are both adjacent to another 4-gon; this 4-gon is adjacent to both b-gons, which is an impossibility since those b-gons can share only one edge.

Actually, the above analysis does not use the fact that the map is a sphere; so it holds for any map. Another, more direct proof can be obtained by using Theorem 14.1.1. $\qquad\square$

14.2 Elementary polycycles

If G is a $(\{5, b\}, 3)$-map that is bR_1, then we remove all its b-gonal faces and obtain a $(5, 3)_{gen}$-polycycle. This $(5, 3)_{gen}$-polycycle is decomposed into elementary $(5, 3)_{gen}$-polycycles along bridges (see Chapter 7 for definition of those notions). In this chapter a *bridge* is an edge, which is not contained in any b-gonal face but whose two vertices are contained into different b-gonal faces.

A $(5, 3)$-polycycle is called 1-*elementary* if it is elementary and its boundary sequence is of the form $2^{n_1} 3^{m_1} \ldots 2^{n_t} 3^{m_t}$, where each n_i is 1 or 2 and, moreover, if $n_i = 1$, then $n_{i-1} = n_{i+1} = 2$. So, every 0-elementary $(5, 3)$-polycycle is also 1-elementary. We obtain that the list of 1-elementary $(5, 3)$-polycycles, which are not 0-elementary, consists of C_2, D, and any E_{2n} with $n \geq 1$.

Theorem 14.2.1 *Given a $(\{5, b\}, 3)$-map that is bR_1, then the set of all bridges, together with edges, incident to b-gonal faces, establish a partition of the set of 5-gonal faces into 1-elementary $(5, 3)$-polycycles.*

Proof. Clearly, snub $Prism_{b'}$ with $b' \geq 2$ cannot occur as component in the decomposition into $(5, 3)_{gen}$-polycycles of the 5-gonal faces of the map. Hence, the set of 5-gonal faces is decomposed into elementary $(5, 3)$-polycycles.

In order to prove that the elementary $(5, 3)$-polycycles, which can appear, are only the 1-elementary ones, we will examine the list (see Figure 7.2) of elementary $(5, 3)$-polycycles.

An admissible $(5, 3)$-polycycle should have the pattern 22 in its boundary sequence, since, otherwise, it would be bounded by a ring of b-gons that are adjacent to at least two b-gons. This eliminate A_i, $1 \leq i \leq 5$.

A pattern $3^{h_1} 23^{h_2} 23^{h_3}$ with $h_i \geq 1$ corresponds to a b-gon with $b = 2 + h_2$; we will prove that this is not possible. Clearly, the pattern $3^{h_1} 23^{h_2} 23^{h_3} 23^{h_4}$ with $h_i \geq 1$ cannot appear, since it would imply that one of the b-gons of the pair has a vertex of degree 2, which cannot be matched by a polycycle. This eliminates B_3.

The $(5, 3)$-polycycle B_2 is not possible, since the closure of two vertices of degree two in 32323 would yield a b-gon with $b = 3$ and such structures do not exist.

Now consider the case of $(5, 3)$-polycycles E_{2n-1}; the closure of two isolated vertices of degree two would yield a $(n + 1)$-gon. But the opposite side of the $(5, 3)$-polycycle E_{2n-1} has the boundary 23^n2; so after uniting with other $(5, 3)$-polycycle it would yield a b-gon with $b > n + 1$, which is impossible. The only remaining admissible $(5, 3)$-polycycles are the 1-elementary ones. □

All 1-elementary $(5, 3)$-polycycles appear in decompositions of $(\{5, b\}, 3)$-maps that are bR_1. See below an example of such a decomposition:

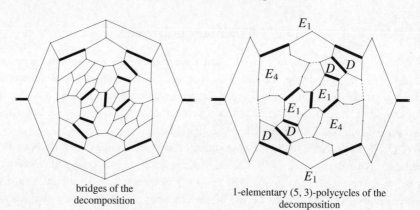

bridges of the
decomposition

1-elementary $(5, 3)$-polycycles of the
decomposition

Theorem 14.2.2 *(i)* $(\{5, b\}, 3)$*-spheres that are* bR_1 *have at most* 32, 52, 92, 212 *vertices for* $b = 6, 7, 8, 9$*, respectively.*
(ii) There are no $(\{5, b\}, 3)$*-tori that are* bR_1*, for* $b \leq 9$*.*

Proof. By Theorem 14.1.1, we obtain $x_3 + (9 - b)p_b = 20$, where x_3 is the number of vertices contained in 3 5-gons. Since the gonality of faces is at most 9, the 1-elementary $(5, 3)$-polycycles, forming its decomposition, are D, C_1, C_2, C_3, E_1, E_2, E_4, E_6, E_8, E_{10}, E_{12}. Denote by p_D, p_{C_1}, ..., the number of those polycycles. By counting the number of interior vertices, we obtain:

$$x_3 = 10p_{C_1} + 7p_{C_2} + 4p_{C_3} + p_{E_1} + \sum_{i=1}^{6} 2i p_{E_{2i}}. \quad (14.1)$$

On the other hand, we have the equations:

$$\begin{cases} e_{b-b} = \frac{1}{2}p_b = \frac{1}{2}p_D + \frac{1}{2}p_{C_2} + \sum_{i=1}^{6} p_{E_{2i}}, \\ e_{5-b} = (b-1)p_b = 3p_D + 10p_{C_1} + 10p_{C_2} + 9p_{C_3} + 6p_{E_1} + \sum_{i=1}^{6}(6+2i)p_{E_{2i}}, \\ p_5 = p_D + 10p_{C_1} + 8p_{C_2} + 6p_{C_3} + 3p_{E_1} + \sum_{i=1}^{6}(2i+2)p_{E_{2i}}. \end{cases}$$
$$(14.2)$$

We then consider the linear programming problem (see, for example, [Chv83]):

> **maximize** p_b
> **subject to** *Equations* (14.1), (14.2) *and* $x_3 + (9 - b)p_b = 20$
> $p_b \geq 0, p_5 \geq 0, x_3 \geq 0, p_D \geq 0, \ldots$

For $b = 6, 7, 8, 9$, the maximal value of p_b is 6, 8, 12, 24 giving the above upper bounds on the number of vertices, thereby proving (i).

Actually, for $b = 9$, we can get a more simple proof. Combining the first two equations of (14.2), we get:

$$e_{5-9} - 6e_{9-9} = 5p_9 = 7p_{C_1} + 10p_{C_2} + 9p_{C_3} + 6p_{E_1} + \sum_{i=1}^{6} 2ip_{E_{2i}}. \qquad (14.3)$$

It is clear, that the linear programming problem:

> **maximize** $\sum_i a_i x_i$
> **subject to** $\sum_i b_i x_i = b$,
> $x_i \geq 0$ with $a_i, b_i > 0$

has the solution $b \max_i \frac{a_i}{b_i}$. Using equation (14.3) for p_9, equation (14.1) for x_3 and the relation $x_3 = 20$, we get $p_9 \leq 20\frac{6}{5} = 24$.

Let us consider (ii). First, we obtain, by Theorem 14.1.1, $x_3 = 0$, then $p_{C_1} = p_{C_2} = p_{E_1} = p_{E_{2i}} = 0$. Subsequently, we obtain the relations:

$$e_{9-9} = \frac{1}{2}p_9 = \frac{1}{2}p_D, \quad \text{and} \quad e_{5-9} = 8p_9 = 3p_D;$$

so, $p_9 = p_D = 0$. Hence, there are no $(\{5, 9\}, 3)$-tori that are $9R_1$. If $b \leq 8$, then the proof comes directly from application of Theorem 14.1.1(ii). $\square$

The enumeration of $(\{5, b\}, 3)$-spheres bR_1 for $b \leq 9$ was done by computer using an exhaustive search scheme. The result are presented on Figures 10.4, 14.1, 14.2, 14.3, and 14.4. It turns out that the upper bound in Theorem 14.2.2 is always attained. The $(\{5, b\}, 3)$-spheres, realizing the upper bound, have their 5-gons organized into elementary $(5, 3)$-polycycles D and E_1. The program, enumerating the $(\{5, b\}, 3)$-spheres, terminates for $b \leq 9$, which is another proof that there is a finite number of $(\{5, b\}, 3)$-spheres bR_1 for $b \leq 9$.

Recall that a special perfect matching of a 6-valent tiling by 3-gons is a perfect matching such that every vertex is contained in exactly one vertex, and every vertex is contained in exactly one 3-gon whose face opposite to it belongs to the perfect matching.

Theorem 14.2.3 *(i) The elementary* $(5, 3)$*-polycycles, appearing in the decomposition of a* $(\{5, 10\}, 3)$*-torus that is* $10R_1$*, are* D *and* E_1*.*

(ii) Every $(\{5, 10\}, 3)$*-torus that is* $10R_1$*, corresponds, in a following way, to a* 6*-valent tiling of the torus by* 3*-gons, together with a special perfect matching:*

- 10*-gonal faces correspond to vertices,*
- $(5, 3)$*-polycycles* D *and* E_1 *correspond to triangular faces,*
- *bridges between* $(5, 3)$*-polycycles make the first part of the edge-set, while edges between two* 10*-gons make the second part. This second part of the edge-set form a special perfect matching.*

Proof. A priori, the $(5, 3)$-polycycles, appearing in the decomposition of such a $(\{5, 10\}, 3)$-torus are D, C_1, C_2, C_3, E_1, and E_{2i}, with $1 \leq i \leq 7$. We use the same notation as in the preceding theorem. By Theorem 14.1.1, we have $x_3 = p_{10}$; so in the same way as in the above theorem we obtain:

$$\begin{cases} p_{10} = x_3 = 10p_{C_1} + 7p_{C_2} + 4p_{C_3} + p_{E_1} + \sum_{i=1}^{7} 2ip_{E_{2i}}, \\ e_{10-10} = \frac{1}{2}p_{10} = \frac{1}{2}p_D + \frac{1}{2}p_{C_2} + \sum_{i=1}^{7} p_{E_{2i}}, \\ e_{5-10} = 9p_{10} = 3p_D + 10p_{C_1} + 10p_{C_2} + 9p_{C_3} + 6p_{E_1} + \sum_{i=1}^{7}(6+2i)p_{E_{2i}}. \end{cases}$$

Subtracting those equations, we obtain successively:

$$\begin{cases} e_{5-10} - 6e_{10-10} = 6p_{10} = 10p_{C_1} + 7p_{C_2} + 9p_{C_3} + 6p_{E_1} + \sum_{i=1}^{7} 2ip_{E_{2i}}, \\ e_{5-10} - 6e_{10-10} - 6x_3 = 0 = -50p_{C_1} - 35p_{C_2} - 15p_{C_3} - \sum_{i=1}^{7} 10ip_{E_{2i}}. \end{cases}$$

The second equation implies $p_{C_1} = p_{C_2} = p_{C_3} = p_{E_{2i}} = 0$; hence, (i) follows.

The above equalities yield $p_{10} = x_3 = p_{E_1} = p_D$. We say that a $(5, 3)$-polycycle is *incident* to a 10-gonal face if it shares a sequence of edges with it. A $(5, 3)$-polycycle D or E_1 is incident to exactly three 10-gonal faces. Hence, every 10-gonal face is incident to exactly three polycycles D and three polycycles E_1. Consider now the torus, whose vertices are 10-gonal faces and faces are $(5, 3)$-polycycles D or E_1. Its edges are bridges, which are common to two adjacent elementary $(5, 3)$-polycycles, or the edges linking two polycycles D and separating two 10-gonal faces.

Clearly, this torus is 6-valent and the set of edges between two adjacent 10-gons define a special perfect matching. From this special perfect matching, we can get the position of polycycles D and E_1 and get the original $(\{5, 10\}, 3)$-torus. So, (ii) follows. $\qquad\square$

Remark 14.2.4 *There is a large number of possibilities for special perfect matchings. See on picture below two such possibilities.*

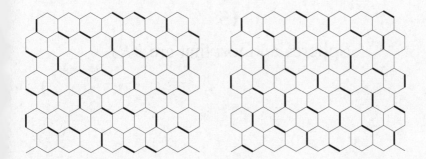

Note also that Theorem 14.2.3 combined with Theorem 9.3.4 implies that there exist a one-to-one correspondence between $(\{5, 10\}, 3)$-torus $10R_1$ and $(\{4, 8\}, 3)$-torus $4R_1$ and $8R_5$.

Theorem 14.2.5 *For any $b \geq 10$, there exist a $(\{5, b\}, 3)$-torus that is bR_1.*

Proof. Such tori can be obtained as quotients of a $(\{5, b\}, 3)$-plane. We will get again such a plane from the graphite lattice $\{6, 3\}$ with added structure on it. On any vertex of the structure below, which is incident to an boldfaced edge, we put a $(5, 3)$-polycycle D, while on other vertices we put a $(5, 3)$-polycycle E_1. The obtained $(\{5, b\}, 3)$-plane is bR_1.

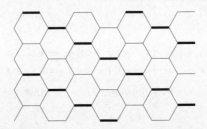

Note that the $(5, 3)$-polycycles D are adjacent between themselves by pairs; every pair of $(5, 3)$-polycycles D can be substituted by a $(5, 3)$-polycycle E_{2n} with $n \geq 1$. It is easy to see that the structure is a $(\{5, 10 + n\}, 3)$-plane that is $(10 + n)R_1$. $\square$

Note also that there are many other possibilities for creating $(\{5, b\}, 3)$-torus bR_1 (for example, substituting E_1 by C_3 or inserting $(5, 3)$-polycycles C_1).

Conjecture 14.2.6 *For any $b \geq 10$, there exist an infinity of $(\{5, b\}, 3)$-spheres that are bR_1.*

15

Spheres and tori that are bR_2

We consider here ($\{a, b\}$, 3)-spheres and tori that are bR_2 (i.e. b-gons are organized into disjoint simple cycles). After some general results, the toroidal case is treated. Then, in Section 15.3, we consider the special case of ($\{a, b\}$, 3)-spheres with a cycle of b-gons. In this chapter, we follow [DGr02, DDS05a, DuDe06].

15.1 ($\{a, b\}$, 3)-maps bR_2

Recall that p_a and p_b denote, respectively, the number of a- and b-gons in a ($\{a, b\}$, 3)-map bR_2.

Theorem 15.1.1 *Let G be a ($\{a, b\}$, 3)-sphere or torus that is bR_2. Then it holds that:*

$$\begin{cases} (6 - a)x + (4 - (4 - a)(4 - b))p_b = 4a & \text{on sphere,} \\ (6 - a)x + (4 - (4 - a)(4 - b))p_b = 0 & \text{on torus,} \end{cases} \tag{15.1}$$

where x is the number of vertices contained either in three a-gonal faces, or in three b-gonal faces.

Proof. Let x_0, x_1, x_2, x_3 be the number of vertices contained in exactly zero, one, two or three a-gonal faces. So, $x = x_0 + x_3$.

There are x_0 cycles of length 3 of b-gonal faces, which correspond to $3x_0$ b-gonal faces.

We have $x_1 = (b - 4)(p_b - 3x_0) + (b - 3)3x_0$ and $x_2 = 2(p_b - 3x_0) + 3x_0$. By counting the number e of edges in two different ways, we get:

$$2e = 3(x_0 + x_1 + x_2 + x_3) = ap_a + bp_b .$$

Euler formula $2 - 2g = (x_0 + x_1 + x_2 + x_3) - e + p_a + p_b$, where g is the genus of the oriented map, can be rewritten as:

$$2 - 2g = -\frac{1}{2}(x_0 + x_1 + x_2 + x_3) + p_a + p_b.$$

234

Eliminating p_a in above two equations, we obtain:

$$(6 - a)(x_0 + x_1 + x_2 + x_3) - 4(1 - g)a = 2(b - p)p_b,$$

which, after substitution of x_1 and x_2, yields the formula of the theorem. $\square$

Corollary 15.1.2 *For $(a, b) = (3, b)$, $(4, b)$, $(5, 6)$, $(5, 7)$, the number of $(\{a, b\}, 3)$-spheres bR_2 is finite.*

Proof. For those cases, Equation (15.1) takes the form $bp_b + 3x = 12$, $4p_b + 2x = 16$, $2p_6 + x = 20$, $p_7 + x = 20$. Hence, p_b is bounded by some constant, which implies that p_a (and so, the number v of vertices) is bounded as well. $\square$

Euler formula implies also, that the number e_{a-a} of edges of adjacency of two a-gons in any $(\{a, b\}, 3)$-sphere bR_2 is:

$$\frac{6a + p_b((a - 3)(b - 4) - 6)}{6 - a}.$$

So, this number is a constant 24, 30 for $(a, b) = (4, 10)$, $(5, 7)$, respectively, and the number of such spheres is finite for $(a, b) = (4, b < 10)$, $(5, b < 7)$.

Theorem 15.1.3 *For any $b \geq 8$, there exist an infinity of $(\{5, b\}, 3)$-spheres that are bR_2.*

Proof. Take two Dodecahedra and remove one vertex from each of them. By merging three pending edges, we obtain a $(\{5, 8\}, 3)$-sphere $8R_2$ with three 8-gonal faces and 18 5-gonal faces partitioned into two $(5, 3)$-polycycles A_3. Conclusion follows from Theorem 11.0.2.

In order to prove the result for $b > 8$, we need to find an initial graph that is bR_2. For any $n \geq 0$, take three $(5, 3)$-polycycles E_{2n} (where E_0 is the gluing of two $(5, 3)$-polycycles D, while, for $n \geq 1$, E_n can be seen on Figure 7.2) and glue them along their open edges. The resulting graph has two boundary sequences of the form $(23^{2+n})^3$; each could be filled by three $(6 + n)$-gons by adding one vertex. Those two vertices can be truncated and replaced by $(5, 3)$-polycycle A_3. The resulting $(\{5, 9 + n\}, 3)$-sphere is $(9 + n)R_2$ and we can apply Theorem 11.0.2, in order to get infiniteness. $\square$

Theorem 15.1.4 *All $(\{a, b\}, 3)$-spheres that are bR_2, with $a = 3, 4$ are (besides Tetrahedron and Cube):*

1 *The $(\{4, b\}, 3)$-spheres with a 4-cycle of b-gons with $p_4 = 2(b - 3)$ and $p_b = 4$, shown, for first values $b = 5, 6, 7, 8$, on Figure 15.1.*

2 *$Prism_3$ and the following spheres:*

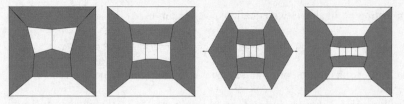

Figure 15.1 Cases $b = 5, 6, 7, 8$ of the infinite series of $(\{4, b\}, 3)$-spheres with a 4-cycle of b-gons

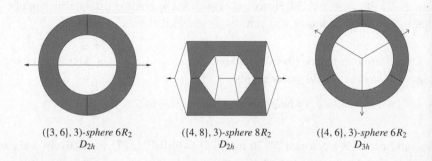

$(\{3, 6\}, 3)$-*sphere* $6R_2$	$(\{4, 8\}, 3)$-*sphere* $8R_2$	$(\{4, 6\}, 3)$-*sphere* $6R_2$
D_{2h}	D_{2h}	D_{3h}

Proof. Equation (15.1) of Theorem 15.1.1 for $(\{4, b\}, 3)$-spheres bR_2 is $4p_b + 2x = 16$. Its possible solutions are $(p_b, x) = (4, 0), (3, 2), (2, 4), (1, 6)$, and $(0, 8)$. The solution $(0, 8)$ corresponds to Cube, while $(1, 6)$ is, clearly, impossible.

On the other hand, all $(4, 3)$-polycycles have been classified. Let G be a $(\{4, b\}, 3)$-sphere bR_2 with at least one b-gon. Then G contains a $(4, 3)$-polycycle P bounded by b-gonal faces.

If P is $P_2 \times P_k$ with $k \geq 2$, then it is adjacent to four b-gonal faces. Hence, $p_b \geq 4$. By the above reasoning, this implies $p_b = 4$, and so there is just one cycle and no interior vertices for the other $(4, 3)$-polycycle. So, this other polycycle is of the form $P_2 \times P_{k'}$. Only possibility is that $b = k + 3$ and $k' = k$.

If P is $\{4, 3\} - v$, then $p_b \geq 3$, and $x \geq 1$. The only possibility is $p_b = 3$ and $x = 2$. Hence, there is just one cycle and the other polycycle contains exactly one interior vertex. So, the sphere is the unique $(\{4, 6\}, 3)$-sphere $6R_2$.

If P is $\{4, 3\} - e$, then $p_b \geq 2$ and $x \geq 2$. If $p_b = 3$, we would get a cycle of length one, which is impossible; hence, $p_b = 2$, $x = 4$. The only possibility for filling with 4-gons is $b = 8$ and we obtain a $(\{4, 8\}, 3)$-sphere with a 2-cycle (i.e. two 2-gons adjacent on disjoint edge) of 8-gons.

For $(\{3, b\}, 3)$-spheres bR_2, Equation (15.1) can be rewritten as $bp_b + 3x = 12$. On the other hand, there exist only two $(3, 3)$-polycycles, that can be part of such a sphere: the 3-gon and a pair of adjacent 3-gons.

If this polycycle is a 3-gon, then we obtain $p_b \geq 3$. $12 \geq bp_b \geq 3b$, which implies $b \leq 4$. The only possibility is $b = 4$ and the sphere is $Prism_3$.

If this polycycle is a pair of adjacent 3-gons, then $p_b \geq 2$, $12 \geq bp_b \geq 2b$, which implies $b \leq 6$. Values lower than 6, are not possible; hence, we obtain a $(\{3, 6\}, 3)$-sphere with a 2-cycle of 6-gons. $\hfill\square$

15.2 $(\{5, b\}, 3)$-tori bR_2

Theorem 15.2.1 *A $(\{5, 8\}, 3)$-torus is $8R_2$ if and only if it is $5R_2$.*

Proof. Take a $(\{5, 8\}, 3)$-torus and assume that it is $8R_2$. We obtain, by Theorem 15.1.1, the relation $x_0 + x_3 = 0$, i.e. $x_0 = x_3 = 0$.

Also, by Euler formula (1.1), we obtain $p_5 = 2p_8$. The total number e of edges is equal to:

$$\frac{1}{2}(5p_5 + 8p_8) = 9p_8 \ .$$

Moreover, the property $8R_2$ implies $e_{8-8} = p_8$, $e_{5-8} = 6p_8$ and $e_{5-5} = 2p_8$.

Denote by $p_{5,i}$ the number of 5-gonal faces, which are adjacent to exactly i 5-gonal faces. By direct counting, it follows:

$$p_5 = \sum_{i \geq 0} p_{5,i} \quad \text{and} \quad 2p_8 = e_{5-5} = \frac{1}{2}\sum_{i \geq 0} i p_{5,i} \ .$$

Subtracting those two equations, we obtain $0 = \sum_{i \geq 0}(\frac{i}{2} - 1)p_{5,i}$. Suppose that $p_{5,i} \neq 0$ for $i > 2$.

Then at least one vertex is contained in exactly three 5-gonal faces, which is impossible. So, the equation reduces to:

$$0 = -p_{5,0} - \frac{1}{2}p_{5,1} \ .$$

This implies $p_{5,i} = 0$ for $i < 2$. So, our $(\{5, 8\}, 3)$-torus is strictly face-regular $5R_i$ with $i = 2$.

If the map is $5R_2$, then, again, Theorem 15.1.1 implies $x_0 + x_3 = 0$, i.e. $x_0 = x_3 = 0$.

Also, by the same computation, we get $e_{8-8} = p_8$, $e_{5-8} = 6p_8$, and $e_{5-5} = 2p_8$.

Take an 8-gonal face F; since $x_0 = x_3 = 0$, the corona sequence of F does not contain the pattern 88. Assume that F contains the pattern 858; by the property $5R_2$, the 5-gonal faces should have the corona 88855. This implies that $x_0 + x_3 > 0$.

So, the corona 858 is impossible and this implies that 8-gonal faces are adjacent to at most two 8-gonal faces. Since, on average, 8-gonal faces are adjacent to two 8-gonal faces, this implies property $8R_2$. $\hfill\square$

Theorem 15.2.2 *A $(\{5, b\}, 3)$-torus bR_2 exists if and only if $b \geq 8$.*

Proof. Consider the graphite lattice $\{6, 3\}$ and put a 5-gon in every vertex of it. The obtained structure is, clearly, a $(\{5, 8\}, 3)$-plane, which is $8R_2$. In order to obtain $(\{5, b\}, 3)$-planes with $b \geq 8$, we need to modify the structure. The 5-gons can be organized into pairs of adjacent ones. Every such pair, which is highlighted in the diagram below, can be changed into a $(5, 3)$-polycycle E_{2n} with $n \geq 1$.

We obtain a $(\{5, 8 + n\}, 3)$-plane that is $(8 + n)R_2$. All above planes are periodic; hence, by taking the quotient (by a translation subgroup of their automorphism group), we obtain $(\{5, b\}, 3)$-tori that are bR_2.

In order to prove the non-existence of $(\{5, b\}, 3)$-torus with $b \leq 7$, it is sufficient to use Theorem 15.1.1; it yields $p_b = 0$, which is impossible. $\square$

See on Figures 15.2, 15.3 some $(\{5, b\}, 3)$-tori that are bR_2 for $b = 9, 10$.

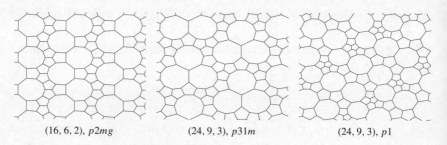

          $(16, 6, 2)$, $p2mg$ $(24, 9, 3)$, $p31m$ $(24, 9, 3)$, $p1$

Figure 15.2 Some $(\{5, 9\}, 3)$-tori that are $9R_2$

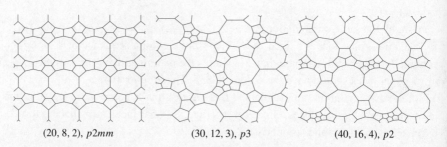

          $(20, 8, 2)$, $p2mm$ $(30, 12, 3)$, $p3$ $(40, 16, 4)$, $p2$

Figure 15.3 Some $(\{5, 10\}, 3)$-tori that are $10R_2$

15.3 $(\{a, b\}, 3)$-spheres with a cycle of b-gons

Consider $(\{a, b\}, 3)$-spheres with all b-gons organized in a single cycle of length, say, n. They form an interesting subset of the $(\{a, b\}, 3)$-spheres bR_2. In this section we drop condition $a < b$ supposing instead only $a, b \geq 3$. In fact, we suppose $b > 3$ since there is no $(\{a, 3\}, 3)$-sphere $3R_2$.

Theorem 15.3.1 *For $b \geq 4$, $a \geq 5$, a $(\{a, b\}, 3)$-sphere with a n-cycle of b-gons may exist only for the following values of the parameters n, a, and b.*

(i) If $b = 4$, then $n = a$ and the unique possibility is $Prism_a$.

(ii) If $b \geq 5$, then $a \leq 7$, and if $a \geq 6$, then $b = 5$. Besides, it holds:

1 if $a = 7$, then $n - (x + x') = 28$; $n \geq 28$. There is a unique $(\{5, 7\}, 3)$-sphere with a 28-cycle of 5-gons;

2 if $a = 6$, then there are four $(\{5, 6\}, 3)$-spheres with a 12-cycle of 5-gons (see Figure 10.8);

3 if $a = 5$, then it holds:

(a) if $b = 6$, then $n \leq 10$, and all existing spheres are presented on Figure 10.5;

(b) if $b = 7$, then $n \leq 20$ and all existing spheres are presented on Figure 15.6.

Proof. In case (i), the only way the 4-gons can arrange themselves into a cycle is by forming $Prism_n$ with $n = a$ (see, for example, the classification of $(4, 3)$- and $(4, 3)_{gen}$-polycycles in Chapter 4).

Denote by I_n, O_n $(a, 3)$-polycycles inside and outside of the n-cycle of b-gons. Denote by x, x' the number of interior vertices of polycycles I_n, O_n, respectively. Formula (15.1) can be rewritten as follows:

$$(a - 6)(x + x') = (4 - (a - 4)(b - 4))n - 4a.$$

Consider the case (ii) and assume, at first, $a \geq 7$. Since $a - 6 > 0$ and $x + x' \geq 0$, we have $4 - (a - 4)(b - 4) > 0$. For $a \geq 7$, this inequality holds only if $b = 5$ and $a = 7$. For these values a and b, Equation (15.1) takes the form:

$$x + x' = n - 28 .$$

So, no $(\{5, 7\}, 3)$-sphere with cycles of 5-gons exists for $n < 28$. For $n = 28$, $x + x' = 0$ and the unique sphere is presented on Figure 15.5.

Now consider the case $a = 6$. In this case, Equation (15.1) takes the form:

$$(12 - 2b)n = 24 .$$

Hence, $12 - 2b > 0$, i.e. $b < 6$. Since $b \geq 5$, we have $b = 5$ and $n = 12$. There exist four $(\{5, 6\}, 3)$-spheres with a cycle of 5-gons: see Figure 10.8 and the proof of Lemma 10.4.1.

The case $a = 5$ is the most interesting one. Now Equation (15.1) takes the form:

$$x + x' = 20 - (8 - b)n .$$ (15.2)

For $b < 8$, this equality restricts n, since $x + x' \geq 0$.   □

Remark 15.3.2

(i) *All $(\{5, 7\}, 3)$-spheres with a n-cycle of 5-gons are enumerated in [HaSo07] for $n \leq 50$; furthermore they also constructed an infinite sequence of such spheres.*

(ii) *Some infinite sequences of $(\{5, b\}, 3)$-spheres with a cycle of b-gons are constructed in [MaSo07] for $b = 10$ and $b \equiv 2, 3 \pmod 5$, $b \geq 13$. They showed the existence of $(\{5, b\}, 3)$-spheres with a cycle of b-gons for $b \equiv 0, 1, 4 \pmod 5$, $b \geq 5$, $b \neq 9$, for $b \equiv 1, 4 \pmod 5$, $b \geq 5$ and for $b \equiv 0 \pmod{10}$, $b \geq 20$.*

We now give two infinite series from [DGr02]:

1 If $t \geq 1$, define a $(\{5, 5t + 3\}, 3)$-sphere D_{2d} with a 4-cycle of $(5t + 3)$-gons by taking two identical $(5, 3)$-polycycles tC_1 as caps.

2 If $t \geq 1$, define a $(\{5, 5t + 2\}, 3)$-sphere D_{2d} by taking two identical $(5, 3)$-polycycles P as caps. If $t = 1$, then $P = A_4$ and, if $t \geq 2$, then $P = B_2 + (t - 2)C_1 + B_2$.

The first examples of those infinite series are shown below:

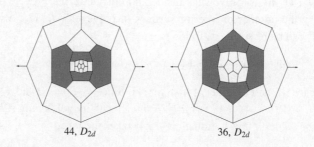

44, D_{2d} 36, D_{2d}

We conjecture that there exist an infinity of $(\{5, b\}, 3)$-spheres with a cycle of b-gons for every $b \geq 8$.

The existence of a $(\{a, b\}, 3)$-sphere with a cycle of b-gons is equivalent to that both, some boundary sequence $\mathbf{c} = 23^{c_1} 23^{c_2} \ldots 23^{c_m}$ and its b-*complement* (i.e. $\mathbf{c}' = 23^{c'_1} 23^{c'_2} \ldots 23^{c'_m}$ with $c'_i = b - 4 - c_i$), can be filled by a-gons. So, unicity of such sphere means that only one such $\mathbf{c}$ exists and it admits unique filling by a-gons. For b-*self-complementary* $\mathbf{c}$ (i.e. its b-complement coincides with $\mathbf{c}$, up to cyclic shift or reversal), the unicity of such a sphere is equivalent to unicity of filling of $\mathbf{c}$.

See below the $(\{a, b\}, 3)$-spheres with a 2-cycle of b-gons:

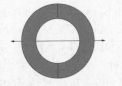 

($\{3, 6\}$, 3)-sphere D_{2h} with a   ($\{4, 8\}$, 3)-sphere D_{2h} with a   ($\{5, 10\}$, 3)-sphere D_{2h} with a
    2-cycle of 6-gons   2-cycle of 8-gons   2-cycle of 10-gons

See below the ($\{a, b\}$, 3)-spheres with a 3-cycle of b-gons:

($\{3, 4\}$, 3)-sphere D_{3h} with a   ($\{4, 6\}$, 3)-sphere D_{3h} with a   ($\{5, 8\}$, 3)-sphere D_{3h} with a
    3-cycle of 4-gons   3-cycle of 6-gons   3-cycle of 8-gons

Theorem 15.3.3 *(i) Given two* ($\{5, 10\}$, 3)-*spheres that are* $10R_2$ *and have a pair of adjacent 5-gons, adjacent only to 5-gons, then we can merge them into a new* ($\{5, 10\}$, 3)-*sphere that is* $10R_2$, *not 3-connected and has a 2-cycle of 10-gons adjacent on two opposite edges.*

(ii) Any ($\{5, 10\}$, 3)-*sphere, having a 2-cycle of 10-gons adjacent on opposite edges, is obtained by the above procedure of merging of two* ($\{5, 10\}$, 3)-*spheres that are* $10R_2$.

Proof. (i) First, split the common edge of two 5-gons of those spheres. Then merge the cut spheres along those half-edges and obtain the requested new ($\{5, 10\}$, 3)-sphere.

(ii) In the case of ($\{5, 10\}$, 3)-sphere having a 2-cycle on opposite edges, the process can, obviously, be reversed by cutting the edges of the cycle. $\square$

It turns out, that every known ($\{5, 10\}$, 3)-sphere $10R_2$, has exactly two such pairs of 5-gons. Hence, we can, for example, "inscribe a Dodecahedron" in them and obtain not 3-connected ($\{5, 10\}$, 3)-spheres that we $10R_2$.

Theorem 15.3.4 *For any* $t \geq 0$, *there exists a* ($\{5, 8\}$, 3)-*sphere with t 3-cycles of 8-gons, which is 3-connected. This sequence corresponds to the* ($\{5, 8\}$, 3)-*spheres* $8R_2$, *containing a triple of mutually adjacent 5-gons, which are not adjacent to 8-gonal faces.*

Proof. Take two Dodecahedra and remove a vertex in each of them. Glue the edges belonging to different Dodecahedra together. We obtain the unique $(\{5, 8\}, 3)$-sphere with a 3-cycle of 8-gons. This construction can be generalized by taking $t + 1$ Dodecahedra, removing one vertex of two Dodecahedra and two opposite vertices of $t - 1$ remaining Dodecahedra. The construction can be glued, in order to form a $(\{5, 8\}, 3)$-sphere with t 3-cycles of 8-gons (see Figure 15.4).

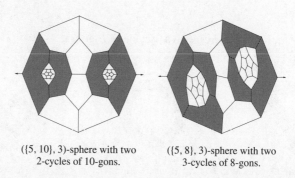

($\{5, 10\}$, 3)-sphere with two   ($\{5, 8\}$, 3)-sphere with two
2-cycles of 10-gons.   3-cycles of 8-gons.

Figure 15.4 Examples of spheres with two cycles from Theorems 15.3.3 and 15.3.4

Take a $(\{5, 8\}, 3)$-sphere $8R_2$ with such a triple of 5-gons. By hypothesis, we can add a cycle of 5-gons around this triple. The obtained $(5, 3)$-polycycle has exactly three vertices of degree 2. It is easy to see that, if we add another 5-gon to it, then we close the structure and obtain a Dodecahedron. If not, then add a 3-cycle of 8-gons. Hence, we should add around the structure a cycle of 5-gons. Then again, we can choose to add a 5-gon, and so three 5-gons, in order to obtain a $(\{5, 8\}, 3)$-sphere with a 3-cycle of 8-gons, or we can continue to add three 8-gons and the argument can be repeated. Hence, since the graph is finite, eventually, we will obtain a $(\{5, 8\}, 3)$-sphere with t 3-cycles of 8-gons. $\qquad\square$

Remark 15.3.5 *The following two methods were used to enumerate* $(\{5, 7\}, 3)$-*spheres with a n-cycle of 7-gons (see Figure 15.6 for the drawings):*

(i) *Harmuth ([Har00]) used* CPF *(see [Har]) to enumerate all such spheres with* $n \leq 16$. *More precisely, he enumerated, using a modification of* CPF, *all* $(\{5, 7\}, 3)$-*spheres that are* $7R_2$, *with at most 84 vertices.*

(ii) *Another method is possible. Any* $(\{5, 7\}, 3)$-*sphere with a n-cycle of 7-gons corresponds to two* $(5, 3)$-*polycycles* $\mathcal{P}$ *and* $\mathcal{P}'$. *Denote by* x *and* x' *the number of interior vertices of those polycycles, by* p_5 *and* p_5' *the number of 5-gons of those polycycles, and by* v_3, v_3' *the number of 3-valent vertices on the boundary. From Theorem 5.2.1, we get* $p_5 = 6 + v_3 - n$ *and* $x \leq 10 + v_3 - 2n$ *and a similar relation holds for* p_5' *and* x'. *But we have* $v_3 + v_3' = 3n$; *therefore, one of the numbers* v_3,

v_3', say, v_3 satisfies to $v_3 \leq \frac{3}{2}n$ and so we have:

$$p_5 \leq 6 + \frac{n}{2} \text{ and } x \leq 10 - \frac{n}{2}.$$

Since $n \leq 20$, this implies that we need to enumerate all $(5, 3)$-polycycles with $p_5 \leq 16$ and less than $16 - p_5$ interior vertices. The result are as follows:

p_5	x_{max}	Nr. of (5,3)-polycycles			
6	10	18	12	4	3147
7	9	35	13	3	6850
8	8	87	14	2	12803
9	7	204	15	1	13448
10	6	518	16	0	4160
11	5	1287			

To every such $(5, 3)$-polycycle, we add, if possible, a cycle of 7-gons and then test if we can fill with 5-gons the obtained $(5, 3)$-boundary (see Section 5.2 for the relevant algorithms). When the fillings is possible, we get $(\{5, 7\}, 3)$-spheres with a cycle of 7-gons.

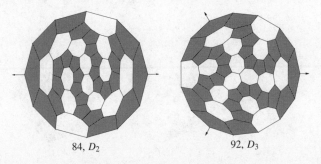

84, D_2 92, D_3

Figure 15.5 ($(\{5, 7\}, 3)$-spheres with a n-cycle of all 5-gons, for $n \leq 31$

The same strategy, as the one used for $(\{5, 7\}, 3)$-spheres with n-cycle of 7-gons, can be used for $(\{5, b\}, 3)$-spheres with n-cycles of b-gons with $b \geq 8$. It yields many $(\{5, b\}, 3)$-spheres for $b = 8, 9$, and 10, which are presented in [DDS05a].

A $(\{5, 7\}, 3)$-sphere with a n-cycle of 5-gons and $n \leq 31$ has at most 19 7-gons. Hence, one of its $(7, 3)$-polycycles has less than nine 7-gons. We enumerated all $(7, 3)$-polycycles up to nine 7-gons, and found only two spheres depicted on Figure 15.5. The first sphere was in [DGr02], while the second coincides with the one found in [HaSo07].

The enumeration of $(\{5, 7\}, 3)$-spheres with more than one cycle of 7-gons was done in the following way. We can assume that the sphere G has at least 88 vertices, because such spheres with at most 84 vertices were found in [Har00]. By the relation $p_7 + x = 20$, this implies that $x \leq 3$. Hence, we will restrict our enumeration to spheres having at most 3 vertices contained only in 5- or only in 7-gonal faces. Moreover, if the sphere G has at least one cycle of 7-gons, then it contains at least two $(5, 3)$-polycycles bounded by cycle of 7-gons. It is easy to see that those $(5, 3)$-polycycles belong to the list described in Remark 15.3.5.ii. Hence, our method is the following. Take one $(5, 3)$-polycycle from this list, add to it a cycle of 7-gons. Then add 5- and 7-gon in every possible way, while making sure that x is never larger than 3 and that every 7-gon is adjacent to exactly two 7-gons.

The above method gave two new $(\{5, 7\}, 3)$-spheres $7R_2$. So, we have the following theorem:

Theorem 15.3.6 *The list of* $(\{5, 7\}, 3)$-*spheres* $7R_2$ *consists (besides of Dodecahedron)* 10 *spheres with a single cycle of 7-gons (see Figure 15.6) and* 16 *spheres with more than one cycle of 7-gons (see Figure 15.7).*

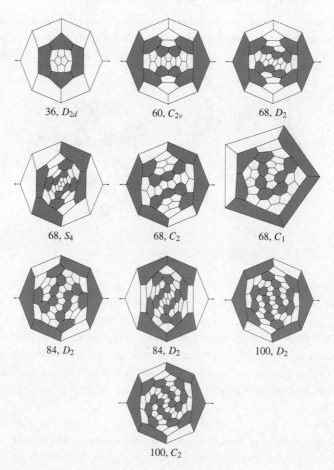

Figure 15.6 All $(\{5, 7\}, 3)$-spheres with a cycle of all 7-gons

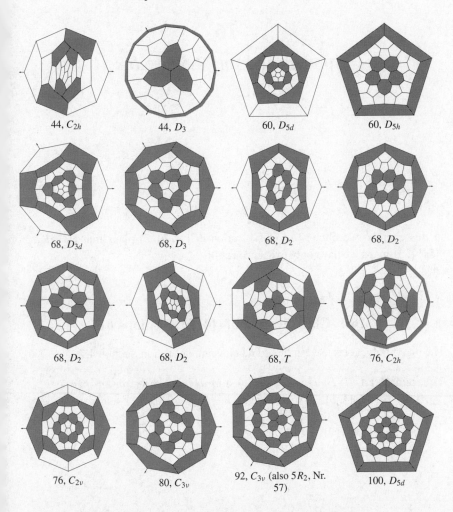

Figure 15.7 All ({5, 7}, 3)-spheres with 7-gons organized into $t \geq 2$ cycles

16

Spheres and tori that are bR_3

In this chapter, we obtain full classification of $(\{4, b\}, 3)$-maps that are bR_3. For $(\{5, b\}, 3)$-maps we have only existence results.

16.1 Classification of $(\{4, b\}, 3)$-maps bR_3

The $(4, 3)$-polycycles used in theorem below are defined in Section 4.2.

Theorem 16.1.1 *(i) There is no $(\{4, b\}, 3)$-torus that is bR_3, for any $b \geq 5$.*
(ii) The list of $(\{4, b\}, 3)$-spheres that are bR_3, consists of:

- *Following $(\{4, 9\}, 3)$-sphere and 2-connected $(\{4, 12\}, 3)$-sphere:*

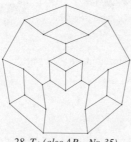　　　　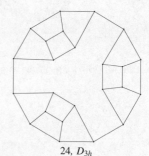

　　　28, T_d (also $4R_2$, Nr. 35)　　　　　　　24, D_{3h}

- *Unique sphere F_b, having two $(4, 3)$-polycycles $\{4, 3\} - e$ and two $(4, 3)$-polycycles $P_2 \times P_{b-6}$, if $b \geq 8$. It is of symmetry D_{2h} if $b = 8$ and D_2 if $b > 8$.*
- *Unique sphere G_b, having two $(4, 3)$-polycycles $\{4, 3\} - v$ and three $(4, 3)$-polycycles $P_2 \times P_{b-5}$, if $b \geq 7$. It is of symmetry D_{3h} if $b = 7$ and D_3 if $b > 7$.*

246

- *Unique sphere H_b, having two points incident to three b-gons and three $(4, 3)$-polycycles $P_2 \times P_{b-3}$, if $b \geq 5$. It is of symmetry D_{3h} if $b = 5$ and D_3 if $b > 5$.*
- *A family of spheres $K_{b,h}$, for $1 \leq h \leq b - 5$, with two $(4, 3)$-polycycles $P_2 \times P_{b-4}$, two polycycles $P_2 \times P_{h+1}$ and two polycycles $P_2 \times P_{b-3-h}$. If $h = 1$, then the symmetry is O_h if $b = 6$ and D_4, otherwise. If $h > 1$, then the symmetry is D_{2d} if $h = \frac{b-4}{2}$ and D_2, otherwise.*

Proof. (i) Take G a $(\{4, b\}, 3)$-map bR_3. The set of 4-gonal faces of G is split into $(4, 3)$-polycycles: $\{4, 3\} - v$, $\{4, 3\} - e$ and $P_2 \times P_k$ for $2 \leq k \leq 7$. Consider the sphere $b(G)$ formed by the b-gons of G with two b-gons adjacent if they share an edge. Since G is bR_3, $b(G)$ is 3-valent.

Every vertex, which is incident to three b-gonal faces, corresponds to a 3-gonal face of $b(G)$. Every $(4, 3)$-polycycle $\{4, 3\} - v$ also corresponds to a 3-gonal face. Every $(4, 3)$-polycycle $\{4, 3\} - e$ corresponds to a 2-gonal face. On the other hand, all $P_2 \times P_k$ correspond to 4-gonal faces. A 3-valent map, whose faces have gonality at most four, does not exist on the torus and, clearly, has at most 8 vertices on the sphere by Euler Formula 1.1.

(ii) Take G a $(\{4, b\}, 3)$-sphere bR_3, consider its associated map $G' = b(G)$. G' has at most 8 vertices and so G has at most 32, 40, 48, and 56 for, respectively, $b = 7, 8, 9, 10$. The enumeration of $(\{4, b\}, 3)$-spheres cover those values and so the result is proved for $b \leq 10$. Assume now $b \geq 11$.

G' is a 3-valent sphere with faces of gonality at most four. There are exactly five such maps: Tetrahedron, $Bundle_3$, $Prism_2$, $Prism_3$, and Cube.

If G' is Tetrahedron, then all its faces are 3-gonal; hence, they all correspond to $(4, 3)$-polycycles $\{4, 3\} - v$ or to vertices. Clearly, in order for a face to be b-gonal, they should be all $\{4, 3\} - v$ or all vertices. This corresponds to strictly face-regular Nr. 35 or to a Tetrahedron.

If G' is $Bundle_3$, then, clearly, the sphere is the 2-connected one, indicated above in the statement of the Theorem.

If G' is $Prism_2$, then its face-set consists of two 2-gons and two 4-gons. Hence, the map G is formed of two $(4, 3)$-polycycles $\{4, 3\} - e$, one $(4, 3)$-polycycle $P_2 \times P_k$, and one $(4, 3)$-polycycle $P_2 \times P_{k'}$ with a priori $k \neq k'$. Given a polycycle $P_2 \times P_k$, which is adjacent to the b-gonal faces F_1, F_2, F_3, and F_4, the sequence of $k - 1$ 4-gons can be adjacent either to F_1 and F_3, or to F_2 and F_4. Hence, we need to fix the orientations of the $(4, 3)$-polycycles $P_2 \times P_k$.

On every 4-gon, there are two possibilities for orienting the polycycles $P_2 \times P_k$. So, we have a total of four possibilities and, after reducing by isomorphism, two possibilities. One possibility corresponds to a b-gonal face having corona

$4b4b444b$, which is impossible. So, the orientation of 4-gonal faces should be done in such a way, that their coronas are of the form $4^{b-7}b4b444b$; hence, $k = k'$ and we obtain the sphere F_b.

If G' is $Prism_3$, then two 3-gonal faces of G' correspond, in the sphere G, to $\{4, 3\} - v$ or vertices, incident only to b-gonal faces. Three 4-gonal faces correspond to the $(4, 3)$-polycycles $P_2 \times P_{k_i}$ with $1 \le i \le 3$. We must find the values of k_i and the orientations of those polycycles.

Denote by $(F_j)_{1 \le j \le 3}$ and $(F'_j)_{1 \le i \le 3}$, two cycles of faces of length 3. Let us consider the faces F_i. Their boundary sequences are of the form either $b\alpha b4b4$, or $b\alpha b4^{k_i-1}b4$, or $b\alpha b4^{k_i-1}b4^{k_{i'}-1}$. Here α is void, if the faces F_i are incident to a common vertex, and $\alpha = 44$, if the faces F_i are adjacent to a common $(4, 3)$-polycycle $\{4, 3\} - v$. Clearly, the first pattern, i.e. $b\alpha b4b4$, is impossible, since it implies gonality 5 or 7. So, the pattern $b\alpha b4^{k_i-1}b4^{k_{i'}-1}$ is also impossible and the faces F_i have their boundary of the form $b\alpha b4^{k_i-1}b4$. This implies $k_1 = k_2 = k_3$.

For faces F'_i, we obtain, by the same argument, that their boundary is of the form $b\alpha' b4^{k_i-1}b4$ with α' being void for a vertex and equal to 44 for a polycycle $\{4, 3\} - v$. So, we have $\alpha = \alpha'$, i.e. either two vertices, or two $(4, 3)$-polycycles $\{4, 3\} - v$. We obtain the series G_b and H_b.

Assume now that G' is Cube. By the previous analysis, all 4-gonal faces are organized in $(4, 3)$-polycycles $P_2 \times P_k$. We need to fix the orientation on every 4-gonal face. Since there are two choices for every one of six faces, this makes a total of 64 choices. Every b-gonal face should be incident to at least one sequence of 4-gons. This reduces to 22 cases. Under symmetry, this reduces to just three cases (a very similar analysis is done in Theorem 9.2.2). The first case is depicted below:

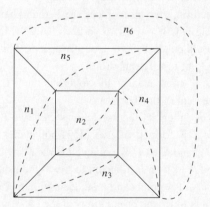

The letters correspond to the length of the $(4, 3)$-polycycles in the following way: n_1 corresponds to $P_2 \times P_{n_1+1}$. Every vertex is incident to three 4-gonal faces. So, three dotted paths appear. If two of those paths are incident to the vertex, i.e. if the corresponding $(4, 3)$-polycycle is incident to an isolated 4-gon to the b-gonal face, then the length of the last dotted path is set. It gives $n_2 = n_5 = n_6 = n_3 = b - 4$. Consider now the case of vertex being incident to just one dotted path. By previous assignment, we get the length $n_1 = n_4 = 1$. This map is $K_{b,1}$.

The second case is depicted below:

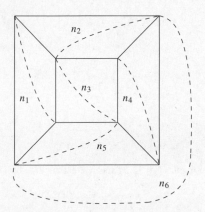

By the same argument, we get $n_1 = n_4 = b - 5$. This reasoning for other vertices yields the equations:

$$n_3 + n_5 = n_3 + n_2 = n_2 + n_6 = n_5 + n_6 = b - 4.$$

It yields $n_5 = n_6$, $n_3 = n_6$ and $n_3 = b - 4 - n_2$. The corresponding map is $K_{b,h}$.

The third case is depicted below:

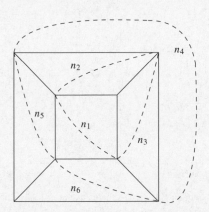

We obtain $n_1 = n_2 = n_3 = n_4 = n_5 = n_6 = b - 5$. Two remaining vertices, which are not incident to any dotted path, yield the equation $3(b - 5) = b - 3$, i.e. $b = 6$, which is already covered.   □

See on Figures 16.1, 16.2, 16.3, and 16.4 the lists of weakly face-regular ($\{4, b\}$, 3)-spheres that are bR_3 for $b = 7, 8, 9$, and 10, respectively.

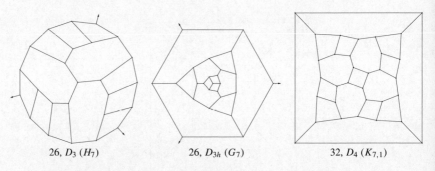

26, D_3 (H_7)	26, D_{3h} (G_7)	32, D_4 ($K_{7,1}$)

Figure 16.1 All face-regular ($\{4, 7\}$, 3)-spheres that are $7R_3$, besides Cube

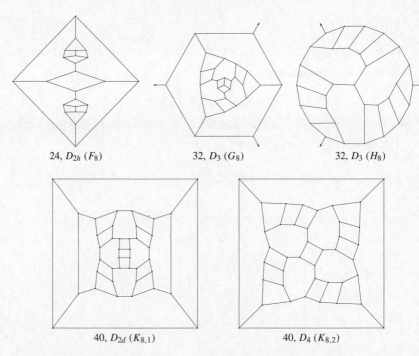

24, D_{2h} (F_8)	32, D_3 (G_8)	32, D_3 (H_8)

40, D_{2d} ($K_{8,1}$)	40, D_4 ($K_{8,2}$)

Figure 16.2 All face-regular ($\{4, 8\}$, 3)-spheres that are $8R_3$, besides Cube

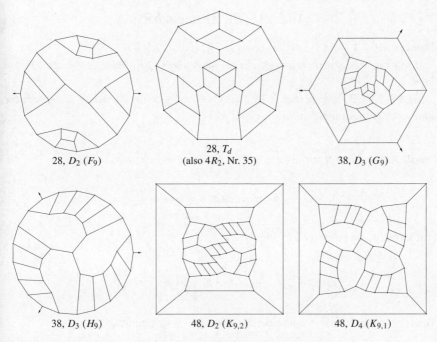

28, D_2 (F_9)

28, T_d
(also $4R_2$, Nr. 35)

38, D_3 (G_9)

38, D_3 (H_9)

48, D_2 ($K_{9,2}$)

48, D_4 ($K_{9,1}$)

Figure 16.3 All face-regular ({4, 9}, 3)-spheres that are $9R_3$, besides Cube

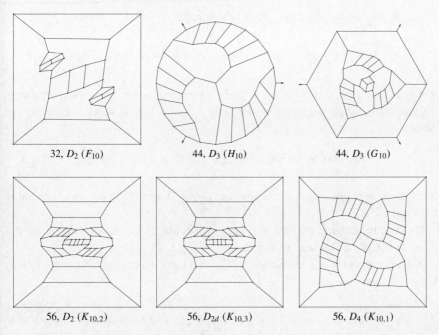

32, D_2 (F_{10})

44, D_3 (H_{10})

44, D_3 (G_{10})

56, D_2 ($K_{10,2}$)

56, D_{2d} ($K_{10,3}$)

56, D_4 ($K_{10,1}$)

Figure 16.4 All face-regular ({4, 10}, 3)-spheres that are $10R_3$, besides Cube

16.2 $(\{5, b\}, 3)$-maps bR_3

Theorem 16.2.1 *(i) A $(\{5, 7\}, 3)$-torus is $7R_3$ if and only if it is $5R_1$.*

Moreover, the corresponding $(\{5, 7\}, 3)$-plane that is $5R_1$ belongs to Case 17 from Table 9.3.

(ii) A $(\{5, 7\}, 3)$-sphere that is $7R_3$ satisfies to $x_0 + x_3 = 20$, where x_i denotes the number of vertices incident to i 5-gonal faces.

Proof. Assume first that the torus is $7R_3$. We have the standard relations:

$$p_5 = p_7 \quad \text{and} \quad e = 6p_7.$$

Furthermore, the property $7R_3$ yields:

$$e_{7-7} = \frac{3}{2}p_7, \quad e_{5-7} = 4p_7 \quad \text{and} \quad e_{5-5} = \frac{1}{2}p_7.$$

Then, by expressing above numbers in terms of x_i, we obtain:

$$2e_{7-7} = 3x_0 + x_1, \quad e_{5-7} = x_1 + x_2 \quad \text{and} \quad 2e_{5-5} = x_2 + 3x_3.$$

By combining above equalities, we get the equality:

$$0 = 3p_7 + p_7 - 4p_7 = 2e_{7-7} - e_{5-7} + 2e_{5-5} = 3x_0 + 3x_3,$$

which yields $x_0 = x_3 = 0$ and so, $x_1 = 3p_7$, $x_2 = p_7$. Denote by $p_{5,k}$ the number of 5-gonal faces, which are adjacent to exactly k 5-gons. Again, by easy counting, we obtain:

$$x_1 = 5p_{5,0} + 3p_{5,1} + p_{5,2} \quad \text{and} \quad 2x_2 = 2p_{5,1} + 4p_{5,2}.$$

Since $x_2 = p_7 = p_5 = p_{5,1} + p_{5,2}$, we get $p_{5,2} = 0$, which implies $p_{5,0} = 0$ and so our torus is $5R_1$.

On the other hand, if the torus is $5R_1$, then it holds, by the same computation, that $x_0 = x_3 = 0$, $x_1 = 3p_7$ and $x_2 = p_7$.

Now, denoting by $p_{7,k}$ the number of 7-gonal faces, which are adjacent to exactly k 7-gonal faces, we obtain:

$$x_2 = 7p_{7,0} + 5p_{7,1} + 3p_{7,2} + p_{7,3} \quad \text{and} \quad 2x_1 = 2p_{7,1} + 4p_{7,2} + 6p_{7,3}.$$

Hence, we get:

$$0 = 6p_7 - 2x_1$$
$$= 6(p_{7,0} + p_{7,1} + p_{7,2} + p_{7,3}) - (2p_{7,1} + 4p_{7,2} + 6p_{7,3})$$
$$= 6p_{7,0} + 4p_{7,1} + 2p_{7,2}.$$

Therefore, $p_{7,0} = p_{7,1} = p_{7,2} = 0$ and the torus is $7R_3$.

In the case of spheres, the proof is very similar; we only indicate the different formula:

$$p_5 = 12 + p_7, \quad \text{and} \quad e = 30 + 6p_7,$$

which yields $3x_0 + 3x_3 = 2e_{7-7} - e_{5-7} + 2e_{5-5} = 60.$ □

Theorem 16.2.2 *For any $b \geq 7$, there exist a $(\{5, b\}, 3)$-torus that is bR_3.*

Proof. Take the following picture of a $(\{5, 7\}, 3)$-torus that is $5R_1$ and $7R_3$ (Case 17)

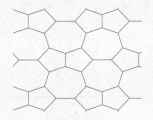

and replace every pair of adjacent 5-gons by an elementary $(5, 3)$-polycycle E_{2n}. The obtained $(\{5, 7 + n\}, 3)$-torus is $(7 + n)R_3$. See below the result for $b = 8, 9$ and 10. □

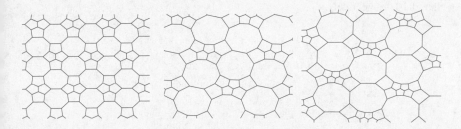

Conjecture 16.2.3 *For any $b \geq 7$, there exist an infinity of $(\{5, b\}, 3)$-spheres that are bR_3.*

Theorem 16.2.4 *There exist an infinity of $(\{5, b\}, 3)$-spheres that are bR_3, for $b = 9$, 10 and 12.*

Proof. The following ($\{5, 9\}$, 3)-, ($\{5, 10\}$, 3)-spheres:

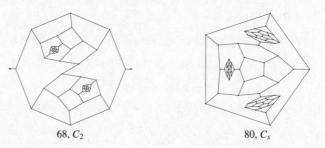

68, C_2 80, C_s

are $9R_3$, $10R_3$ and contains two (5, 3)-polycycles A_2; so, Theorem 11.0.2 solves the cases $b = 9$ and 10. By truncating two opposite vertices of the 3-fold axis of the following ($\{5, 9\}$, 3)-sphere $9R_3$:

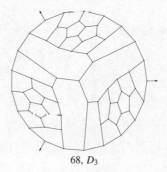

68, D_3

and filling those truncations by (5, 3)-polycycles A_3, we obtain a ($\{5, 12\}$, 3)-sphere $12R_3$, which, by Theorem 11.0.2, solves the case $b = 12$. $\square$

See on Figures 16.5, 16.6, and 16.7 some ($\{5, b\}$, 3)-tori that are bR_3, for $b = 8$, 9, and 10, respectively.

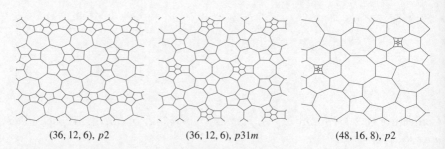

(36, 12, 6), $p2$ (36, 12, 6), $p31m$ (48, 16, 8), $p2$

Figure 16.5 Some ($\{5, 8\}$, 3)-tori that are $8R_3$

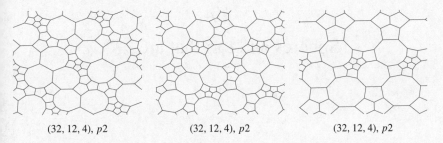

(32, 12, 4), *p*2   (32, 12, 4), *p*2   (32, 12, 4), *p*2

Figure 16.6 Some ({5, 9}, 3)-tori that are 9R_3

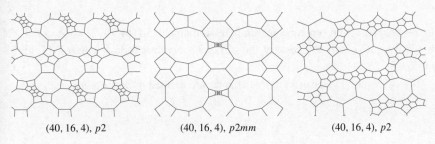

(40, 16, 4), *p*2   (40, 16, 4), *p*2*mm*   (40, 16, 4), *p*2

Figure 16.7 Some ({5, 10}, 3)-tori that are 10R_3

17

Spheres and tori that are bR_4

In this chapter we present a conjectural classification of all $(\{4, b\}, 3)$-spheres bR_4 for $b \leq 8$. We also obtain a necessary and sufficient condition for existence of $(\{4, b\}, 3)$-tori bR_4.

The $(4, 3)$-polycycles used in this chapter are defined in Section 4.2.

17.1 $(\{4, b\}, 3)$-maps bR_4

Lemma 17.1.1 *Let G be a $(\{4, b\}, 3)$-sphere or torus that is bR_4, and let x_i denote the number of vertices, which are contained into i 4-gonal faces. Then we have:*

$$\begin{cases} x_0 + x_3 = 8 & \text{on sphere,} \\ x_0 + x_3 = 0 & \text{on torus.} \end{cases}$$

Proof. Denote by $G' = b(G)$ the map formed by b-gonal faces and their adjacencies; so, it is a 4-valent map. The set of 4-gonal faces of G is partitioned into $(4, 3)$-polycycles $\{4, 3\} - v$, $\{4, 3\} - e$ and $P_2 \times P_k$ with $k \geq 2$. Denote by $p_{\{4,3\}-v}, \dots$ the corresponding number of such polycycles.

Those polycycles correspond, respectively, to 3-, 2- and 4-gonal faces of the map G'. The other faces of G' are 3-gonal ones, corresponding to vertices of G, which are incident to three b-gonal faces.

Hence, the map G' is a 4-valent one, whose faces are 2-, 3-, or 4-gonal. Euler formula (1.1) for those maps is $4\chi = 2p_2 + p_3$ with p_i being the number of i-gonal faces and $\chi = 2$ for spheres and 0 for tori. We have $p_2 = p_{\{4,3\}-e}$ and $p_3 = p_{\{4,3\}-v} + x_0$. So, it holds that:

$$4\chi = 2p_{\{4,3\}-e} + p_{\{4,3\}-v} + x_0.$$

It turns out that $x_3 = 2p_{\{4,3\}-e} + p_{\{4,3\}-v}$. $\qquad\square$

The above theorem also admits a more standard proof, which does not make use of the classification of $(4, 3)$-polycycles. It is very useful for classifying the corresponding maps.

Theorem 17.1.2 *(i) $(\{4, b\}, 3)$-tori that are bR_4, exist if and only if $b \geq 8$.*

(ii) A $(\{4, 8\}, 3)$-torus is $8R_4$ if and only if it is $4R_0$.

(iii) Any $(\{4, b\}, 3)$-torus that is bR_4, has no $(4, 3)$-polycycles $\{4, 3\} - v$ and $\{4, 3\} - e$ and has no vertices incident only to b-gons. Such a torus is described by orienting the $(4, 3)$-polycycles $P_2 \times P_k$ in a 4-valent tiling of the torus.

Proof. (i) By standard arguments (double counting and Euler formula), we obtain:

$$p_4 = \frac{b - 6}{2} p_b \quad \text{and} \quad e_{4-4} = \frac{b - 8}{2} p_b.$$

This implies that a $(\{4, b\}, 3)$-torus that is bR_4, exists if and only if $b \geq 8$. On the other hand, there exists a unique $(\{4, 8\}, 3)$-plane that is $8R_4$ (Case 14 in Table 9.1, see Figure 9.5); we need to modify it so as to obtain a $(\{4, b\}, 3)$-plane. In order to do this, we use the following drawing:

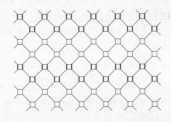

and transform every 4-gon with two boldfaced edges into a $(4, 3)$-polycycle $P_2 \times P_{b-6}$.

(ii) If $b = 8$, then $e_{4-4} = 0$, i.e. all 4-gons are isolated. On the other hand, if a $(\{4, 8\}, 3)$-torus is $4R_0$, then each 8-gon is adjacent to at most four 4-gons and we conclude easily.

(iii) For any $(\{4, b\}, 3)$-torus that is bR_4, we obtain, by Lemma 17.1.1, the equality $x_0 + x_3 = 0$. Therefore, $x_0 = x_3 = 0$, then it holds $p_{\{4,3\}-e} = p_{\{4,3\}-v} = 0$. So, all $(4, 3)$-polycycles are of the form $P_2 \times P_k$. On the other hand, the map $b(G)$ is, clearly, a 4-valent tiling of the torus. Each vertex corresponds to a b-gonal face and each 4-gon corresponds to a $(4, 3)$-polycycle $P_2 \times P_k$. We just need to fix the orientation of this polycycle and the value of k in order to define the torus. $\qquad\square$

Theorem 11.1.4 implies the finiteness of the number of ($\{4, 7\}$, 3)-spheres $7R_4$. They are shown on Figure 17.1.

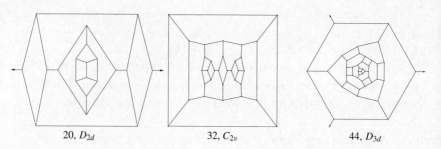

20, D_{2d} 32, C_{2v} 44, D_{3d}

Figure 17.1 All face-regular ($\{4, 7\}$, 3)-spheres that are $7R_4$, besides Cube and strictly face-regular ones Nrs. 27, 28, 30 and 31

Conjecture 17.1.3 *All* (($\{4, 8\}$, 3)-*spheres that are* $8R_4$, *belong to the following list of v-vertex spheres:*

1 *Two infinite series, with* $v = 32 + 8t$ *(*$t \geq 0$*), of spheres containing exactly one* $(4, 3)$-*polycycle* $\{4, 3\} - e$. *If* t *is odd, they are isomorphic and of symmetry* C_2; *if* t *is even positive, they are not isomorphic and one is of symmetry* C_{2h}, *the other of symmetry* C_{2v}. *For* $t = 0$, *the sphere of symmetry* C_{2h} *gains higher symmetry* D_{2h} *(see Figure 17.9).*

2 *Two infinite series, with* $v = 32 + 16t$ *(*$t \geq 0$*), of spheres containing exactly four* $(4, 3)$-*polycycles* $\{4, 3\} - v$. *Their symmetry is* D_{2d}, D_{2h}. *If* $t = 0$, *the sphere of symmetry* D_{2d} *gains the higher symmetry* T_d *(see Figure 17.10). First members of those series are Nrs. 33 and 32.*

3 *Four infinite series, with* $v = 80 + 24t$ *(*$t \geq 0$*), of spheres containing two* $(4, 3)$-*polycycles* $\{4, 3\} - v$ *and six* $(4, 3)$-*polycycle* $P_2 \times P_3$. *Two series are of symmetry* C_2, *one of symmetry* C_{2v}, *and one of symmetry* C_{2h} *(see Figure 17.11).*

4 *Two infinite series, with* $v = 144 + 16t$ *(*$t \geq 0$*) of symmetry* D_2, *of spheres containing twelve* $(4, 3)$-*polycycles* $P_2 \times P_3$ *(see Figure 17.12).*

5 *Three infinite series, with* $v = 128 + 32t$ *(*$t \geq 0$*) of symmetry* D_2, D_{2d}, *and* D_{2h}, *respectively, containing twelve* $(4, 3)$-*polycycles* $P_2 \times P_3$ *(see Figure 17.13).*

6 *A list of sporadic examples, given on Figures 17.2–17.8 (including Nr. 34).*

We indicate below why this list is likely to be complete. The enumeration was done by computer, using following two relations (see Theorems 11.1.4 and 17.1.1):

$$e_{4-4} = 12, \qquad x_0 + x_3 = 8,$$

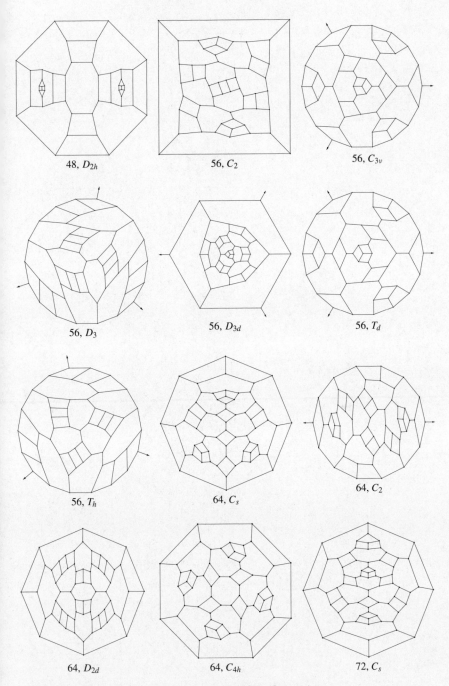

48, D_{2h}

56, C_2

56, C_{3v}

56, D_3

56, D_{3d}

56, T_d

56, T_h

64, C_s

64, C_2

64, D_{2d}

64, C_{4h}

72, C_s

Figure 17.2 Sporadic ({4, 8}, 3)-spheres that are 8R_4 (first part)

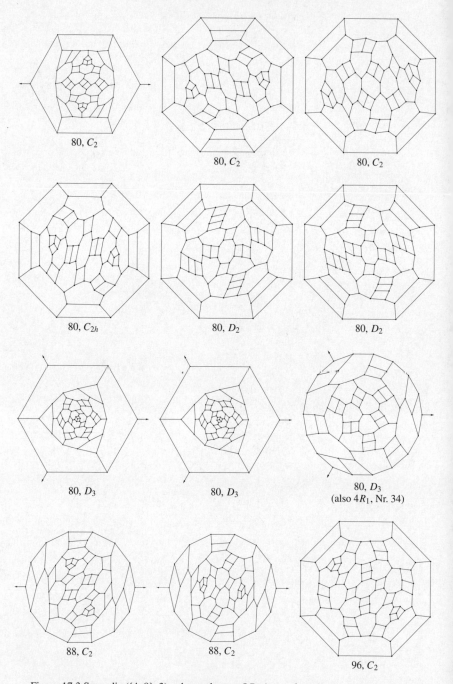

80, C_2

80, C_2

80, C_2

80, C_{2h}

80, D_2

80, D_2

80, D_3

80, D_3

80, D_3
(also $4R_1$, Nr. 34)

88, C_2

88, C_2

96, C_2

Figure 17.3 Sporadic ({4, 8}, 3)-spheres that are $8R_4$ (second part)

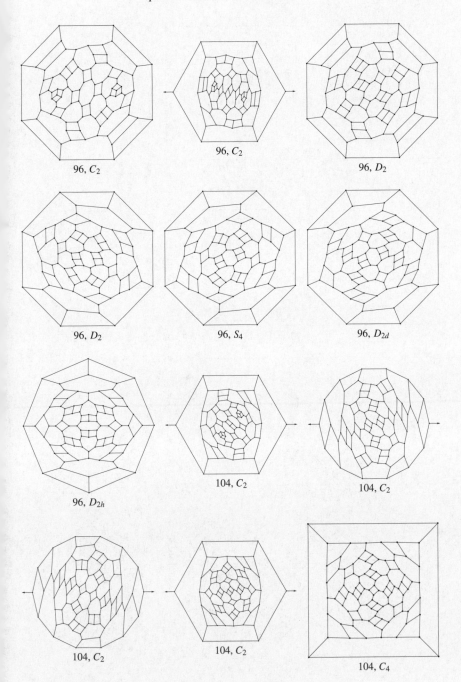

96, C_2

96, C_2

96, D_2

96, D_2

96, S_4

96, D_{2d}

96, D_{2h}

104, C_2

104, C_2

104, C_2

104, C_2

104, C_4

Figure 17.4 Sporadic ($\{4, 8\}$, 3)-spheres that are $8R_4$ (third part)

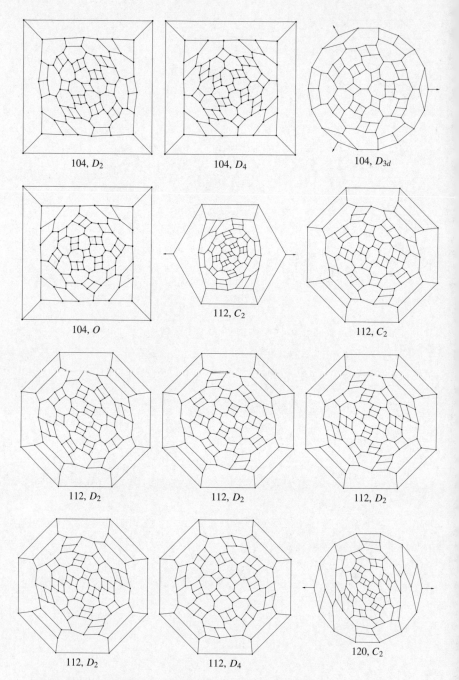

104, D_2

104, D_4

104, D_{3d}

104, O

112, C_2

112, C_2

112, D_2

112, D_2

112, D_2

112, D_2

112, D_4

120, C_2

Figure 17.5 Sporadic ($\{4, 8\}$, 3)-spheres that are $8R_4$ (fourth part)

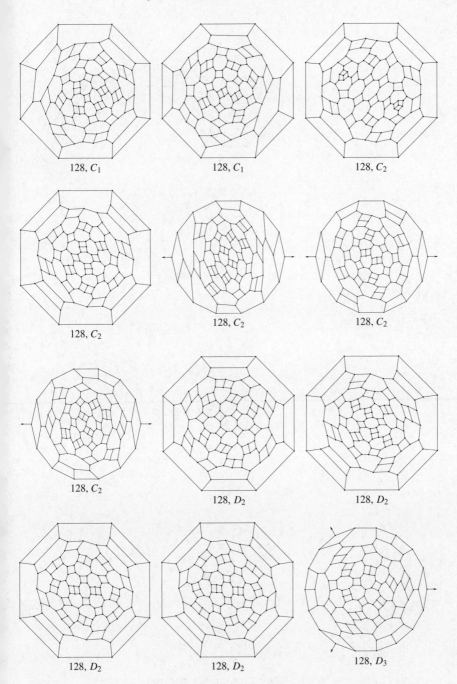

128, C_1 128, C_1 128, C_2

128, C_2 128, C_2 128, C_2

128, C_2 128, D_2 128, D_2

128, D_2 128, D_2 128, D_3

Figure 17.6 Sporadic ({4, 8}, 3)-spheres that are 8R_4 (fifth part)

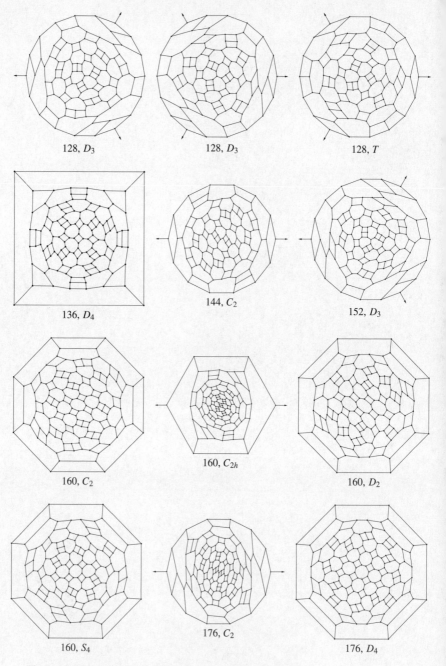

128, D_3 128, D_3 128, T

136, D_4 144, C_2 152, D_3

160, C_2 160, C_{2h} 160, D_2

160, S_4 176, C_2 176, D_4

Figure 17.7 Sporadic ($\{4, 8\}$, 3)-spheres that are $8R_4$ (sixth part)

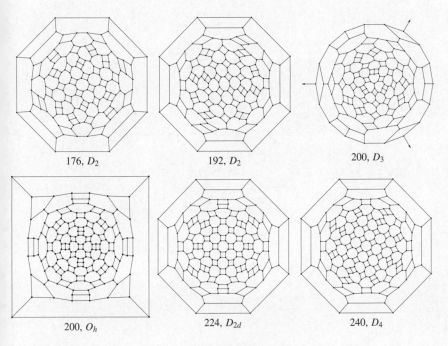

176, D_2 192, D_2 200, D_3

200, O_h 224, D_{2d} 240, D_4

Figure 17.8 Sporadic $(\{4, 8\}, 3)$-spheres that are $8R_4$ (seventh part)

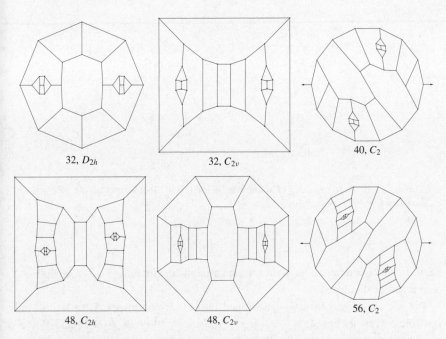

32, D_{2h} 32, C_{2v} 40, C_2

48, C_{2h} 48, C_{2v} 56, C_2

Figure 17.9 First members of infinite series of $(\{4, 8\}, 3)$-spheres that are $8R_4$ and contain two $(4, 3)$-polycycles $\{4, 3\} - e$

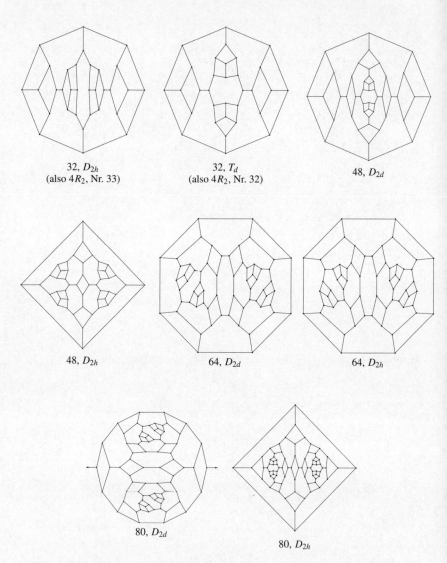

Figure 17.10 First members of infinite series of ({4, 8}, 3)-spheres that are $8R_4$ and contain four (4, 3)-polycycles $\{4, 3\} - v$

i.e. if, in the enumeration, we found some partial map with $e_{4-4} > 12$ or $x_0 + x_3 > 8$, then it can be discarded.

The set of 4-gonal faces is partitioned into (4, 3)-polycycles. Clearly, the only possible polycycles are $\{4, 3\} - v$, $\{4, 3\} - e$, $P_2 \times P_2$, $P_2 \times P_3$, $P_2 \times P_4$ and $P_2 \times P_5$.

The enumeration consisted of the following progressive steps:

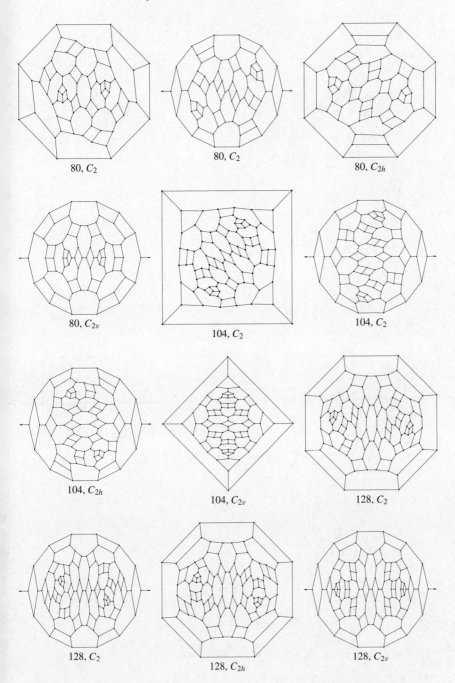

Figure 17.11 First members of infinite series of ({4, 8}, 3)-spheres that are $8R_4$ and contain two (4, 3)-polycycles {4, 3} − v

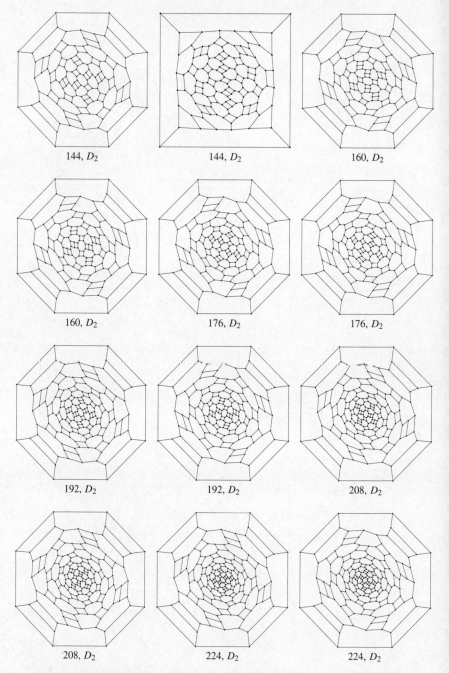

144, D_2 144, D_2 160, D_2

160, D_2 176, D_2 176, D_2

192, D_2 192, D_2 208, D_2

208, D_2 224, D_2 224, D_2

Figure 17.12 First members of infinite series of ($\{4, 8\}$, 3)-spheres that are $8R_4$, contain twelve (4, 3)-polycycles $P_2 \times P_3$ and have symmetry D_2

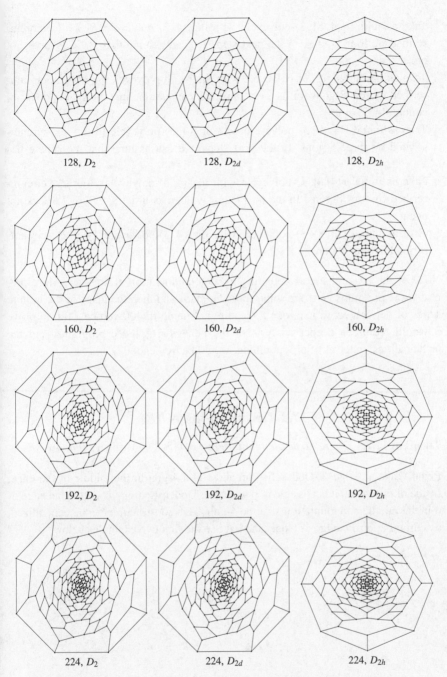

128, D_2 128, D_{2d} 128, D_{2h}

160, D_2 160, D_{2d} 160, D_{2h}

192, D_2 192, D_{2d} 192, D_{2h}

224, D_2 224, D_{2d} 224, D_{2h}

Figure 17.13 First members of infinite series of ($\{4, 8\}$, 3)-spheres that are $8R_4$, contain twelve (4, 3)-polycycles $P_2 \times P_3$ and have symmetry D_2, D_{2d} or D_{2h}

1 Take as initial $(\{4, 8\}, 3)$-polycycle with the $(4, 3)$-polycycle $P_2 \times P_5$, circum-scribed by four 8-gons. In the next steps we can assume that $P_2 \times P_5$ does not occur.

2 Take as initial $(\{4, 8\}, 3)$-polycycle with the $(4, 3)$-polycycle $\{4, 3\} - e$, circum-scribed by two 8-gons. In the next steps, we can assume that the sphere does not contain $\{4, 3\} - e$.

3 Take as initial $(\{4, 8\}, 3)$-polycycle with the $(4, 3)$-polycycle $\{4, 3\} - v$, circum-scribed by three 8-gons. In the next steps, we can assume that the sphere has $x_3 = 0$.

4 Take as initial $(\{4, 8\}, 3)$-polycycle with the $(4, 3)$-polycycle $P_2 \times P_4$, circum-scribed by four 8-gons. In the next steps, we can assume that $P_2 \times P_4$ does not occur.

5 Take as initial $(\{4, 8\}, 3)$-polycycle with the $(4, 3)$-polycycle $P_2 \times P_3$, circum-scribed by four 8-gons.

None of the programs, corresponding to such enumeration, terminates. More precisely, the programs generate some sporadic maps and then the maps of the infinite series of our conjecture. So, our conjecture is true up to 400 vertices. But to prove it would require a deeper analysis of the behavior of those programs, and the technicalities, in terms of programming, would be tremendous.

17.2 $(\{5, b\}, 3)$-maps bR_4

Theorem 17.2.1 *There exists an infinite series of* $(\{5, 8\}, 3)$-*spheres that are* $8R_4$.

Proof. Take an edge of Dodecahedron and add a 4-gon in the middle of the edge. Inside of this 4-gon it is possible to put another Dodecahedron, by cutting an edge of it in the middle and gluing it inside the 4-gon. This construction can be generalized, by cutting opposite edge of Dodecahedra. See the first two examples below.   □

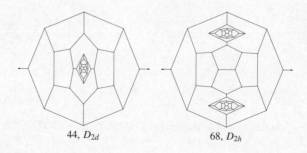

44, D_{2d} 68, D_{2h}

In above procedure, cutting opposite edges of Dodecahedra is just one of several possibilities. In fact, given one edge of Dodecahedron, there are five other edges, which can be cut, in order to obtain $(\{5, 8\}, 3)$-spheres $8R_4$.

Theorem 17.2.2 *A $(\{5, 7\}, 3)$-sphere that is $7R_4$ and contains, in its set of 5-gonal faces, a $(5, 3)$-polycycle with boundary sequence containing at most three 2, belongs to an infinite series of such spheres, having $v = 20 + 24t$ vertices with $t \geq 1$ and symmetry D_{3d} (Figure 17.14).*

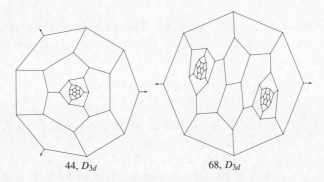

                   44, D_{3d} 68, D_{3d}

Figure 17.14 Both $(\{5, 7\}, 3)$-spheres that are $7R_4$ and have at most 76 vertices; they are the cases $t = 1, 2$ of an infinite series of those having $20 + 24t$ vertices

Proof. Denoting by v_3, v_2 the number of vertices of degree 3, 2 on the boundary, we obtain $v_3 \leq 2v_2$ and $v_2 \leq 3$.

The formula $p_5 = 6 - v_2 + v_3$, expressing the number of 5-gonal faces in a $(5, 3)$-polycycle (see Theorem 5.2.1), yields $p_5 \leq 9$. An exhaustive enumeration amongst all $(5, 3)$-polycycles with nine 5-gons yields the elementary $(5, 3)$-polycycle A_3 as the only possibility.

So, now we extend this polycycle by adding a ring of 7-gons around it. Since every 7-gon is adjacent to four 7-gons, we need to add another ring of 7-gons around them. Every 7-gonal face in those rings is adjacent to four 7-gonal faces. So, we are forced to add another ring of 5-gons. If we add another 5-gon, then there is only one possibility for filling the structure and one adds two more 5-gons and obtains a sphere.

If not, then we add a ring of 7-gons and the argument can be repeated. Since the graph is finite, eventually, we will obtain a sphere belonging to the infinite series. $\square$

$(\{5, 7\}, 3)$-spheres that are $7R_4$ and which are not obtained by the above Theorem 17.2.2, probably, exist since there is a lot of $(\{5, 7\}, 3)$-tori that are $7R_4$.

Theorem 17.2.3 *For any $b \geq 7$, there exists a $(\{5, b\}, 3)$-torus that is bR_4.*

Proof. The proof consists of taking the following initial $(\{5, 7\}, 3)$-plane $7R_4$ that is also $5R_2$ (the sporadic subcase of Case 18 of Table 9.3):

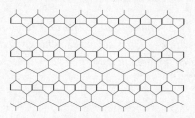

In order to obtain a $(\{5, b\}, 3)$-torus that is bR_4, we need to modify the structure of the 5-gons. The 5-gons can be paired to form the polycycle E_0 in the following way:

Every one of those pairs of 5-gons can be replaced by an E_{2n} with $n \geq 1$. We obtain a periodic $(\{5, 7+n\}, 3)$-plane that is $(7+n)R_4$. Then needed torus is obtained as its quotient.   □

Theorem 17.2.4 *There is an infinity of* $(\{5, b\}, 3)$*-spheres that are* bR_4, *for* $b = 7$, 10, 13, 16.

Proof. If we take Cube, truncate some of its vertices and replace them by polycycles A_3, so that every 4-gon is incident to exactly t truncated vertices, then it is easy to see that the obtained graph is a $(\{5, 4 + 3t\}, 3)$-sphere that is $(4 + 3t)R_4$. Clearly, for any $1 \leq t \leq 4$, such sets of vertices exist. The conclusion follows from Theorem 11.0.2.   □

See on Figures 17.15, 17.16, 17.17, and 17.18 some examples of $(\{5, b\}, 3)$-tori that are bR_4, for $b = 7, 8, 9$, and 10.

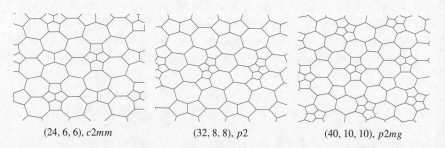

(24, 6, 6), $c2mm$ (32, 8, 8), $p2$ (40, 10, 10), $p2mg$

Figure 17.15 Some $(\{5, 7\}, 3)$-tori that are $7R_4$

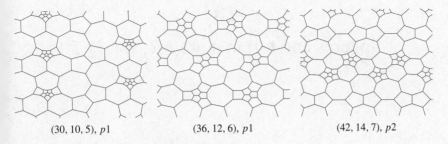

(30, 10, 5), *p1*   (36, 12, 6), *p1*   (42, 14, 7), *p2*

Figure 17.16 Some ({5, 8}, 3)-tori that are $8R_4$

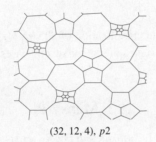

(32, 12, 4), *p2*

Figure 17.17 A ({5, 9}, 3)-torus that is $9R_4$

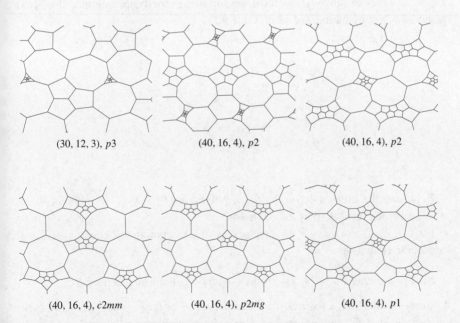

(30, 12, 3), *p3*   (40, 16, 4), *p2*   (40, 16, 4), *p2*

(40, 16, 4), *c2mm*   (40, 16, 4), *p2mg*   (40, 16, 4), *p1*

Figure 17.18 Some ({5, 10}, 3)-tori that are $10R_4$

18
Spheres and tori that are bR_j for $j \geq 5$

In this chapter we present the classification of all $(\{a, b\}, 3)$-spheres bR_j for $j \geq 5$. We also obtain the minimal such spheres for several cases.

18.1 Maps bR_5

Theorem 18.1.1 *(i) A $(\{4, b\}, 3)$-torus that is bR_5, exists if and only if $b \geq 7$.*
(ii) Any $(\{4, 7\}, 3)$-torus that is $7R_5$, is also $4R_0$.

Proof. In order to prove (i), we take the following strictly face-regular $(\{4, 7\}, 3)$-plane $7R_5$ (belonging to Case 13 of Table 9.3):

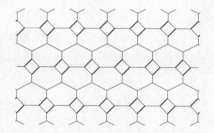

Replacing the isolated 4-gons by $(4, 3)$-polycycles $P_2 \times P_{2+n}$, we obtain $(\{4, 7 + n\}, 3)$-torus that is $(7 + n)R_5$. So, (i) holds.

For $(\{4, 7\}, 3)$-tori that are $7R_5$, we get, by direct counting, the equality $e_{4-4} = 0$; hence, the conclusion. $\qquad \square$

Note that there exist a $(\{4, 7\}, 3)$-torus that is $4R_0$ but which is not $7R_5$.

Lemma 18.1.2 *Take a map G, such that its set $\mathcal{F}$ of faces is partitioned into two classes, $\mathcal{F}_1$ and $\mathcal{F}_2$, so that any face F in $\mathcal{F}_1$ is 6-gonal and adjacent to exactly five other faces of $\mathcal{F}_1$. Then it holds that:*

(i) *A face $F \in \mathcal{F}_2$ is adjacent only to faces in $\mathcal{F}_1$.*

(ii) *There exists a 3-valent map G', such that $G = GC_{2,1}(G')$ or $G = GC_{1,2}(G')$ (see Section 2.1).*

Proof. (i) follows from direct analysis of possible corona of faces. Given a face $F \in \mathcal{F}_2$, denote by $N(F)$ the neighborhood of F in $\mathcal{F}_1$ (i.e. the set of all faces from $\mathcal{F}_1$, which are adjacent to F). Clearly, the set $\mathcal{F}_1$ is partitioned into $N(F_1), \ldots, N(F_k)$. Suppose that two sets $N(F_i)$ and $N(F_j)$ have an adjacency. Then the following two cases are possible:

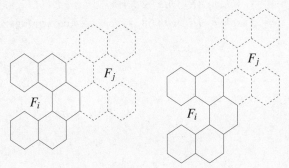

Both of those cases correspond to the local configuration arising in the Goldberg-Coxeter construction (see Chapter 2). Moreover, the choice of a local configuration determines the whole structure completely, i.e. there is only one choice globally.

Now, define the map G' with faces corresponding to the set $\mathcal{F}_2$, edges corresponding to pairs $N(F_i)$, $N(F_j)$, having some adjacencies, and vertices corresponding to triples $N(F_i)$, $N(F_j)$, $N(F_k)$, having pairwise adjacencies. G' is a 3-valent plane graph and $GC_{2,1}(G')$ or $GC_{1,2}(G')$ is isomorphic to G. $\square$

Lemma 18.1.3 *The set of $(5, 3)$-polycycles, having boundary sequence $(23^h)^g$, consists of:*

- *elementary $(5, 3)$-polycycle A_2 with $h = 3$, $g = 2$,*
- *elementary $(5, 3)$-polycycle A_3 with $h = 2$, $g = 3$,*
- *elementary $(5, 3)$-polycycle A_5 with $h = 1$, $g = 5$,*
- *elementary $(5, 3)$-polycycle D with $h = 0$, $g = 5$.*

Proof. Take a $(5, 3)$-polycycle P with boundary sequence $(23^h)^g$ and assume that it consists of several elementary $(5, 3)$-polycycles put together.

The major skeleton $Maj(P)$ formed by the elementary components of P is a tree (see Theorem 8.0.1). This tree has at least one vertex of degree 1. Such a vertex corresponds to an elementary $(5, 3)$-polycycle, say P_{el}.

If P_{el} has at least twice the pattern 22 in its boundary sequence, then $h = 0$ and we are done. Hence, we can assume that it has the pattern 22 only once in its boundary sequence; so, it is B_2 or B_3 (see Figure 7.2). But those do not fit. Hence, the only possible cases are the elementary $(5, 3)$-polycycles indicated above. □

Theorem 18.1.4 *Take a* $(\{5, b\}, 3)$-*map that is* bR_5, *such that the corona of each* b-*gon is* $b^5 5^{b-5}$ *and whose graph* $b(G)$ *is connected.*
Then it is one of the following:

- $GC_{2,1}(Dodecahedron)$, *i.e. the strictly face-regular Nr. 55*,
- *the unique* $(\{5, 7\}, 3)$-*sphere* $7R_5$ *with* 260 *vertices, depicted in Figure 18.2*,
- *the unique* $(\{5, 8\}, 3)$-*sphere* $8R_5$ *with* 92 *vertices, depicted in Figure 18.1*,
- *the unique* $(\{5, 9\}, 3)$-*sphere* $9R_5$ *with* 68 *vertices, depicted in Figure 18.1*.

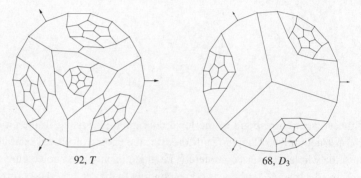

$$92, T \qquad\qquad\qquad 68, D_3$$

Figure 18.1 The smallest $(\{5, b\}, 3)$-sphere that is bR_5 for $b = 8$ and 9

Proof. The connectedness of $b(G)$ implies that the set of 5-gonal faces is partitioned into $(5, 3)$-polycycles. The corona condition implies that the boundary sequence is of the form $(23^{b-6})^g$. Lemma 18.1.3 gives that the number g depends only on b.

After replacing those $(5, 3)$-polycycles by b-gons, we get a map G with 6- and g-gons only and with $g \le 5$. Hence, this map is on the sphere; it satisfies the condition of Lemma 18.1.2 and we get the result. □

Theorem 18.1.5 *The smallest* $(\{5, 8\}, 3)$-*sphere that is* $8R_5$, *is depicted on Figure 18.1.*

Proof. Euler formula (1.1) for a $(\{5, 8\}, 3)$-sphere gives $12 = p_5 - 2p_8$; hence, $e = 30 + 9p_8$ and $v = 20 + 6p_8$. Since the example, given above, has 92 vertices, we can assume in the following that $p_8 \le 12$. Further enumeration yields:

$$e_{5-5} = 3p_8, \quad e_{8-8} = 5p_8 \quad \text{and} \quad e_{5-5} = 30 + \left(3 + \frac{1}{2}\right) p_8 .$$

The set of 5-gonal faces is partitioned into elementary $(5, 3)$-polycycles by bridges.

Since the sphere is $8R_5$, every 8-gon is adjacent to exactly three 5-gons; hence, the only elementary $(5, 3)$-polycycles, which can appear in the decomposition, are A_3, A_4, A_5, D, B_3, C_3, E_1, E_2, E_3, E_4. Denote by $p_{A_3}, \ldots,$ their numbers and by t the number of adjacencies of elementary $(5, 3)$-polycycles along their open edges.

By direct counting, we get the equalities:

$$\begin{cases} e_{5-8}=9p_{A_3} + 10p_{A_4} + 10p_{A_5} + 5p_D + 11p_{B_3} + 12p_{C_3} \\ \quad +9p_{E_1} + 10p_{E_2} + 11p_{E_3} + 12p_{E_4} - 2t \\ e_{5-5}=18p_{A_3} + 15p_{A_4} + 10p_{A_5} + 12p_{B_3} + 6p_{C_3} + 3p_{E_1} \\ \quad +5p_{E_2} + 7p_{E_3} + 9p_{E_4} + t. \end{cases}$$

Furthermore, the number t of adjacencies between $(5, 3)$-polycycles satisfies the inequality:

$$2t \leq 2p_D + 3p_{E_1} + 2p_{E_2}.$$

By combining above equalities and inequalities we get:

$$\begin{cases} 2e_{5-8} - e_{5-5}=\frac{5}{2}p_8 - 30 = \frac{5}{2}(p_8 - 12) \leq 0 \\ 2e_{5-8} - e_{5-5}=5p_{A_4} + 10p_{A_5} + 10p_D + 10p_{B_3} + 18p_{C_3} \\ \quad +15p_{E_1} + 15p_{E_2} + 15p_{E_3} + 15p_{E_4} - 5t \\ \quad \geq 5p_{A_4} + 10p_{A_5} + 5p_D + 10p_{B_3} + 18p_{C_3} + \frac{15}{2}p_{E_1} \\ \quad +10p_{E_2} + 15p_{E_3} + 15p_{E_4} \\ \quad \geq 0. \end{cases}$$

This implies $2e_{5-8} - e_{5-5} = 0$; hence, $p_8 = 12$ and $p_{A_4} = p_{A_5} = p_D = p_{B_3} = p_{C_3} = p_{E_i} = 0$. So, the only polycycle, appearing in the decomposition, is A_3. So, by Theorem 18.1.4, the sphere is obtained by $GC_{2,1}(Tetrahedron)$ and replacing all 3-gons by the polycycles A_3. $\qquad\qquad\square$

Theorem 18.1.6 *There is an infinity of $(\{5, b\}, 3)$-spheres that are bR_5, for $b = 8$,* 11, 14, 17, 20.

Proof. The example for $b = 8$, which is given in Theorem 18.1.5, has four $(5, 3)$-polycycles A_3. The conclusion follows by Theorem 11.0.2.

In order to get a proof for other values of b, we need an initial example. The graph $GC_{2,1}(Tetrahedron)$ can be interpreted in the following way. The triangular faces can be shrunken to just one point. The obtained graph is Dodecahedron with a set S of four special vertices, corresponding to those faces. For any $1 \leq t \leq 5$, there exist a set S_t of $4t$ vertices of Dodecahedron, such that every face is incident to t vertices of this set:

1 For $t = 0$ or 5 (Dodecahedron or strictly face-regular Nr. 14), there is one possible set and it has symmetry I_h.

2 For $t = 1$ or 4 (strictly face-regular Nrs. 6 or 13), there is one possible set and it has symmetry T.

3 For $t = 2$ or 3 (strictly face-regular Nrs. 8, 9 or 11, 12), there are two possibilities, one of symmetry D_3, the other of symmetry T_h.

By doing 5-triakon of Dodecahedron on those sets S_t (i.e. truncate the vertices and replace the obtained 3-gons by $(5, 3)$-polycycle A_3), we get a $(\{5, 5 + 3t\}, 3)$-sphere that is $(5 + 3t)R_5$. The proof of infiniteness is then identical to the case $b = 8$. □

In the above construction, we got a $(\{5, 8\}, 3)$-sphere by putting the spheres together in a path. If we create cycle, then we obtain higher genus surfaces that are $8R_5$. So, for every $g \geq 0$, there is an infinity of oriented $(\{5, 8\}, 3)$-maps of genus g that are $8R_5$.

Also, above operation of removing a vertex leave us with six 5-gons in a circuit; hence, $5R_2$ holds. Clearly, if we manage to eliminate all $(5, 3)$-polycycles A_3, so as to obtain a cycle, then we get an oriented $(\{5, 8\}, 3)$-map of genus g that is $8R_5$ and $5R_2$. Such a structure can be obtained for any g, $g \geq 2$. But for $g = 1$ it does not exist.

Theorem 18.1.7 *There is an infinity of $(\{5, b\}, 3)$-spheres that are bR_5, for $b = 9$, 12, 15, 18, 21.*

Proof. The sphere, shown on Figure 18.1, is such a sphere. It can be obtained by taking $GC_{2,1}(Bundle_3)$ (see Section 10.1) and replacing every 2-gon by a $(5, 3)$-polycycle A_2. The conclusion for $b = 9$ follows from Theorem 11.0.2.

On the other hand, if we take the graph $GC_{2,1}(Bundle_3)$ and remove all 2-gons, then the obtained graph is Cube. For every $0 \leq t \leq 4$, there exist sets S_t of $2t$ vertices, such that every face of Cube is incident to exactly t vertices of S_t. More precisely:

1 For $t = 0$ or 4 (Cube and strictly face-regular Nr. 10), there is one such set and it has symmetry O_h.

2 For $t = 1$ or 3 (strictly face-regular Nrs. 2, 7), there is one such set and it is of symmetry D_{3d}.

3 For $t = 2$ (strictly face-regular Nrs. 4, 5), there are two such sets, one of symmetry T_d and the other of symmetry D_{2h}.

So, take Cube, truncate vertices belonging to the set S_t and fill them by $(5, 3)$-polycycles A_3. Also, insert three 2-gons on the edges corresponding to the 2-gons of the original $(\{5, 9\}, 3)$-sphere that is $9R_5$. We get infiniteness by Theorem 11.0.2. □

Theorem 18.1.8 *If G is a $(\{5, b\}, 3)$-sphere that is bR_{b-2}, then it holds:*

(i) *The graph $b(G)$ is connected.*

(ii) *The set of 5-gonal faces belongs to the following set $\{D, D + D, E_1, E_2, A_5\}$ of $(5, 3)$-polycycles:*

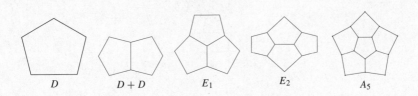

D $D + D$ E_1 E_2 A_5

(iii) $b \leq 7$.

(iv) *If $b = 6$, then such spheres are enumerated in [DeGr01] and presented on Figure 10.7.*

(v) *If $b = 7$, then such sphere has at least 260 vertices. If it has 260 vertices, then it is unique and has symmetry I (see Figure 18.2); otherwise, it has at least 280 vertices.*

Proof. Suppose that the graph $b(G)$ is not connected. This means that there exists a set of 5-gonal faces, on which at least two (say, t) connected components of $b(G)$ meet.

Since every b-gon is adjacent to exactly two 5-gons, in the $(5, 3)$-polycycle, the runs of 3 (i.e. sequence of 3 bounded by 2) of the boundary sequence have length at most one, i.e. every 3-valent vertex is bounded by two 2-valent vertices. This implies $v_3 \leq v_2$ and so, $p_5 \leq 0$, by Theorem 11.0.3. Hence, (i) holds.

Now, the connectedness of $b(G)$ implies that the set of 5-gonal faces form $(5, 3)$-polycycles, i.e. they have $t = 1$ and $p_5 = 6 + v_3 - v_2 \leq 6$. The set of $(5, 3)$-polycycles, satisfying the condition that every run of 3 is of length at most one, is the presented one.

The set of faces of $b(G)$ comes from vertices of G, which are incident to three b-gonal faces (so, they are of gonality 3) and the $(5, 3)$-polycycles (so, they are of gonality 5 or 6). Hence, by Theorem 1.2.3(ii), $b(G)$ has a vertex of degree at most 5; so, $b - 2 \leq 5$ and (iii) holds.

The following formulas are easy:

$$p_5 = 12 + p_7, \quad e = 30 + 6p_7,$$
$$e_{7-7} = \tfrac{5}{2}p_7, \quad e_{5-7} = 2p_7 \quad \text{and} \quad e_{5-5} = 30 + \tfrac{3}{2}p_7.$$

Denote by $n_D, n_{D+D}, n_{E_1}, n_{E_2}, n_{A_5}$ the respective number of such $(5, 3)$-polycycles, depicted in (ii).

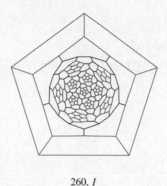

260, *I*

Figure 18.2 The only ({5, 7}, 3)-sphere that is $7R_5$, besides Dodecahedron and has less than 280 vertices

By direct counting, we obtain:

$$\begin{cases} 12 + p_7 = p_5 = n_D + 2n_{D+D} + 3n_{E_1} + 4n_{E_2} + 6n_{A_5}, \\ 30 + \frac{3}{2}p_7 = e_{5-5} = n_{D+D} + 3n_{E_1} + 5n_{E_2} + 10n_{A_5}. \end{cases}$$

By eliminating n_{A_5}, we obtain also:

$$60 - p_7 = -10n_D - 14n_{D+D} - 12n_{E_1} - 10n_{E_2},$$

which implies $p_7 \geq 60$, i.e. that a ({5, 7}, 3)-sphere that is $7R_5$, has at least 260 vertices. Furthermore, such a sphere with exactly 260 vertices has only the elementary (5, 3)-polycycle A_5 and we conclude by Theorem 18.1.4. If it has more than 260 vertices, then $n_i > 0$ for some $1 \leq i \leq 4$, which implies $60 - p_7 \leq -10$ and, hence, the needed result.   □

See below the only known ({4, 8}, 3)-torus that is $8R_5$ and not $4R_1$:

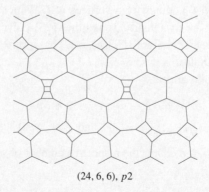

(24, 6, 6), *p2*

See below two examples of $(\{5, 10\}, 3)$-tori that are $10R_5$:

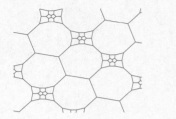 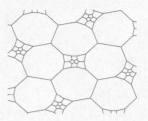

18.2 Maps bR_6

Theorem 18.2.1 *(i) Any $(\{4, b\}, 3)$-torus that is bR_6 and 3-connected, is also $4R_2$.*
(ii) All $(\{4, 8\}, 3)$-tori that are $8R_6$, are also $4R_2$.

Proof. (i) The hypothesis of 3-connectedness of such a map G excludes the existence of $(4, 3)$-polycycle $\{4, 3\} - e$ in the set of 4-gons of this torus.

This means that, if we consider the corresponding 6-valent map $b(G)$, then all its faces are 3-gonal (vertices or $(4, 3)$-polycycles $\{4, 3\} - v$) or 4-gonal ($(4, 3)$-polycycles $P_2 \times P_k$). Clearly, by Euler formula, 4-gons are excluded, i.e. only $\{4, 3\} - v$ exist and the map is $4R_2$.

(ii) In a $(\{4, 8\}, 3)$-torus, no 4-gon can be adjacent only to 4-gons, since it would imply the structure of Cube. If a $(\{4, 8\}, 3)$-torus contains a 4-gon, adjacent to three 4-gons, then it contains a 8-gon, adjacent to at least three 4-gons and, hence, to at most five 8-gons. So, in a $(\{4, 8\}, 3)$-torus, the 4-gons are adjacent to at most two other 4-gons. The result then follows by usual double counting and positivity. $\quad\square$

Theorem 18.2.2 *There are no $(\{4, b\}, 3)$-spheres that are bR_{b-2}, for $b \geq 8$.*

Proof. Every b-gonal face of such a sphere G would be adjacent to exactly two 4-gons. This means that the $(4, 3)$-polycycle $\{4, 3\} - e$ cannot appear in the decomposition of the set of 4-gonal faces. Hence, all faces of the sphere $b(G)$ are 3- or 4-gonal. So, we can conclude using Theorem 1.2.3(ii). $\quad\square$

Theorem 18.2.3 *There are no $(\{5, 8\}, 3)$-tori that are $8R_6$.*

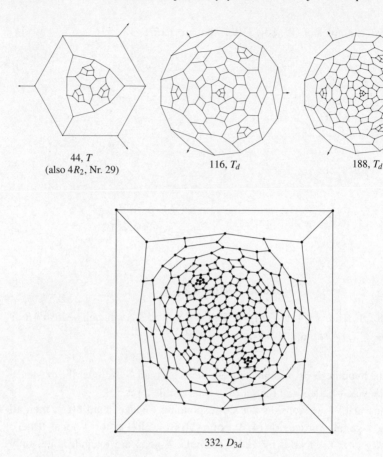

44, *T*
(also 4R_2, Nr. 29)   116, T_d 188, T_d

332, D_{3d}

Figure 18.3 All known ({4, 7}, 3)-spheres that are 7R_5, besides Cube

Proof. By re-doing computations of Theorem 18.1.8 for $b = 8$, we would get that a
({5, 8}, 3)-torus G that is 8R_6, has connected 8(G) and the set of its 5-gonal faces is
partitioned into the (5, 3)-polycycles, depicted in Theorem 18.1.8(ii).

Hence, the torus 8(G) is 6-valent and its faces are 3-gons (intersection of three
8-gons), 5-gons and 6-gons.

Euler formula (1.1) for 6-valent torus is $\sum_i (3 - i)p_i = 0$. Hence, there are no 5-
and 6-gonal faces in 8(G). So, G has no 5-gons, which is impossible.   □

Theorem 18.2.4 *If a* ({5, b}, 3)-*sphere is* bR_j *with* $j \geq 6$ *and* $b(G)$ *is connected,*
then $j \leq b - 4$.

Proof. By Theorem 1.2.3(ii), the map $b(G)$ of such a sphere G contains at least one
2-gon. It is easy to see that the only (5, 3)-polycycle with two vertices of degree 2 on

the boundary is A_2. So, the b-gonal faces, which are adjacent to this $(5, 3)$-polycycle A_2, are adjacent to at least four 5-gonal faces. □

Note that we can construct some $(\{5, 10\}, 3)$-spheres that are $10R_6$. Take Dodecahedron and select a set S of its edges, such that every 5-gon is incident to exactly one edge of this set. Replacing those edges by the $(5, 3)$-polycycles A_2, we obtains such spheres. Up to isomorphism, there exist five such sets in Dodecahedron and they yield five $(\{5, 10\}, 3)$-spheres $10R_6$ with 140 vertices and symmetry groups D_{3d}, C_2, D_2, D_3, T_h.

19
Icosahedral fulleroids

In this chapter, which is an adaptation of [DeDe00], are considered *icosahedral fulleroids* (or *I-fulleroids*, or, more precisely, $I(5, b)$-*fulleroids*, i.e. ($\{5, b\}, 3$)-spheres of symmetry I or I_h). For some values of b, the smallest such fulleroids are indicated and their unicity is proved. Also, several infinite series of them are presented.

The case $b = 6$ is the classical fullerene case. Theorem 2.2.2 gives that all $I(5, 6)$-fulleroids, i.e. fullerenes of icosahedral symmetry, are of the form $GC_{k,l}(Dodecahedron)$. See on Figure 19.1 the first three of the following smallest icosahedral fullerenes besides Dodecahedron:

- $C_{60}(I_h)$, buckminsterfullerene,
- $C_{80}(I_h)$, chamfered Dodecahedron,
- $C_{140}(I)$, smallest chiral one,
- $C_{180}(I_h)$.

Here $C_v(G)$ stands for a ($\{5, 6\}, 3$)-sphere with v vertices and symmetry group G. Although this notation is not generally unique, it will suffice for our purpose.

Both smallest $I(5, 7)$-fulleroids are described in [DrBr96]; see them on Figure 19.2.

All I-fulleroids known so far and simple ways to describe them are given in Section 19.1; based on that some infinite series are introduced. In Section 19.2, a necessary condition for the p-vectors, which implies that five of the new I-fulleroids are minimal for their respective values of v, is derived.

Table 19.1 shows the smallest possible p-vectors for $7 \leq b \leq 20$, accordingly see Lemma 19.2.3 below. The first four columns of the table show the quantities b, p_5, p_b (the number of 5-, b-gons) and the number of vertices v. The invariants m_5, m_b, k_2, k_3, and k_5 are described in Section 19.2. In the 5th column, the I-fulleroids are indexed by their construction, while in the last column their names are given. These names are of the form $F_{5,b}(G)$, where G is the symmetry group, or $F_b(1)$, using the notation given in Theorem 19.1.1. See on Figure 19.3 six $F(5, b)(I_h)$ with $b \geq 8$.

Note that for $b = 12$, the smallest p-vector, which fulfills the condition of Lemma 19.2.3, is not realizable.

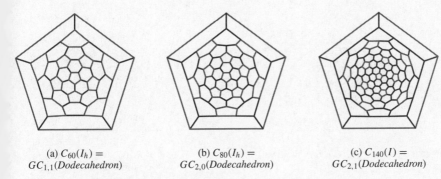

(a) $C_{60}(I_h) =$
$GC_{1,1}(Dodecahedron)$

(b) $C_{80}(I_h) =$
$GC_{2,0}(Dodecahedron)$

(c) $C_{140}(I) =$
$GC_{2,1}(Dodecahedron)$

Figure 19.1 The smallest $I(5, 6)$-fulleroids apart from Dodecahedron

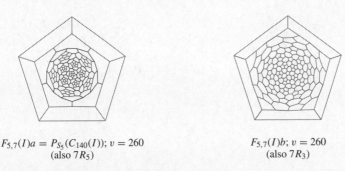

$F_{5,7}(I)a = P_{S_5}(C_{140}(I)); v = 260$
(also $7R_5$)

$F_{5,7}(I)b; v = 260$
(also $7R_3$)

Figure 19.2 Both smallest $I(5, 7)$-fulleroids

19.1 Construction of I-fulleroids and infinite series

Eight I-fulleroids, shown in Figures 19.2 and 19.3, have been found by a systematic investigation of all possible ways to assemble 5-gons and, say, 8-gons into a structure with the desired properties. This will be explained in more detail in Section 19.3. However, it turns out that all eight structures can be conveniently described as the results of decorating small icosahedral fullerenes with $(5, 3)$-polycycles. The pentacon operation P (see Section 1.6), is applied to the set S_5 of twelve 5-gons with 5-fold axis of rotation. The 5-triakon (see Section 1.6) is applied to the set S_3 of 20 vertices belonging to 3-fold axis of rotation.

The E_1- *and* C_3-*replacement* of a 6-gonal face F of a map M consist of replacing a 6-gon by $(5, 3)$-polycycles E_1 or C_3, according to the following schemes:

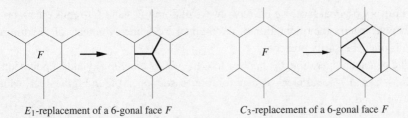

E_1-replacement of a 6-gonal face F

C_3-replacement of a 6-gonal face F

$F_{5,9}(I_h) = P_{S_5}(C_{60}(I_h)); \; v = 180$
(also $9R_3$)

$F_{5,8}(I_h) = P_{S_5}(C_{80}(I_h)); \; v = 200$
(also $8R_4$)

$F_{5,10}(I_h); \; v = 140$
(also $5R_3, 10R_0$)

$F_{5,15}(I_h); \; v = 260$
(also $15R_0$)

$F_{5,12}(I_h) = T_{S_3}(C_{80}(I_h)); \; v = 440$
(also $12R_4$)

$P_{S_5}(T_{S_3}(C_{80}(I_h))) = T_{S_3}(P_{S_5}(C_{80}(I_h))$
$= F_{5,14}(I_h); \; v = 560$ (also $14R_4$)

Figure 19.3 Smallest $I(5, b)$-fulleroids for $b = 8, 9, 10, 12, 14, 15$; each is unique for its number v of vertices

For every 6-gonal face, we have two ways of doing E_1- and C_3-replacement on it. Since, we specify the orientation of the 6-gonal faces, the symmetry of the original map is not necessarily preserved.

The pentacon operation P_{S_5} applied to $GC_{k,l}(Dodecahedron)$ yields a polyhedron with 5-, 6- and 7-gonal faces, except for three cases appearing in Figure 19.1, which are two-faced.

Table 19.1 *Potential p-vectors and invariants for certain I(5, b)-fulleroids*

b	p_5	p_b	v	Constr.	m_5	m_b	k_2	k_3	k_5	Nr.	Name
7	72	60	260	$A_{7,1}$	1	1	3	2	1	2	$F_{5,7}(I)a$
											$F_{5,7}(I)b$
8	72	30	200	$A_{8,1}$	1	1	4	2	1	1	$F_{5,8}(I_h)$
9	72	20	180	$A_{9,1}$	1	0	3	3	1	1	$F_{5,9}(I_h)$
10	60	12	140	$B_{10,1}$	1	0	3	2	2	1	$F_{5,10}(I_h)$
11	312	60	740	$A_{11,5}$	5	1	3	2	1	≥ 1	$F_{11}(1)$
12	132	20	300	$A_{12,2}$	2	0	3	4	1	–	
12	192	30	440	$A_{12,3}$	3	0	6	3	1	1	$F_{5,12}(I_h)$
13	432	60	980	$A_{13,7}$	7	1	3	2	1	≥ 1	$F_{13}(1)$
14	252	30	560	$A_{14,4}$	4	0	7	2	1	1	$F_{5,14}(I_h)$
15	120	12	260	$B_{15,2}$	2	0	3	2	3	1	$F_{5,15}(I_h)$
16	312	30	680	$A_{16,5}$	5	0	8	2	1	?	
17	672	60	1460	$A_{17,11}$	11	1	3	2	1	≥ 1	$F_{17}(1)$
18	252	20	540	$A_{18,4}$	4	0	3	6	1	?	
19	792	60	1700	$A_{19,13}$	13	1	3	2	1	≥ 1	$F_{19}(1)$
20	180	12	380	$B_{20,3}$	3	0	3	2	4	?	

An infinite series of $I(5, 7)$-fulleroids $(H_n)_{n>0}$ of symmetry I is obtained from the series $(GC_{2n+1,0}(Dodecahedron))_{n>0}$ by doing E_1-replacement, in a regular pattern, to one fourth of all 6-gons (instead of just to those 6-gons, which contain a 3-fold rotational axis). The pattern is shown in Figure 19.4, together with fundamental domains for the first four of these $I(5, 7)$-fulleroids. The fundamental domains are kite-shaped pieces of different sizes from the decorated $\{6, 3\}$. The axes of the 5- and the 3-fold rotations will go through the leftmost and rightmost vertex, respectively, of the kite. Note that the first $I(5, 7)$-fulleroid of this series, H_1, is just $F_{5,7}(I)b$. To prove that this method works, it suffices to notice that if one does a 5-fold or 3-fold rotation around the leftmost and rightmost vertices, then the local coherency is preserved and so the global one as well (see [Dre87]).

A similar series $(O_n)_{n>0}$ of $I(5, 8)$-fulleroids of symmetry I is obtained from the series $(GC_{2n+1,0}(Dodecahedron))_{n>0}$ by replacing 6-gons by $(5, 3)$-polycycles C_3 in the same pattern as shown in Figure 19.4.

The following general result was proved in [JeTr01]:

Theorem 19.1.1 *Let $b \geq 8$ and $m \geq 1$ be integers. Then there exists an $I(5, b)$-fulleroid $F(m)$ having $p_b = 60m$ (so, with $v = 20 + 120m(b - 5)$ vertices).*

Clearly, this theorem, together with Theorem 19.2.2 below, implies, in the special case when b is not divisible by 2, 3, or 5, that $F(1)$ has the smallest possible, for an $I(5, b)$-fulleroid, number of vertices. This number, $20(6b - 29)$, appeared, for cases $b = 7, 11, 13, 17, 19$ in Table 19.1.

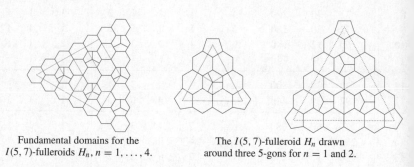

Fundamental domains for the
$I(5, 7)$-fulleroids $H_n, n = 1, \ldots, 4$.

The $I(5, 7)$-fulleroid H_n drawn
around three 5-gons for $n = 1$ and 2.

Figure 19.4 Partial drawing of the $I(5, 7)$-fulleroids H_n

19.2 Restrictions on the *p*-vectors

Recall first the Euler formula for the p-vector of a 3-valent plane graph (see Theorem 1.2.3):

$$0 = 12 + \sum (i - 6)p_i.$$

For tilings with specified symmetries, much stronger conditions on the potential p-vectors are obtained by applying some elementary group theory. Let T be some tiling of the sphere $\mathbb{S}^2$ and let G be a subgroup of its symmetry group $Aut(G)$. The image of an arbitrary face of T by an arbitrary element of G is again a face of T. Therefore, G can be interpreted as acting by permutations on the set of faces. By the same argument, G also acts on the set of edges and on the set of vertices. Recall that the stabilizer of some element x (face, edge or vertex) is the set of elements in G, which map it onto itself. The stabilizer of x is a subgroup of G and is denoted by $Stab_G(x)$. The orbit Gx of an element x is the set of images of x under all the operations in G. Obviously, the set of orbits forms a partition of the set of elements, thus, for example, the set of all face orbits under G forms a partition of the set of faces, and so on. The following elementary statement from group theory tells us how many elements these orbits can have.

Lemma 19.2.1 *If G is a finite group acting on a finite set S and x is an arbitrary element of S, then $|(Gx)| = \frac{|G|}{|Stab_G(x)|}$.*

Now, consider a tiling T with symmetry group I. All the elements of I are rotations. If some rotation maps a face f onto itself, then all the powers of this rotation do so too. Moreover, no two rotations of different orders can fix the same face. Thus, the possible face stabilizers are exactly the groups generated by the rotations of orders 2, 3, and 5.

Another elementary fact is that the stabilizers of two elements in the same orbit are conjugate to each other, and that, moreover, if some subgroup H of G is conjugate to a stabilizer $Stab_G(x)$, then H is the stabilizer of some element in the same orbit as x. Now, since in the icosahedral group I, any two subgroups generated by a rotation of the same order are conjugate, the following statement holds.

Theorem 19.2.2 *In an I-fulleroid, there is at most one orbit of faces with rotational symmetries of orders 2, 3, and 5 respectively. These orbits, if existing, contain exactly 30, 20, and 12 faces, respectively. All the other orbits have trivial stabilizers and contain exactly 60 faces each.*

Table 19.2 gives an overview of the possible combinations of face orbits.

Table 19.2 *The possible face orbits of an I-fulleroid*

orbit size	60	30	20	12
number of orbits	any	≤ 1	≤ 1	1
face gonality	any	$2t$	$3t$	$5t$

Now, a rotation of order 2 can never fix a face of odd gonality or a vertex of odd degree. So, in this case, it has to fix either an edge, or a face of even gonality. Similarly, a rotation of order 3 must fix either a vertex of degree divisible by 3 or a face of gonality divisible by 3. On the other hand, a rotation of order 5 must fix a face, since there is no other element it could fix. We can distinguish two broad classes of $I(5, b)$-fulleroids depending on whether or not the 5-fold rotations fix 5-gons or larger faces. In the first case, the p-vector will be of the form:

$$\mathbf{A}_{b,k}: (p_5, p_b) = \left(12 + 60k, \frac{60k}{b-6}\right),$$

where $k \geq 1$, $b \geq 7$. In the second case, it will be of the form:

$$\mathbf{B}_{b,k}: (p_5, p_b) = \left(60k, 12\frac{5k-1}{b-6}\right),$$

where $k \geq 1$, $b \geq 10$, 5 divides b. In the following, we will give a finer parametrization of potential p-vectors.

For a given integer i, let m_i denotes the number of orbits of faces of gonality i that are not mapped onto themselves by any symmetry but the identity. The total number of such faces is then $60m_i$ by the above. For $j = 2, 3, 5$, set $k_j = k$, if a kj-gon exists which is fixed by a rotation of order j. If no such face exists for any k, then we set $k_j = \frac{6}{j}$.

Lemma 19.2.3 *For each I-fulleroid, the equality:*

$$\sum_i m_i(i - 6) + \sum_{j=2,3,5} k_j = 6$$

holds.

Proof. Each face orbit has either a trivial stabilizer or is fixed by a rotation of order 2, 3, or 5. If some face f of gonality m has a stabilizer of order $j = 2, 3, 5$, then we have $m = k_j j$, and the orbit of f contains exactly $\frac{60}{j}$ faces. Together with Euler formula (1.1), this implies:

$$0 = \sum_i 60m_i(i - 6) + \sum_{j=2,3,5} \frac{60}{j}(k_j j - 6) + 12.$$

Note that this equation remains true if, for some j, there is no face fixed by a rotation of order j, because in that case we have $k_j = \frac{6}{j}$, thus $k_j j - 6 = 0$.
Divide both sides by 60 and simplify the second sum, to obtain:

$$0 = \sum_i m_i(i - 6) + \sum_{j=2,3,5} k_j - \sum_{j=2,3,5} \frac{6}{j} + \frac{1}{5}$$

$$= \sum_i m_i(i - 6) + \sum_{j=2,3,5} k_j - \frac{6}{2} - \frac{6}{3} - \frac{6}{5} + \frac{1}{5}$$

$$= \sum_i m_i(i - 6) + \sum_{j=2,3,5} k_j - 6. \qquad \square$$

Now we give formulas to calculate the p-vector and the number of vertices v from the invariants m_i and k_j. Since the invariants are insensitive to p_6, the following equations hold if and only if the I-fulleroid in question does not contain 6-gons. For the p-vector, we have:

$$p_i = 60 \left(m_i + \sum_{jk_j=i} \frac{1}{j} \right).$$

The number of vertices is:

$$v = \frac{1}{3} \sum_i i p_i = 20 \sum_i \left(i m_i + \sum_{j k_j = i} \frac{i}{j} \right) = 20 \sum_i \left(i m_i + \sum_{j k_j = i} k_j \right),$$

thus:

$$v = 20 \left(\sum_i i m_i + \sum_{j k_j \neq 6} k_j \right).$$

19.3 From the p-vectors to the structures

In the following, we will present a general method for the classification of I-fulleroids with given parameters m_i and k_j. Remaining proofs can be found in [DeDe00].

Theorem 19.3.1 *The I-fulleroid $F_{5,10}(I_h)$ is the only I-fulleroid with p-vector* $(p_5, p_{10}) = (60, 12)$ *and the smallest $I(5, 10)$-fulleroid.*

Proof. First note that 12 is the smallest possible non-zero value of p_{10}, corresponding to the parameter values $m_{10} = 0$ and $k_5 = 2$. This means that there is exactly one orbit of 10-gons, each of which is fixed by a rotation of order 5, and one orbit of 5-gons, which have trivial stabilizers. So, all the rotation axes of order 2 meet edges and all the axes of order 3 meet vertices. Let v_0 be such a vertex and let F be a face adjacent to it.

Assume F is a 10-gon. Because of the 5-fold rotation mapping of F onto itself, every second vertex of F must meet a 3-fold axis. The three faces containing such a vertex must then all be 10-gons, because they are rotated onto each other. This means that F is completely surrounded by 10-gons, each of them in the same orbit as F. By induction, all the faces in the connected component, which contain F, are 10-gons. There is only one connected component; so, there are only 10-gons, a contradiction.

We conclude that v_0 must be incident only to 5-gons. Let v_1 be a vertex incident to any two of these faces. The third face, incident to v_1, cannot be a 5-gon, since this 5-gon would be in the same orbit as F and thus would contain a symmetric image v_2 of v_0, i.e. a vertex meeting a 3-fold rotation axis. There are three essentially different possible positions for v_2. In each case, by applying all symmetries we find that the resulting structure must be Dodecahedron.

Consequently, the third face incident to v_2 is a 10-gon, thus the sought I-fulleroid must contain the substructure depicted in Figure 19.5, where the 3- and the 5-fold rotation centers are indicated by a small triangle and a 5-gon, respectively. By applying these rotations in a systematic way, we obtain the structure shown in Figure 19.3. $\qquad\square$

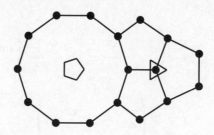

Figure 19.5 A forced substructure for an I-fulleroid with p-vector $(p_5, p_{10}) = (60, 12)$

Obviously, the argumentation in the above proof is straightforward but rather tedious. It is much more convenient to work in the *orbit space* of the group I. The orbit space of some group G is defined as the image of a continuous function, which maps two vertices p and q of the sphere onto the same image vertex if and only if there is some element of G which maps p to q. The orbit space of I is, topologically, just a sphere with three special vertices, namely the images of the three types of rotational centers of order 2, 3, and 5. A simple way to obtain the orbit space is to take a fundamental domain and glue together the boundary vertices, which are mapped onto each other by some group element.

The gonality of the image of a face in the orbit space depends on the stabilizer of that face. In general, an ij-gon with a rotation center of order j is mapped to one i-gon. Likewise, vertices with non-trivial stabilizers are mapped to vertices of, accordingly, smaller degree. Therefore, in the orbit space, we have to consider special features such as 1-gons and vertices of degree 1. Still more strangely, the image of an edge with 2-fold stabilizer is a *half-edge* with only one end-vertex. A loop edge, one which has two identical end-vertices, must be count twice when determining the vertex degree. Likewise, an edge adjacent to the same face on both sides has to be counted twice when determining the face gonality. In both cases, however, a half-edge is counted only once.

Despite these difficulties, we can simplify proofs considerably by working in the orbit space. We will call the image of a tiling in the orbit space its *orbit tiling*.

Theorem 19.3.2 *The I-fulleroid $F_{5,8}(I_h)$ is the only I-fulleroid with p-vector $(p_5, p_8) = (72, 30)$ and the smallest $I(5, 8)$-fulleroid.*

Proof. Again, $p_8 = 30$ is the smallest possible value of p_8, since 2 is the largest rotational order occurring in I which divides 8. The p-vector $(p_5, p_8) = (72, 30)$ corresponds to the parameter values:

$$(m_5, m_8, k_2, k_3, k_5) = (1, 0, 4, 2, 1).$$

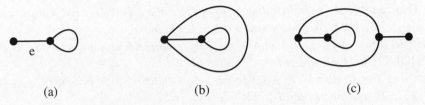

(a)   (b)   (c)

Figure 19.6 Forced substructures for the smallest $I(5, 8)$-fulleroid

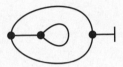

Figure 19.7 The orbit space of $F_{5,9}(I_h)$

The orbit tiling consists of a 1-gon (the image of a 5-gon with a 5-fold symmetry axis), a 5-gon (the image of an asymmetric 5-gon), and a 4-gon (the image of a 2-fold symmetric 8-gon). Exactly one degree 1 vertex and no half-edge occurs, because there are 2-fold symmetric, but no 3-fold symmetric, faces.

The 1-gon gives rise to the subgraph depicted in Figure 19.6(a), because of the degree 3 at its vertex. The edge e is adjacent to the same face at both sides. Thus, if that face is a 4-gon, then the configuration of Figure 19.6(b) occurs. Now, both vertices have degree 3; so, there is no possible continuation. We conclude that the 1-gon must be adjacent to the 5-gon, leading to the configuration of Figure 19.6(c).

Now, the outer face is a 4-gon and there is exactly one vertex of degree 1, as required. The configuration is unique, as shown, and corresponds to the I-fulleroid $F_{5,8}(I_h)$. □

Theorem 19.3.3 *The I-fulleroid $F_{5,9}(I_h)$ is the only I-fulleroid with p-vector $(p_5, p_9) = (72, 20)$ and the smallest $I(5, 9)$-fulleroid.*

Proof. As above, $p_9 = 20$ is smallest possible and the parameter values follow uniquely from the p-vector. The orbit tiling consists of a 1-gon and a 5-gon as above and a 3-gon corresponding to an orbit of 3-fold symmetric 9-gons. One has no vertex of degree 1, but a half-edge. As above, the 1-gon must be adjacent to the 5-gon, which in turn must be adjacent to the 3-gon. The only possible configuration is shown in Figure 19.7, where the half-edge is shown as a "T-shape"; note that a half-edge counts only once and so the outer face is indeed a 3-gon. □

Theorem 19.3.4 *(i) The I-fulleroid $F_{5,12}(I_h)$ is the only I-fulleroid with p-vector $(p_5, p_{12}) = (192, 30)$ and the smallest $I(5, 12)$-fulleroid.*

(ii) The I-fulleroid $F_{5,15}(I_h)$ is the only I-fulleroid with p-vector $(p_5, p_{15}) = (120, 12)$ and the smallest $I(5, 15)$-fulleroid.

(iii) The I-fulleroid $F_{5,14}(I_h)$ is the only I-fulleroid with p-vector $(p_5, p_{14}) = (252, 30)$ and the smallest $I(5, 14)$-fulleroid.

Proof. See [DeDe00] for the case-by-case proof based on the theory of Delaney symbols, which is explained in [Del80, Dre84, DrBr96]. □

References

[Ale48] A. D. Alexandrov, *Vnutrennyaya Geometriya Vypuklyh Poverhnostĕ* 1948 OGIZ, Moscow-Leningrad. Translated in English as *A. D. Alexandrov Selected Works, Part II: Intrinsic Geometry of Convex Surfaces*, ed. by S. S. Kutateladze, Chapman & Hall, FL, 2006.

[Ale50] A. D. Alexandrov, *Vypuklie mnogogranniki*, 1950 GITL, Moscow. Translated in German as *Convexe Polyheder*, Akademie-Verlag, Berlin, 1958. Translated in English as *Convex Polyhedra*, Springer-Verlag, 2005.

[Alt73] A. Altshuler, Construction and enumeration of regular maps on the torus, *Discrete Mathematics* **4** (1973) 201–217.

[ArRi92] D. Archdeacon and R. B. Bruce, The construction and classification of self-dual spherical polyhedra, *Journal of Combinatorial Theory*, Series B **54**-1 (1992) 37–63.

[ArPe90] D. Archdeacon and M. Perkel, *Constructing* polygonal graphs of large girth and Degree, *Congressus Numerantium* **70** (1990) 81–85.

[Bal95] A. T. Balaban, Chemical graphs: looking back and glimpsing ahead, *Journal of Chemical Information and Computer Science* **35** (1995) 339–350.

[BeSi07] M. Beck and R. Sinai, Computing the continuous discretly, integer-point enumeration in polyhedra, *Undergraduate Texts in Mathematics*, Springer Verlag, 2007.

[BoLi95] C. P. Bonnington, C. H. C. Little, and Charles H. C., *The Foundations of Topological Graph Theory*, Springer-Verlag, 1995.

[BBG03] J. Bornhöft, G. Brinkmann, and J. Greinus, Pentagon-hexagon-patches with short boundaries, *European Journal of Combinatorics* **24**-5 (2003) 517–529.

[BGOR99] M. Bousquet-Mélou, A. J. Guttman, W. P. Orrick, and A. Rechnitzer, Inversion Relations, Reciprocity and Polyominoes, *Annals of Combinatorics* **3** (1999) 223–249.

[BrWi93] U. Brehm and J. M. Wills, Polyhedral manifolds, chapter 16 in *Handbook of Convex Geometry*, ed. by P. M. Gruber and J. M. Wills, Elsevier Science Publishers, 1993.

[BCH02] G. Brinkmann, G. Caporossi, and P. Hansen, A constructive enumeration of fusenes and benzenoids, *Journal of Algorithms* **45**-2 (2002) 155–166.

[BCH03] G. Brinkmann, G. Caporossi, and P. Hansen, A survey and new results on computer enumeration of polyhex and fusene hydrocarbons, *Journal of Chemical Information and Computer Science* **43**-3 (2003) 842–851.

[BDDH97] G. Brinkmann, O. Delgado-Friedrichs, A. Dress, and T. Harmuth, CaGe – a virtual environment for studying some special classes of large molecules, MATCH **36** (1997) 233–237.

[BDvN06] G. Brinkmann, O.Delgado-Friedrichs, and U. von Nathusius, Numbers of faces and boundary encoging of patches, *Graphs and Discovery*, ed. by S. Fajtlowicz, P. W.

295

Fowler, P. Hansen, M. F. Janowitz, and F. S. Roberts, American Mathematical Society, 2006.

[BrDe99] G. Brinkmann and M. Deza, Tables of face-regular polyhedra, *Journal of Chemical Information and Computer Science* **40**-3 (1999) 530–541.

[BGGMTW05] G. Brinkmann, S. Greenberg, C. Greenhill, B. D. McKay, R. Thomas, and P. Wollan, Generation of simple quadrangulations of the sphere, *Discrete Mathematics* **305**-1&3 (2005) 33–54.

[BHH03] G. Brinkmann, T. Harmuth, and O. Heidemeier, The construction of cubic and quartic planar maps with prescribed face degrees, *Discrete Applied Mathematics* **128**-2&3 (2003) 541-554.

[BrMK] G. Brinkmann and B. McKay, The Plantri program, http://cs.anu.edu.au/people/bdm/plantri/.

[BrMK06] G. Brinkmann and B. D. McKay, Fast generation of planar graphs, *MATCH* **58**-2 (2007) 333–367.

[BCC92] J. Brunvoll, B. N. Cyvin, and S. J. Cyvin, Enumeration of benzenoid systems and other polyhexes, *Topics in Curr. Chem.* **162** (1992) 65–180.

[BCC96] J. Brunvoll, S. J. Cyvin, and B. N. Cyvin, Azulenoids, *MATCH* **34** (1996) 91–108.

[CaHa98] G. Caporossi and P. Hansen, Enumeration of polyhex hydrocarbons up to h = 21, *Journal of Chemical Information and Computer Science* **38** (1998) 610–619.

[CaKl62] D. Caspar and A. Klug, Physical principles in the construction of regular viruses, *Cold Spring Harbor Symp. Quant. Biol.* **27** (1962) 1–24.

[Cha89] D. Chavey, Tilings by regular polygons – 2; a catalog of tilings, *Computers and Mathematical Applications* **17** (1989) 147–165. Reprinted in *SYMMETRY* 2, ed. by I. Hargittai, Vol. 18, *International Series Modern Applied Mathematics and Computer Science*, Pergamon Press, 1989.

[CDG97] V. Chepoi, M. Deza, and V. P. Grishukhin, Clin d'oeil on ℓ_1-embeddable planar graphs, *Discrete Applied Mathematics* **80** (1997) 3–19.

[CDV06] V. Chepoi, F. Dragan, and Y. Vaxes, Distance and routing labelling schemes for non-positively curved plane graphs, *Journal of Algorithms* **61**-2 (2006) 60–88.

[Chv83] V. Chvátal, *Linear Programming*, W. H. Freeman & Company, 1983.

[Cla] B. Clair, Frieze groups, http://euler.slu.edu/~clair/escher/friezehandout.pdf

[CoDo01] M. Conder and P. Dobcsányi, Determination of all regular maps of small genus, *Journal of Combinatorial Theory*, Series B **81**-2 (2001) 224–242.

[CoDo02] M. Conder and P. Dobcsányi, Trivalent symmetric graphs on up to 768 vertices, *Journal of Combinatorial Mathematics and Combinatorial Computing* **40** (2002) 41–63.

[CoDo05] M. Conder and P. Dobcsányi, Applications and adaptations of the low index subgroups procedure, *Math. Comp.* **74** (2005) 485–497.

[CoLa90] J. H. Conway and J. C. Lagarias, Tilings with polyominoes and combinatorial group theory, *Journal of Combinatorial Theory*, Series A **53** (1990) 183–208.

[Coo04] S. B. Cooper, *Computability Theory*, Chapman & Hall/CRC, 2004.

[Cox71] H. S. M. Coxeter, Virus macromolecules and geodesic domes, in *A Spectrum of Mathematics*, ed. by J. C. Butcher, 98–107, Oxford University Press/Auckland University Press, 1971.

[Cox73] H. S. M. Coxeter, *Regular Polytopes*, 3rd edition, Dover, 1973.

[Cox98] H. S. M. Coxeter, *Non-Euclidean Geometry*, 6th edition, MAA Spectrum. Mathematical Association of America, 1998.

[CBCL96] V. H. Crespi, L. X. Benedict, M. L. Cohen, and S. G. Louie, Prediction of a pure-carbon covalent metal, *Journal of Physics Review*, Series B **53** (1996) R13303-R13305.

[Cro97] P. R. Cromwell, *Polyhedra*, Cambridge University Press, 1997.

[CCBBZGT93] C. J. Cyvin, B. N. Cyvin, J. Brunvoll, E. Brendsdal, F. Zhang, X. Guo, and R. Tosic, Theory of polypentagons, *Journal of Chemical Information and Computer Science* **33** (1993) 466–474.

[DeDe00] O. Delgado-Friedrichs and M. Deza, More icosahedral fulleroids, *DIMACS Series in Discrete Mathematics and Theoretical Computer Science* **51** (2000) 97–115.

[Del80] M. S. Delaney, Quasisymmetries of space group orbits, *MATCH* **9** (1980) 73–80.

[DGMŠ07] M. DeVos, L. Goddyn, B. Mohar, and R. Šámal, Eigenvalues of (3, 6)-polyhedra, submitted.

[Dez02] M. Deza, *Face-regular polyhedra and tilings with two combinatorial types of faces*, in "Codes and Designs", Ohio State University Mathematical Research Institute **10** (2002) 49–71.

[DDG98] A. Deza, M. Deza, and V. P. Grishukhin, Embeddings of fullerenes and coordination polyhedra into half-cubes, *Discrete Mathematics* **192** (1998) 41–80.

[DeDu05] M. Deza and M. Dutour, *Zigzag structure of simple two-faced polyhedra, Combinatorics, Probability and Computing* **14**-1&2 (2005) 31–57.

[DDS05] M. Deza, M. Dutour, and S. Shpectorov, Graphs 4_n that are isometrically embeddable in hypercubes, *Bulletin of South-East Asian Mathematical Society* **29**-3 (2005) 469–484.

[DDS03] M. Deza, M. Dutour and M. I. Shtogrin, 4-*valent plane graphs with 2-, 3- and 4-gonal faces, Advances in Algebra and Related Topics*, (73–97, in memory of B. H. Neumann; Proceedings of ICM Satellite Conference on Algebra and Combinatorics, Hong Kong 2002), World Scientific, 2003.

[DDS05a] M. Deza, M. Dutour, and M. I. Shtogrin, Filling of a given boundary by p-gons and related problems, to appear in Proceedings of International Conference General Theory of Information Transfer and Combinatorics (Bielefeld, 2004), ed. by L. Baumer, in *Electronic Notes in Discrete Mathematics* **21** (2005) 279.

[DDS05b] M. Deza, M. Dutour, and M. I. Shtogrin, Elliptic polycycles with holes, *Uspechi Mat. Nauk.* **60**-2 (2005) 157–158 (in Russian). English translation in Russian *Mathematical Surveys* **60**-2, 349–351.

[DDS05c] M. Deza, M. Dutour, and M. I. Shtogrin, Elementary elliptic (R,q)-polycycles, chapter in Analytic, Computational and Statistical Approaches for Network Analysis, ed. by M. Dehmer and F. Emmert-Streib, Wiley, to appear.

[DFG01] M. Deza, P. W. Fowler, and V. P. Grishukhin, Allowed boundary sequences for fused polycyclic patches, and related algorithmic problems, *Journal of Chemical Information and Computer Science* **41**-2 (2001) 300–308.

[DFRR00] M. Deza, P. W. Fowler, A. Rassat, and K. M. Rogers, Fullerenes as tilings of surfaces, *Journal of Chemical Information and Computer Science* **40**-3 (2000) 550–558.

[DFSV00] M. Deza, P. W. Fowler, M. I. Shtogrin, and K. Vietze, Pentaheptite modifications of the graphite sheet, *Journal of Chemical Information and Computer Science* **40**-6 (2000) 1325–1332.

[DeGr99] M. Deza and V. P. Grishukhin, Hexagonal sequences, in *Proceedings of the Conference on General Algebra and Discrete Mathematics*, Potsdam, 1998, ed. by K. Denecke and H.-J. Vogel, Shaker-Verlag, 1999, 47–68.

[DeGr01] M. Deza and V. P. Grishukhin, Face-regular bifaced polyhedra, *Journal of Statistical Planning and Inference* **95**-1&2, Special issue in honor of S. S. Shrikhande (2001) 175–195.

[DGr02] M. Deza and V. P. Grishukhin, Maps of p-gons with a ring of q-gons, *Bulletin of Institute of Combinatorics and its Applications* **34** (2002) 99–110.

[DGS04] M. Deza, V. P. Grishukhin, and M. I. Shtogrin, *Scale-Isometric Polytopal Graphs in Hypercubes and Cubic Lattices*, World Scientific and Imperial College Press, 2004.

[DHL02] M. Deza, T. Huang, and K-W. Lih, Central circuit coverings of octahedrites and medial polyhedra, *Journal of Mathematical Research and Exposition* **22**-1 (2002) 49–66.

[DeLa97] M. Deza and M. Laurent, *Geometry of Cuts and Metrics*, Springer-Verlag, 1997.

[DeSh96] M. Deza and S. Shpectorov, Recognition of ℓ_1-graphs with complexity O(nm), or football in a hypercube, *European Journal of Combinatorics* **17**-2&3 (1996) 279–289.

[DSS06] M. Deza, S. Shpectorov, and M. I. Shtogrin, Non-extendible elliptic (3, 5)-polycycles, *Isvestia of Russian Academy of Sciences, Sec. Math.* **70**-7 (2006) 3–17.

[DeSt98] M. Deza and M. I. Shtogrin, *Polycycles*, Voronoi Conference on Analytic Number Theory and Space Tilings (Kyiv, September 7–14, 1998), Abstracts, Kyiv, 19–23.

[DeSt99a] M. Deza and M. I. Shtogrin, Primitive polycycles and helicenes, *Uspechi Mat. Nauk.* **54**-6 (1999) 159–160 (in Russian). English translation in Russian *Mathematical Surveys* **54**-6, 1238–1239.

[DeSt99b] M. Deza and M. I. Shtogrin, Three, four and five-dimensional fullerenes, *Southeast Asian Bulletin of Mathematics* **23** (1999) 9–18.

[DeSt00a] M. Deza and M. I. Shtogrin, Infinite primitive polycycles, *Uspechi Mat. Nauk.* **55**-1 (2000) 179–180 (in Russian). English translation in Russian *Mathematical Surveys* **55**-1, 169–170.

[DeSt00b] M. Deza and M. I. Shtogrin, Polycycles: symmetry and embeddability, *Uspechi Mat. Nauk.* **55**-6 (2000) 129–130 (in Russian). English translation in Russian *Mathematical Surveys* **55**-6, 1146–1147.

[DeSt00c] M. Deza and M. I. Shtogrin, Uniform partitions of 3-space, their relatives and embedding, *European Journal of Combinatorics* **21**-6 (2000) 807–814.

[DeSt01] M. Deza and M. I. Shtogrin, Clusters of cycles, *Journal of Geometry and Physics* **40**-3&4 (2001) 302–319.

[DeSt02a] M. Deza and M. I. Shtogrin, Criterion of embedding of (r,q)-polycycles, *Uspechi Mat. Nauk.* **57**-3 (2002) 149–150 (in Russian). English translation in Russian *Mathematical Surveys* **57**-3, 589–591.

[DeSt02b] M. Deza and M. I. Shtogrin, Extremal and non-extendible polycycles, *Proceedings of Steklov Mathematical Institute* **239** (2002) 117–135. (Translated from Trudy Mathematicheskogo Instituta imeni V. A. Steklova) **239** (2002) 127–145.

[DeSt02c] M. Deza and M. I. Shtogrin, Mosaics, embeddable into cubic lattices, *Discrete Mathematics* **244**-1&3 (2002) 43–53.

[DeSt03] M. Deza and M. I. Shtogrin, Octahedrites, *Symmetry* **11**-1,2,3,4, Special Issue *Polyhedra in Science and Art* (2003) 27–34.

[DeSt05] M. Deza and M. I. Shtogrin, Metrics of constant curvature on polycycles, *Mathematical Notes* **78**-2 (2005) 223–233.

[DeSt06] M. Deza and M. I. Shtogrin, Types and boundary unicity of polypentagons, *Uspechi Mat. Nauk* **61**-6 (2006) 183–184 (in Russian). English translation in Russian *Mathematical Surveys* **61**-6, 1170–1172.

[Dia88] J. R. Dias, *Handbook of Polycyclic Hydrocarbons. Part B: Polycyclic Isomers and Heteroatom Analogs of Benzenoid Hydrocarbons*, Elsevier, 1988.

[Diu03] M. Diudea, *Capra – a leapfrog related map transformation*, Studia Universitatis Babes-Bolyai, Chemia **48** (2003) 3–21.

[Dre84] A. W. M. Dress, Regular polytopes and equivariant tessellations from a combinatorial point of view, in *Algebraic Topology*, SLN 1172, Göttingen, 1984, 56–72.

[Dre87] A. W. M. Dress, Presentations of discrete groups, acting on simply connected manifolds, in terms of parametrized systems of Coxeter matrices – a systematic approach, *Advances in Mathematics* **63**-2 (1987) 196–212.

[DrBr96] A. W. M. Dress and G. Brinkmann, Phantasmagorical fulleroids, *MATCH* **33** (1996) 87–100.

[DDE96] M. S. Dresselhaus, G. Dresselhaus, and P. C. Eklund, *Science of Fullerenes and Carbon Nanotubes*, Academic Press, 1996.

[Dut] S. Dutch, Symmetry, crystals and polyhedra, www.uwgb.edu/dutchs/symmetry/symmetry.htm

[Dut02] M. Dutour, PlanGraph, a GAP package for planar graphs, www.liga.ens.fr/~dutour/PlanGraph/, 2002.

[Dut03] M. Dutour, l1emb, a program for testing graphical l^1-embedding, www.liga.ens.fr/~dutour/L1emb/, 2003.

[Dut04a] M. Dutour, Point groups, www.liga.ens.fr/~dutour/PointGroups/, 2004

[Dut04b] M. Dutour, TorusDraw, a Matlab program for toroidal maps, www.liga.ens.fr/~dutour/TorusDraw, 2004.

[Du07] M. Dutour, Programs for this book, www.liga.ens.fr/~dutour/BOOK_Polycycle/, 2007.

[DuDe03] M. Dutour and M. Deza, Goldberg–Coxeter construction for 3- and 4-valent plane graphs, *Electronic Journal of Combinatorics* **11**-1 (2004) R20.

[DuDe06] M. Dutour and M. Deza, Face-regular 3-valent two-faced spheres and tori, Research Memorandum Nr. 976 of Institute of Statistical Mathematics, Tokyo, 2006 (see also www.liga.ens.fr/~dutour/).

[Ebe1891] V. Eberhard, *Zur Morphologie der Polyeder*, Teubner, 1891.

[Egg05] A. Egging, Kuerzeste Randlaengen von Patches mit Drei-, Vier-, Fuenf- und Sechsecken, Diploma thesis, Bielefeld, 2005.

[Eps00] D. Epstein, Hyperbolic tilings, www.ics.uci.edu/~eppstein/junkyard/hypertile.html

[Fa48] I. Fary, On straight-line representation of planar graphs, *Acta Univ. Szeged. Sect. Sci. Math.* **11** (1948) 229–233.

[FeKuKu98] G. Fejes Tóth, G. Kuperberg and W. Kuperberg, Highly saturated packings and reduced coverings, *Monatsh. Math.* **125**-2 (1998) 127–145.

[FeKu93] G. Fejes Tóth and W. Kuperberg, *Packings and coverings with convex sets*, Handbook of Convex Geometry, Vol. A, B, North-Holland, 1993, 799–860.

[Fow93] P. W. Fowler, Systematics of fullerenes and related clusters, *Phil. Trans. R. Soc. London*, Series A (1993) 39–51.

[FCS88] P. W. Fowler, J. E. Cremona, and J. I. Steer. Systematics of bonding in non-icosahedral carbon clusters, *Theor. Chim. Acta* **73** (1988) 1–26.

[FoCr97] P. W. Fowler and J. E. Cremona, Fullerenes containing fused triples of pentagonal rings, *Journal of Chemical Society Faraday. Trans.* **93**-13 (1997) 2255–2262.

[FoMa95] P. W. Fowler and D. E. Manolopoulos, *An Atlas of Fullerenes*, Clarendon Press, 1995.

[FYO95] M. Fujita, M. Yoshida, and E. Osawa, Morphology of new fullerene families with negative curvature, *Fullerene Science and Technology* **3**-1 (1995) 93–105.

[GaHe93] Y-D. Gao and W. C. Herdon, Fullerenes with four-membered rings, *Journal of American Chemical Society* **115** (1993) 8459–8460.

[GaSo01] E. Gallego and G. Solanes, Perimeter, diameter and area of convex sets in the hyperbolic plane, *Journal London Mathematical Society* **64**-2 (2001) 161–178.

[GAP02] The GAP Group, GAP — groups, algorithms, and programming, www.gap-system.org, Version 4.3, 2002.

[God71] C. Godbillon, *Éléments de topologie algébrique*, Hermann, 1971.

[Gol35] M. Goldberg, An isoperimetric problem for polyhedra, *Tohoku Mathematical Journal* **40** (1935) 226–236.

[Gol37] M. Goldberg, A class of multi-symmetric polyhedra, *Tohoku Mathematical Journal* **43** (1937) 104–108.

[Gra03] J.E. Graver, The (m,k)-patch boundary code problem, *MATCH* **48** (2003) 189–196.

[Gra05] J.E. Graver, Catalog of all fullerenes with ten or more symmetries, *Graphs and Discovery, DIMACS Ser. Discrete Math. Theoret. Comput. Sci.*, **69**, American Mathematical Society, 2005, 167–188.

[Gre01] J. Greinus, Patches mit minimaler Randlänge, Diploma thesis, Bielefeld, 2001.

[GrTu87] J.L. Gross amd T.W. Tucker, *Topological Graph Theory*, Hamilton, 1987.

[Grü67] B. Grünbaum, *Convex Polytopes*, Interscience, 1967.

[Grü03] B. Grünbaum, *Are your polyhedra the same as my polyhedra?*, Discrete and Computational Geometry: The Goodman-Pollack Festschrift, ed. by B. Aronov, S. Basu, J. Pach, and M. Sharir, Springer Verlag, 2003, 461–488.

[GLST85] B. Grünbaum, H.-D. Löckenhoff, G.C. Shephard, and A.H. Temesvari, The enumeration of normal 2-homeohedral tilings, *Geom. Dedicata* **19** (1985) 109–174.

[GrSh83] B. Grünbaum and G.C. Shephard, The 2-homeotoxal tilings of the plane and the 2-sphere, *Journal of Combinatorial Theory*, Series B **34** (1983) 113–150.

[GrSh87a] B. Grünbaum and G.C. Shephard, *Tilings and Patterns*, W.H. Freeman, 1987.

[GrSh87b] B. Grünbaum and G.C. Shephard, Edge-transitive planar graphs, *Journal of Graph Theory* **11**-2 (1987) 141–155.

[GrMo63] B. Grünbaum and T.S. Motzkin, The number of hexagons and the simplicity of geodesics on certain polyhedra, *Canadian Journal of Mathematics* **15** (1963) 744–751.

[GrZa74] B. Grünbaum and J. Zaks, The existence of certain planar maps, *Discrete Mathematics* **10** (1974) 93–115.

[GHZ02] X. Guo, P. Hansen, and M. Zheng, Boundary uniqueness of fusenes, *Discrete Applied Mathematics* **118**-3 (2002) 209–222.

[HaSo07] R. Hajduk, F. Kardos, and R. Soták, On the azulenoids with large ring of 5-gons, manuscript, 2007.

[HLZ96] P. Hansen, C. Lebatteux, and M. Zheng, The boundary-edges code for polyhexes, *Journal of Molecular Structure* (Theochem) **363** (1996) 237–247.

[HaHa76] F. Harary and H. Harborth, Extremal animals, *Journal of Combinatorics, Information and System Sciences* **1** (1976) 1–8.

[Har90] H. Harborth, Some mosaic polyominoes, *Ars Combinatoria* **29A** (1990) 5–12.

[HaSl96] R.H. Hardin and N.J.A. Sloane, McLaren's improved snub cube and other new spherical designs in three dimensions, *Discrete and Computational Geometry* **15** (1996) 429–441.

[Hat01] A. Hatcher, Algebraic topology, www.math.cornell.edu/~hatcher/AT/ATpage.html

[Har00] T. Harmuth, *Home page*, http://people.freenet.de/thomas.harmuth/

[Har] T. Harmuth, http://people.freenet.de/thomas.harmuth/

[Hei98] O. Heidemeier, Die Erzeugung von 4-regulären, planaren, simplen, zusammenhängenden Graphen mit vorgegebenen Flächentypen, Diploma thesis, Bielefeld (1998).

[HeBr87] W. C. Herndon and A. J. Bruce, Perimeter code for benzenoid aromatic hydrocarbons, in *Graph Theory and Topology in Chemistry*, ed. by R. B. King and D. H. Rouvray, *Studies in Physical and Theoretical Chemistry* **51** (1987) 491–513.

[HoSh93] D. A. Holton and J. Sheehan, *The Petersen Graph*, Cambridge University Press, 1993.

[Hum90] J. E. Humphreys, *Reflection Groups and Coxeter Groups*, Cambridge University Press, 1990.

[Hum96] J. F. Humphreys, *A Course in Group Theory*, Oxford Science Publications, Oxford University Press, 1996.

[Jen90] S. Jendrol, Convex 3-polytopes with exactly two types of edges, *Discrete Mathematics* **84** (1990) 143–160.

[JeTr01] S. Jendrol and M. Trenkler, More icosahedral fulleroids, *Journal of Mathematical Chemistry* **29** (2001) 235–243.

[Joh66] N. W. Johnson, Convex polyhedra with regular faces, *Canadian Journal of Mathematics* **18** (1966) 169–200.

[Jos06] J. Jost, *Compact Riemann Surfaces: An Introduction to Contemporary Mathematics*, 3rd edition. Universitext, Springer Verlag, 2006.

[Kar07] F. Kardos, Symmetry of fulleroids, Ph.D. thesis, Kosice University, 2007.

[Kep1619] J. Kepler, *Harmonice Mundi*, 1619.

[Ki00] R. B. King, Unusual permutation groups in negative curvature carbon and boron nitride structures, *Croatia Chemica Acta* **73**-4 (2000) 993–1015.

[Kir94] E. C. Kirby, On toroidal azulenoids and other shapes of fullerene cage, *Fullerene Science and Technology* **2**-4 (1994) 395-404.

[Kir97] E. C. Kirby, *Recent works on toroidal and other exotic fullerene structures*, chapter 8 in *From Chemical Topology to 3-Dimensional Geometry*, ed. by A. T. Balaban, Plenum Press, 1997, 263–296.

[KoFi87] W. Fischer and E. Koch, On 3-periodic minimal surfaces, *Z. Krist.* **179**-1&4 (1987) 31–52.

[KoFi96] W. Fischer and E. Koch, Spanning minimal surfaces, *Philos. Trans. Roy. Soc. London*, Series A **354** (1996) 2105–2142.

[KlZh97] D. J. Klein and H. Zhu, All-conjugated carbon species, chapter 9 in *From Chemical Topology to 3-Dimensional Geometry*, ed. by A. T. Balaban, Plenum Press, 1997, 297–341.

[Löc68] H.-D. Löckenhoff, Über die Zerlegung der Ebene in zwei Arten topologisch verschiedener Flächen, Inaugural Dissertation, Marburg, 1968.

[LySc77] R. C. Lyndon and P. E. Schupp, Combinatorial group theory, *Ergebnisse der Mathematik und ihrer Grenzgebiete*, 89, Springer-Verlag, 1977

[MaSo07] T. Madaras and R. Soták, More maps of p-gons with a ring of q-gons, to appear in *Ars Combinatorica*.

[Mag74] W. Magnus, *Non-Euclidean Tesselations and Their Groups*, Academic Press, 1974.

[Mag07] The MAGMA Group, *Magma Computer Algebra System*, Sydney University, 2007.

[Mal70] J. Malkevitch, Properties of planar graphs with uniform vertex and face structure, *Memoirs of the American Mathematical Society*, American Mathematical Society 1970.

[Man71] P. Mani, Automorhismen von Polyhedrischen Graphen, *Mathematische Annalen* **192** (1971) 279–303.

[MaSh07] M. Marcusanu and S. Shpectorov, The classification of l_1-embeddable fullerenes, manuscript, 2007.

[McK98] B. D. McKay, Isomorph-free exhaustive generation, *Journal of Algorithms* **26** (1998) 306–324.

[Moh97] B. Mohar, Circle packing of map in polynomial time, *European Journal of Combinatorics* **18** (1997) 785–805.

[MoTh01] B. Mohar and C. Thomassen, *Graphs on Surfaces*, Johns Hopkins University Press, 2001.

[Mor97] J. F. Moran, The growth rate and balance of homogeneous tilings in the hyperbolic plane, *Discrete Mathematics* **173** (1997) 151–186.

[Mun71] J. R. Munkres, *Topology*, 2nd edition, Prentice-Hall, 2000.

[Nak90] M. Nakahara, *Geometry, Topology and Physics*, Graduate Student Series in Physics, Adam Hilger, 1990.

[Nak96] A. Nakamoto, Triangulations and quadrangulations of surfaces, Diploma thesis, Keio University, 1996.

[Neg85] S. Negami, Uniqueness and faithfulness of embedding of graphs into surfaces, Ph.D. thesis, Tokyo Institute of Technology, 1985.

[New72] M. Newman, *Integral Matrices*, Academic Press, 1972.

[OKHy96] M. O'Keeffe and B. G. Hyde, *Crystal Structures I: Pattern and Symmetry*, Mineral. Soc. of America, 1996.

[Oss69] R. Osserman, *A Survey of Minimal Surfaces*, van Nostrand, 1969.

[Par06] R. Parker, Short theorems with long proofs in finitely presented groups, Nikolaus Conference, 2006, www.math.rwth-aachen.de:8001/Nikolaus2006/prog.html

[PSC90] K. F. Prisacaru, P. S. Soltan, and V. D. Chepoi, On embeddings of planar graphs into hypercubes, *Proceedings of Moldavian Academy of Sciences, Mathematics* **1** (1990) 43–50 (in Russian).

[Rot88] J. J. Rotman, An introduction to algebraic topology, *Graduate Texts in Mathematics*, Springer Verlag, 1988.

[Sah94] C. H. Sah, A generalized leapfrog for fullerene structures, *Fullerenes Science and Technology* **2**-4 (1994) 445–458.

[ScSw95] P. Schröder and W. Sweldens, *Spherical Wavelets: Efficiently Representing Functions on the Sphere*, Proceedings of ACM Siggraph (1995) 161–172.

[SeSe94] B. Servatius and H. Servatius, Self-dual maps on the sphere, *Discrete Mathematics* **134**-1&3 (1994) 139–150.

[SeSe95] B. Servatius and H. Servatius, The 24 symmetry pairings of self-dual maps on the sphere, *Discrete Mathematics* **140**-1&3 (1995) 167–183.

[SeSe96] B. Servatius and H. Servatius, Self-dual graphs, *Discrete Mathematics* **149**-1&3 (1996) 223–232.

[Sht99] M. I. Shtogrin, Primitive polycycles: criterion, *Uspechi Mat. Nauk.* **54**-6 (1999) 177–178 (in Russian). English translation in Russian *Mathematical Surveys* **54**-6, 1261–1262.

[Sht00] M. I. Shtogrin, Non-primitive polycycles and helicenes, *Uspechi Mat. Nauk.* **55**-2 (2000) 155–156 (in Russian). English translation in Russian *Mathematical Surveys* **55**-2, 358–360.

[Ste22] E. Steinitz, Polyeder und Raumeinteilungen, *Enzykl. Math. Wiss.* **3** (1922) 1–139.

[Thu98] W. P. Thurston, Shapes of polyhedra and triangulations of the sphere, in *Geometry and Topology Monographs 1*, The Epstein Birthday Schrift, ed. by J. Rivin, C. Rourke, and C. Series, Geom. Topol. Publ., 1998, 511–549.

[Tut63] W. T. Tutte, How to draw a graph, *Proc. London Math. Soc.* **13** (1963) 743–768.

[Wa36] K. Wagner, Bemerkungen zum Vierfarbenproblem, *Jahresbericht. German. Math.-Verein* **46** (1936) 26–32.

[Zal69] V. A. Zalgaller, Convex polyhedra with regular faces, Seminar in Mathematics of Steklov Mathematical Institute, Leningrad **2** Consultants Bureau, 1969.

[Zie95] G. Ziegler, Lectures on polytopes, *Graduate Texts in Mathematics*, 152, Springer Verlag, 1995.

Index